Clinical
Bacteriology

Clinical
Bacteriology

E. Joan Stokes

Elizabeth

M.B., F.R.C.P., F.R.C.Path.
Clinical Bacteriologist, University College Hospital, London

Foreword by A. A. Miles, Kt., C.B.E.,
M.D., F.R.C.P., F.R.C.Path., F.R.S.

Fourth Edition

AN EDWARD ARNOLD PUBLICATION
Distributed by
YEAR BOOK MEDICAL PUBLISHERS, INC.
35 E. Wacker Dr., Chicago

First published 1955
by Edward Arnold (Publishers) Ltd.
25 Hill Street, London W1X 8LL
Second Edition 1960
Reprinted 1962
Third Edition 1968
Reprinted 1970
Fourth Edition 1975

Distributed in the United States of America, South and Central America, Puerto Rico and the Philippines by
Year Book Medical Publishers, Inc.
by arrangement with
Edward Arnold (Publishers) Ltd.

ISBN: 0–8151–8263–5
Library of Congress Catalog Number: 75–35302

to T.R. and E.F.R.

Printed in Great Britain by
Butler & Tanner Ltd., Frome and London

Foreword

Of the many disciplines in clinical pathology, bacteriology is perhaps the least easy to codify in fixed routines. Every patient with a suspected infection is a new biological problem that both the clinician and the bacteriologist can solve only by following their noses wherever their investigations lead them ; and any attempted identification of the infecting microbe in the laboratory may lead the bacteriologist along unexpected paths. Clinical bacteriology is an exploratory art that demands flexibility of mind and technique, and the latest methods are less important to the (would-be) practitioner than a set of uniformly good procedures with which to explore the common and the not-too-rare infections of man, and with which he stands a reasonable chance of discovering new ones.

The methods in this book will go far to meet these requirements. They are well tried. Many indeed had their beginning in 1940–44 in the work of Dr. Stokes and other former colleagues of mine in the Emergency Pathological Services of the London Sector 4, and much has been added to keep pace with advances in post-war medicine ; and they have an added merit in being subordinated to an all round picture of the practice of clinical bacteriology in a hospital department.

The work of those war years, when the treatment of war wounds and the difficulties of epidemic control in the emergency hospitals stimulated the study of cross-infection, also bears fruit in an innovation—Chapter on Hospital Epidemiology. The inclusion of such a chapter very properly implies that the hospital bacteriologist today must be more than the explorer of each single clinical problem as it arises ; as a watchdog in hygiene and as a hospital epidemiologist, he has a function that is almost equally important.

Dr. Stokes insists a little on the academic approach. If it is academic to bear in mind the better established principles of bacteriology and immunology and to define the limits of, and justifications for, the hodge-podge of techniques that make up the practice of those sciences, then in attempting to do these things the book may be said to have an academic background ; but to my mind they are equally the hallmarks of good practice.

A. A. MILES.

Preface to the Fourth Edition

The main challenge which hospital laboratories have had to face since the last edition of this book is how to maintain a reliable and rapid service despite an ever increasing work load ; automation is not, at present, the answer in our subject. A computer can help with records but the logistics of work in bacteriology, which differs from that in other disciplines, must be understood by those devising the system. A brief account of the kind of information needed is therefore included in Chapter 2.

Estimation of the reliability and speed of methods is now possible by examining simulated specimens sent by the Quality Control Laboratory and the use of this service is discussed. Time must not be wasted on inessentials and some of the methods, sputum culture for example, have been modified with this in mind.

Serology for the diagnosis of syphilis has changed completely ; new methods, one of them mechanised, are described. It has not been easy to decide which old methods to retain and whether to include any virology. Although complement fixation tests are no longer essential for diagnostic bacteriology they are essential for virology and such tests are undertaken increasingly in hospital laboratories, the method is therefore retained but described in general terms. When virological examination is worth attempting a note is made of how to obtain and send the sample but no purely virological tests are described.

Identification methods are not much changed but nomenclature has been brought in line with Wilson and Miles, *Topley and Wilson's Principles of Bacteriology, Virology and Immunity* (1975) 6th edition.

Only methods which have proved satisfactory in my department are included and references are limited to publications which support their choice.

I am very grateful to the Association of Clinical Pathologists who have allowed me to use material from my broadsheets on Blood Culture and Antibiotic Sensitivity Tests and to Miss P. M. Waterworth, co-author of the latter, the revision of Chapter 7 owes much to her criticism and advice. She has also read the proofs and revised the index.   Mr. A. W. Cremer has checked some of the proofs and provided much practical information.   I am very grateful to him and to other members of the staff who have helped to introduce new methods.   Finally I would like to thank Mr. P. Luton for the new photographs and Mr. V. K. Asta of U.C.H. Medical School for the new diagrams.

Preface to the First Edition

Examination of bacteriological specimens differs from other investigations in clinical pathology in that no complete set of rules can be laid down to guide the laboratory worker.   The value of the result depends as much on the knowledge and judgement of the person who examines the cultures at each stage as on his technical accuracy.   If it were possible to insist on the isolation in pure culture and identification of each type of colony appearing on every culture the task of the pathologist responsible for maintaining a high standard of work would be simple.   For reasons of speed and economy, however, this is impossible, with the result that it is very difficult for the pathologist to ensure that each specimen receives adequate attention and to decide how far each examination shall proceed. This book is an account of how these problems, and others encountered in the routine laboratory of a large general hospital, are met.

Most of the methods recommended are well known.   They have been tested both experimentally and in routine use in this laboratory.   They are described dogmatically for the sake of clearness and reasons for preferring them are given, but no claim is made that they are the best available.   The reader will have no difficulty in finding descriptions of alternative methods elsewhere.   If he so wishes he can compare the various culture media by the methods described in Chapter 10.

The approach to the investigation of infection is frankly academic. I make no apology for this because I believe that by regarding each investigation as a separate problem to be pursued, as far as is required, in essentially the same manner as a research project, knowledge of infectious disease will grow and the best interests of the individual patient are served.   It is possible to do this without elaborate laboratory equipment, numerous staff or delayed reports, by keeping the purpose of each investigation in mind and limiting it to essentials.

I hope the book will prove useful not only to pathologists and bacteriological technicians but also to clinicians, who wish to know what help they can reasonably expect from the laboratory, to those who teach nurses how to collect specimens for culture and to Resident Medical Officers and others responsible for the protection of patients from pathogenic bacteria in the hospital environment.

It is impossible to acknowledge properly the originators of all the methods quoted. Many of them are unknown to me; those that are known are mentioned in the text.

I am greatly indebted to Professor A. A. Miles; first for his teaching in the application of academic bacteriology to routine investigation; second for allowing me to use his *Practical Notes on Elementary and Clinical Bacteriology*, written in 1944 for University College Hospital Medical School, and finally for his kindness in writing the foreword. For his encouragement and for reading part of the manuscript I wish to thank Professor Wilson Smith. I am also very grateful to Dr. R. W. Riddell and Dr. J. R. May, for reading the proofs; to Dr. J. H. Hale and Dr. J. H. Humphry for reading parts of the manuscript, and to Dr. Robert Blowers, Dr. G. A. James, Mr. R. A. Bono and Mr. L. R. Jeffries for help in testing methods in the laboratory.

I would also like to thank H.M. Stationery Office for permission to reproduce Table 38, and the photographic department of U.C.H. Medical School for the photographs.

Contents

The practice of clinical bacteriology

Clinical bacteriology is the study of specimens taken from patients suspected of infectious disease to find, first, if there is any change in kind or distribution of the normal flora and, second, if the abnormal bacteria found are the cause of the disease. In most cases it is fairly easy to answer the first of these questions. The second is often difficult and sometimes impossible to solve ; it may be approached in two ways. The question is asked, " We have found microbe A, is it causing disease D in this patient ? " We may use the statistical argument that in many previous cases A has been found, to everyone's satisfaction, to be the cause of D ; therefore the chance of it being so in this case is very great and the assumption may safely be made. There are many pitfalls in the use of this argument for individual cases because no two patients are exactly alike and there is also wide variation in virulence between strains of the same species of microbe. For example, it has been established beyond reasonable doubt that *Strept. pyogenes* causes sore throat, and if we use this argument we shall assume that in all cases of sore throat when this microbe is found it is the cause of the infection. But this is not so ; healthy people are sometimes carriers of *Strept. pyogenes* and such a person may develop adenovirus infection, in which case if we say that microbe A (*Strept. pyogenes*) is the cause of disease D (sore throat) we may be at fault.

" Pathogen " and " saprophyte "

It is clear from this that the terms pathogen and saprophyte are not precise. *Strept. pyogenes* is undoubtedly a pathogen when it causes fatal septicaemia, but it is apparently a saprophyte when it is found in the throat of a healthy person. Similarly achromobacteria, normally regarded as saprophytes, have been known to cause a fatal septicaemia.

The presence of a pathogenic microbe in human tissues results in a variety of conditions, ranging from healthy carriage to the moribund state ; the factors which determine whether clinical signs of infection will develop and, when this happens, at what stage the relation of host to parasite will reach equilibrium are not well understood. Some acute infectious diseases produce lasting specific immunity and in these it is easy to explain the lack of signs of infection which follows subsequent contamination by the causal microbe. In others it is known that the pathogen needs special conditions before infection can be established, which explains why the presence of virulent *Cl. welchi* in a wound is not always followed by signs of gas gangrene. Specific immunity and special growth conditions are not however the only factors concerned, and often when there is an unusual response to the presence of bacteria in the tissues we have to assume without evidence, other than the condition we are attempting to explain, that the resistance of the patient or the virulence of the strain is abnormal.

It follows then that by pathogen we mean a microbe which is often dangerous and by saprophyte one which is seldom or never dangerous. This leads us in the practice of clinical bacteriology to adopt the following rules :

1. Never without good reason dismiss a microbe as a contaminant because it is not an accepted pathogen.
2. Never without good reason accept a microbe as the *necessary* cause of a disease merely because it is an accepted pathogen.

Koch's postulates in clinical bacteriology

Another way of tackling the problem " Is microbe A causing disease D " ? is to apply Koch's postulates :

1. The microbe should be found in all cases of the disease, distributed in the body according to the lesions observed.
2. The microbe should be grown artificially in pure culture for several subcultures.
3. The pure culture should reproduce the disease in a susceptible animal.

Postulate 1 must be modified for the single case to read, " The microbe should be constantly found associated with the lesions during the course of the disease."

We may add that the causal role of a microbe is strengthened if it can be shown that the serum of an infected animal contains a high antibody level which is specifically protective against infection by the microbe, particularly if a rise in titre during infection and a fall in titre after recovery can be demonstrated. It may also be strengthened by showing the development of a specific allergic reaction to a preparation of the microbe injected intradermally.

This method of approach is much more satisfactory than the statistical argument, but it is often impossible to carry out since many human infections cannot be recognizably reproduced in animals. It is worth while to test the antibody level in the patient's serum but a negative result does not exclude its causal role. A positive result indicates that the microbe, or another with similar antigens, has been present in sufficient quantity to stimulate antibody formation. Even if the titre can be shown to rise during infection and fall afterwards it does not finally prove that A is causing D. Remember, for example, the positive proteus agglutination with serum from patients suffering from rickettsial diseases.

Two types of evidence

There are two types of evidence which may be gained from laboratory investigations. First, simple tests may be made which are in themselves of little significance but which when taken into consideration with the history and physical signs may help to establish the diagnosis. If urine from a patient with symptoms of renal tract infection contains acid-fast bacilli it is probable that the patient has renal tuberculosis and the demonstration of the bacilli adds weight to the diagnosis. Similarly, if Gram-negative diplococci are found in a smear from the urethra of a woman with symptoms of acute infection, the finding adds weight to the diagnosis of gonorrhoea. But neither of these findings when considered alone has much significance because they are occasionally seen in specimens taken from healthy people. The diagnosis in these cases is essentially a clinical one.

The second type of evidence is based on the isolation in pure culture and identification of the organism concerned. In the case just considered this could be reported not as " acid-fast bacilli seen " but as " *Myco. tuberculosis* present ". This is a piece of evidence which is significant by itself without the support of clinical findings and if the patient has no other signs of tuberculosis the presence of the organism still needs an explanation.

Circular arguments

It is very important that there should be no confusion between the two types of evidence; " acid-fast bacilli seen " is never synonymous with " *Myco. tuberculosis* present ".

The first type of evidence is often useful in the diagnosis of infection in individual patients but can never lead to advance in our knowledge of infectious diseases, and unless its limitations are clearly understood its use may lead to much confusion of thought. The clinician says, " I think this is disease D which we know is caused by microbe A. Can this microbe be found ? " The bacteriologist sees a microbe resembling A and says, " The patient has disease D, therefore this must be A." This is an obvious example of a circular argument which is quite invalid. The same type of argument is very common in more subtle forms. For example, a patient may have symptoms suggesting glandular fever and when a Paul-Bunnell test is made a low titre of heterophile agglutinins is present which disappears after absorption. This result is insignificant because it occurs from time to time in normal serum. The bacteriologist may be tempted in borderline cases of this kind to stretch a point in view of the clinical findings and say that in this case the result can be considered to be weakly positive ; if he does so he falls into the trap of a circular argument. The clinical signs are an indication that the test should be repeated later to see if it becomes positive, but they must not be allowed to influence its interpretation. Laboratory tests are often repeated when the results fail to fit the clinical findings but very seldom when they satisfy clinical expectations. It is illogical to use a test for diagnosis of a particular infection and *at the same time* use the clinical findings to gauge its reliability.

Before tests are adopted for routine use they should be extensively tried out on *known* positive and negative material and their value and limitations established. When this has been done the experimental stage is over and it then remains for the bacteriologist to see that the conditions of the test are observed and that proper controls are included. If there is any doubt about its reliability when handled in a routine laboratory, duplicate tests should be set up in *all* cases irrespective of the clinical findings, and if this reveals variation of results which cannot be overcome the test is useless and must be discarded.

Another false argument frequently encountered is, " This colony resembles A which is commonly found in site S. The specimen came from S therefore the microbe must be A." Part of the work of a clinical bacteriologist is to recognize changes in the distribution of the human flora. He must not therefore assume that there are no changes and use this as evidence in the identification of microbes.

Legitimate use of clinical findings

It may now seem that it would be better if the bacteriologist had no knowledge at all of the clinical condition so that he is not in a position to be biased ; but information about the patient's clinical state is very valuable if legitimately used. When laying down rules for routine investigation they should be made to cover as far as possible all known eventualities and, no matter what the clinical findings are, nothing should be omitted from this routine. If, however, the patient's condition suggests infection by a certain group of organisms special methods may be employed from the outset, which may make it possible to isolate and identify the bacteria more quickly than by the routine method alone. For example, a wound swab may be sent from a patient suspected of gas-gangrene ; the routine method will be followed, but in addition a Nagler plate will be inoculated and if *Cl. welchi* is present it will be possible to identify it by this method next day, whereas if the routine method alone had been used the bacterial diagnosis would not have been possible for 48 hours at the earliest.

Avoidance of false reasoning

False reasoning along the lines indicated may be avoided either by fully identifying all the microbes found, which is not usually possible in a hospital laboratory, or by adopting the following procedure. All bacteria isolated from sites which are normally sterile are assigned to their genus on laboratory evidence alone and if it is useful to proceed further the species is also named. If the laboratory identification has stopped short at the genus this is made plain in the report. For example, a microbe identified by colonial and microscopic morphology and by a positive satellitism test is reported as an " haemophilus species " ; when necessary, tests for the utilization of X and V growth factors may be made and the species finally identified, but in most cases this delays the report and serves no useful purpose. Haemolytic streptococci on the other hand must be fully identified and grouped because the result is important both in the prognosis and treatment of the individual case and from the point of view of hospital epidemiology ; they are therefore reported by name.

Specimens from sites which have a normal flora are treated somewhat differently. It is rarely necessary to identify all the microbes cultured. The routine procedure is to exclude all known pathogens using the best selective methods available and reports are sent which make it clear that this has been done.   For example, the report of a faeces culture reads, " No organisms of the Salmonella or dysentery groups isolated."   This tells the clinician what he needs to know without giving him any misleading information and is to be preferred to the type of report which states, " Cultures yield *Esch. coli, Strept. faecalis* and *Proteus vulgaris*."   Naming the species is at first sight impressive, but a full identification of them cannot be made without delay and it misleads the clinician who may be led to believe that those are the only viable bacteria in the specimen. " No pathogens isolated " or " Cultures yield normal flora only " are further common variations ; they assume that we know which of the faecal flora are pathogens under all conditions and are therefore undesirable.

The amount of work necessary for satisfactory identification of the microbes frequently encountered in clinical bacteriology and the way in which different investigations may be reported will be considered later, but the guiding principles for avoiding unsound reasoning in diagnostic work are the same in all cases and may be stated thus :

1. The report must be based on laboratory evidence only.
2. All microbes named must be isolated in pure culture and identified by biochemical or serological tests.
3. When identification has proceeded as far as the genus only this must be made clear.
4. If identification falls short of the genus in the examination of cultures from sites with a normal flora, the microbe is considered to be insufficiently important for identification and is not mentioned in the report which concerns known pathogens only.

The need for speed

It may appear that the second of these rules is incompatible with speed and that it is more valuable for the experienced bacteriologist to use his " clinical " judgement of the appearance of microbes to send a quick report because, it may be argued, he will only fail to identify atypical strains which are infrequent, and from a practical point of view do not matter, since they are probably of low virulence. Speed in clinical bacteriology is of course very important and it is perfectly legitimate for the bacteriologist to use his experience of morphological appearances to give the clinician a quick *preliminary* report, which may be very valuable in indicating lines of treatment or the need for isolation; but the investigation must never be permitted to stop at this stage. The idea that strains which appear atypical are comparatively harmless is unsound and epidemics have arisen because such strains have not been recognized. The preliminary report must therefore be checked to avoid this risk and also to ensure that both the clinician and the bacteriologist do not put too much faith in it. The great variation in appearance that occurs among different strains of the same species, and the unreliability of colonial and microscopic morphology as a final test of identity, can only be appreciated if it is checked by biochemical and serological methods; atypical strains will continue to be thought rare if these are omitted.

If bacteriological reports based on inadequate tests are entered in the case records much harm may be done. Since the report came from the laboratory it may be regarded by the clinician as accurate in the same way that a chemical estimation is accurate and the records with the bacteriological findings may be included in a survey of cases for research purposes with very misleading results. If the principles of reporting investigations listed above are followed, there is no danger of a misunderstanding about the value of the findings.

" Academic " and " clinical " bacteriology

Since the discovery of antimicrobial drugs a large proportion of routine work has been devoted to testing bacterial sensitivity to them, to find which is most likely to be successful in treatment. Assay of the drugs in body fluids is sometimes required to find if dosage is sufficient or potentially toxic. Much of this work is new and there is still considerable difference of opinion about the value of different antibiotics in different infections and the best methods of giving them. These drugs are capable of causing gross morphological variation in bacteria growing in their presence which makes preliminary identification from colonial and microscopic appearance hopelessly unreliable. It is therefore more important than ever before that routine methods should be based on sound scientific reasoning and a knowledge of the factors which influence bacterial growth. It is sometimes thought that academic and clinical bacteriology are totally different and that the clinical bacteriologist need only know a few tricks for the rapid identification of pathogens and may safely leave a more fundamental knowledge of bacterial behaviour and host-parasite relation-ship to his more learned colleague.   If the bacteriologists in day-to-day contact with the material from human infections are insufficiently trained or fail to apply conscientiously the scientific method, they will fail to recognize departures from the accepted behaviour of host and parasite.   Such observations may not be of great value in the treatment of individual patients but it is on them that the growth of our knowledge of infectious disease depends.

2

General procedure

Delay in the diagnosis of acute infections may have disastrous consequences both for the patient himself, whose chance of recovery diminishes with every hour that the bacteria are allowed to flourish in his tissues, and also for other patients and staff who are thus exposed for an unnecessarily long time to the risk of infection.

The results of investigations of infected material and the speed with which they are obtained depend not only on laboratory methods but also on the manner in which the specimens are taken and the promptness with which they are transmitted to the laboratory. When sampling fluids such as urine and cerebrospinal fluid, which are normally sterile, extreme care must be taken to avoid contamination ; greater precautions are necessary than are used in the operation theatre. Aseptic surgical technique prevents contamination of wounds by pathogens but takes little account of airborne saprophytes, many of which fall into the wound during the operation without harm to the patient ; but the presence of a single contaminating organism in a culture may completely ruin the investigation. Specimens from sites such as fauces, vagina, or alimentary canal which have a normal flora may be taken with less stringent precautions but extraneous material should be excluded as far as possible.

Pathogens may not be cultivated if there is delay in sending the specimen to the laboratory. Delicate bacteria may die from lack of nutrition, lowered temperature and the action of enzymes. When saprophytes are present, either originally in the specimen, or introduced accidentally during sampling, they may survive and multiply at room temperature before cultures are made, gaining advantage in numbers over the pathogens which are subsequently outgrown.

Collection of specimens

If reliable results are to be obtained the following rules for taking specimens should be observed :

1. All specimens must be labelled with the patient's name and the date and time of sampling, and must be accompanied by a signed form which states the name and age of the patient, the ward or

9

department, the nature of the specimen and the site from which it was taken, the clinical diagnosis and duration of the illness, the examination required and the nature of any antibacterial treatment.

2. Specimens for culture must never be in contact with antiseptics or disinfectants. If the site must be cleaned, a dry sterile swab moistened with sterile saline and held in sterile forceps should be used, and the site should be dried before sampling.

3. All specimens for culture must be sent to the laboratory on the day of collection and with as little delay as possible; speed is essential in the following investigations :

(*a*) *N. meningitidis* and *N. gonorrhoeae* are very sensitive to exposure to air in the cold. Cerebrospinal fluid from patients with signs of meningitis must arrive warm and be incubated immediately. Negative results from specimens which have cooled are valueless. Swabs for the culture of *N. gonorrhoeae* should be seeded on warm blood agar at the bedside unless special medium for transport is available (page 362).

(*b*) *Eye swabs.* Lachrymal secretion contains lysozyme, an enzyme which rapidly kills bacteria, therefore swabs should be plated at the bedside.

4. *Urine.* Catheterization is no longer considered justifiable simply for urine culture, therefore most samples are clean midstream specimens inevitably contaminated with a few bacteria, among which are likely to be common urinary pathogens which will multiply in warm urine. It is therefore essential that either the specimen be cultured within an hour of voiding or that it be refrigerated forthwith to prevent bacterial growth.

5. *Faeces.* Bedpans for collecting specimens for culture should be sterile to prevent accidental contamination by bacteria from another sample of faeces. A part of the specimen containing any abnormal material such as pus or mucus is placed in a clean screwcapped bottle or waxed carton. A warm specimen freshly passed into a warmed bedpan is necessary for examination for amoebæ. Rectal swabs are taken while the patient " bears down " by passing the swab through the anal canal. They are very satisfactory for the culture of dysentery bacilli provided they arrive in the laboratory while still moist. It is helpful to moisten the swab in sterile peptone water before sampling.

6. *Sputum*, for culture, should be collected into a sterile screwcapped bottle when the patient first wakes in the morning. If there

is very little of it, it may be possible to collect a satisfactory sample by placing the patient comfortably so that his head and shoulders are lower than his chest. If he remains thus for about ten minutes sputum may drain into the trachea and then he will be able to cough it into a conveniently placed container. Specimens for microscopic examination must be taken either into a container which has never previously been used (waxed cartons are suitable) or into one specially treated to destroy any dead acid-fast bacilli still adherent to it from previous specimens. It should be clean and dry; it need not be sterile.

7. *Serous fluids :* pleural, pericardial, synovial, ascitic. Two specimens are sent, one in a sterile screw-capped bottle for culture, the other citrated for a cell count. If culture for *Myco. tuberculosis* is required, a large volume of fluid, about 200 ml. if possible, is sent in a sterile bottle containing citrate to prevent clotting.

8. *Cerebrospinal fluid.* Two samples are sent in sterile bottles, about 3 ml. each, one for culture (see rule 2) the other for cell count and chemical analysis.

9. *Blood.*

(*a*) For culture; the samples are best taken by a member of the laboratory staff (see page 37).

(*b*) For serology; although serological tests can be made on unsterile serum, samples should be taken with a dry sterilized syringe and needle and the blood should be delivered into a sterile container. This course avoids all danger of accidentally infecting the patient, and it prevents lysis of the red cells which makes the serum unsuitable for the tests; this can be often caused by haemolytic aerobic sporebearers which multiply in the blood at room temperature. Blood should not be allowed to spill on the operator's skin because of the risk of hepatitis.

10. *Tissue.* Biopsy or autopsy specimens should be sent in a *dry* sterile container. They must not be allowed to dry, therefore small samples are sent without delay; saline and water are to some extent bactericidal and should not be added.

11. Whenever possible specimens should be sent early in the day so that there is time to examine them during normal working hours. When an investigation is required which needs immediate attention notice should be given at least an hour before the specimen is taken.

12. All specimens for culture are particularly dangerous to the nurses and laboratory staff who handle them. Leaking containers

should be incinerated or autoclaved with their contents and another sample requested. If the examination is continued, not only is there danger of laboratory infection but the presence of contaminating organisms may lead to a false diagnosis and so indirectly harm the patient.

Macroscopic examination

Unsuitable specimens. When a specimen is received in the laboratory it is usually assumed that the rules listed above have been followed ; nevertheless, it is as well to look out for signs of a badly taken specimen ; the swab that smells of antiseptic, the leaking urine jar and the small watery specimen of " sputum " that is really saliva. Information about some specimens can be gained from the presence of normal flora. Clearly something is wrong if a throat swab or a sample of faeces is sterile, and the examination must be repeated. It is important that at least one original culture medium should show the normal flora. When faeces or swabs for diphtheria are seeded on highly selective media only it is impossible to judge if the specimen has been contaminated with disinfectant, because sterile cultures may be due to the efficiency of the method of selection. When large numbers of specimens from an epidemic are to be examined it is reasonable to take one or two plates containing non-selective medium, mark them in small squares and seed each specimen first on one of these squares and then on to the appropriate medium for the selection of pathogens. In this way unsatisfactory specimens will be recognized without an extravagant use of medium.

Selection of material for examination. It is possible, as a routine, to examine only a very small part of the specimen ; the chance of a positive result is increased by careful selection. Note any pus or blood in the specimen and whenever possible choose a purulent part for culture.

Microscopic examination

Examination of stained films or of a wet preparation should precede culture. The purpose of the examination is twofold ; first it is a guide to further procedures, for example, if fungi are seen in the film a culture is made in conditions favourable for their growth in addition to the routine method for the cultivation of bacteria ; second it is the most valuable indication of the proportion of different species in the specimen. Sometimes bacteria seen in the film fail to grow;

this may be either because they are dead or because they have not been cultured under favourable conditions. It must be remembered that death of bacteria frequently alters their staining properties so that Gram-positive organisms appear Gram-negative, and also that the morphology of a bacterium in a pathological fluid may undergo a profound change when cultured on artificial media.

Urine deposits, faeces, and mucous discharges are examined wet for cells, fungi, protozoa and helminths. Dark-ground or phase-contrast microscopy is needed to find spirochaetes in blood, urine, or tissue fluids.

Specimens may be concentrated before examination. This is easily done by centrifuging liquid specimens such as cerebrospinal fluid, urine, and citrated serous fluids, but the procedure is rather more complex for specimens such as faeces and sputum which must be homogenized before they can be satisfactorily concentrated (see Chapters 4 and 6).

Routine and selective methods of culture

It is impossible to lay down a strict routine for all specimens because each investigation develops individually according to the findings at each stage. Broadly, however, all specimens from sites normally sterile are seeded on blood agar and incubated in air plus 5 to 10 per cent CO_2, and anaerobically plus about 10 per cent CO_2, because the majority of medically important bacteria will grow under these conditions within 48 hours. A negative result may be as important as a positive one; therefore, when primary cultures are sterile and yet from the clinical findings and microscopy infection seems likely, another sample, as large as possible, is concentrated by centrifugation and cultured immediately in air, air plus 5 to 10 per cent CO_2 and anaerobically plus about 10 per cent CO_2. A variety of highly nutrient liquid and solid media are employed, suitable for the growth of all known human pathogens. Incubation of solid media should be continued for at least a week and liquid media are subcultured at frequent intervals for a month before they are discarded.[1]

Animal inoculation is sometimes of value, but unless the animal happens to be extremely susceptible to small numbers of the microbe, success is more likely with cultures. Bacteria inoculated into an animal have to run the gauntlet of its antibacterial defences; if

[1] A more detailed account of the method will be found in Chapter 3.

there are few bacteria it is probable that they will perish before they have a chance to establish themselves and take advantage of the animal's tissues for growth.

It is not practicable in a clinical laboratory to incubate plates longer than 48 hours as a routine ; therefore, if no colonies are visible after this time they are reported as " sterile after 48 hours' incubation " and the prolonged investigation outlined above is made only when the establishment of a negative result is of prime importance.

Primary cultures should not be discarded until the end of the investigation when the report has been sent. It is convenient at the end of each day to collect plate cultures which are not to be reincubated in a wire basket with a dated label so that they can easily be found when necessary.

Routine primary seeding on MacConkey's bile salt agar is not essential but it is often time-saving. Ability to grow on this medium is often a guide to identity and, since swarming is inhibited, it may be possible to obtain pure cultures more quickly from colonies on the MacConkey plate when swarming obscures all growth on blood agar.[1]

Selective culture greatly increases the chance of recovering known pathogens from sites with a normal flora. There are several methods of selection. The atmosphere, the composition of the medium, and the incubation temperature may all be varied to favour the growth of the species to be selected, and antibiotics may be spread on the surface of solid media or incorporated in it to inhibit the growth of contaminating organisms. An example of selection by atmosphere is the isolation of haemolytic streptococci from throat swabs by incubating the cultures anaerobically. Although haemolytic streptococci are aerobes, they grow well under anaerobic conditions and many strains produce larger colonies and more marked haemolysis when cultured thus. Moreover, the growth of many commensals, staphylococci, neisseria, and diphtheroids is inhibited so that they form very small colonies. This is an efficient method because not only are many unwanted bacteria inhibited, but the growth of the selected species may be enhanced.

Most selective media contain chemicals which inhibit the growth of contaminating organisms efficiently but also to some extent inhibit the growth of the species selected. Such media may be invaluable

[1] For prevention of swarming on blood agar see page 149.

for the recovery of pathogens from contaminated specimens but should be used solely for this purpose. If a bacterium is sought in a body fluid which is normally sterile, for example, blood or urine, it is better to use good nutrient media such as blood agar and broth, which have no inhibitory effect. Media which select by differential inhibition are tellurite blood agar for *C. diphtheriae*, Wilson and Blair's bismuth sulphite medium for *S. typhi* and desoxycholate citrate agar for the shigella and salmonella groups. The inhibiting effect of a given batch of medium can be demonstrated by comparing surface viable counts of a pure culture of the selected organism on blood agar and on the medium (see page 346). After 24 hours' incubation the count on blood agar will be higher, sometimes as much as an hundredfold higher, and although after further incubation more colonies may appear on the selective medium the count never reaches that on blood agar. Selection by these media can, therefore, never approach perfection because they are incapable of supporting the growth of very small inocula. Tellurite and bismuth sulphite cultures are always incubated for 48 hours before they are discarded as negative. Occasionally colonies appear on the second day when the overnight culture was negative. Tetrathionate broth, which selects paratyphoid bacilli and other salmonellae, differs from these because the selected bacteria can metabolize tetrathionate, and most strains grow in the medium at least as well as they grow in a good nutrient broth; it is therefore reasonable to use it for urine cultures.

In later chapters various selective media will be recommended for particular purposes; they are by no means the only ones available. It is better to become technically expert with a good medium rather than to change continually to media which are said to be superlative. It is therefore advisable to choose a medium, not necessarily the one described, and gain experience with it until there is good reason for making a change. Most highly selective media are variable and each new batch of the inhibiting substance should be tested before it is used to make media for routine work (see page 343).

Penicillin was first used to inhibit Gram-positive bacteria and neisseria in cultures of cough plates and naso-pharyngeal swabs for the isolation of haemophilus (Fleming, 1929). It can be employed alone or in combination with other antibiotics, whenever resistant bacteria, yeasts, fungi or viruses are to be isolated from a mixture containing sensitive contaminants; the amount to be added to the medium must be calculated, because many relatively resistant bacteria will not grow in very high concentrations of antibiotics.

Selection by alteration of incubation temperature is not often practicable in medical bacteriology. A raised temperature will enhance the enrichment of selenite broth for some Salmonella and is also used for selecting faecal coli from vegatative types in water testing ; a low temperature will select *Listeria monocytogenes* from commensal bacteria.

Antagonists of antibacterial substances

Sometimes an infection must be investigated during antibacterial treatment. In such cases the chance of obtaining positive cultures is reduced, but if the anti-bacterial drugs in the specimen are neutralized by antagonists, a positive result may yet be obtained. Sulphonamides act by depriving the bacteria of an essential growth factor, para-aminobenzoic acid. The effect is easily neutralized by adding 0·1 per cent of a saturated solution of para-aminobenzoic acid to broth. It may as a routine be incorporated in all media containing broth. The medium is improved by the addition of the acid and there are sufficient buffering substances in it to make readjustment of the pH unnecessary.

Beta-lactamases are enzymes produced by many bacteria which neutralize penicillins and cephalosporins, staphylococcal penicillinase was the first to be recognized. They can be spread on the surface of media or incorporated in it to neutralize the effect of these drugs on sensitive bacteria. A suitable sterile preparation can be obtained commercially.[1]

Sampling before these drugs are given is much better than relying on neutralization because the preparation, being a product of bacterial metabolism, may contain small amounts of substances inhibitory to some bacteria. Moreover, there is a risk of contamination from the preparation which cannot be filtered without very considerable loss in potency (Hamilton-Miller and Brumfitt, 1974).

No satisfactory antagonists have so far been described for the newer antibiotics.

[1] From Whatman Biochemicals Ltd., Springfield Mill, Maidstone, Kent, England.

Records

It is impossible to lay down hard and fast rules for the treatment of specimens because each one must be considered separately. It is, however, very valuable to have a rigid routine for recording results so that anyone working in the same laboratory can take over an investigation at a moment's notice and follow clearly what has already been done. Good records have the added advantage that they are available for research, and if there is an inquiry about the validity of bacterial diagnosis in a court of law or elsewhere it is possible to refer to the actual methods of identification and the results obtained. Without them it is impossible for the bacteriologist to keep in touch with the work done in his laboratory. The following method has been found satisfactory :

Each specimen as it comes into the laboratory is given a number which is stamped with a duplicate numbering machine on the front of the two-part request form where the report will be entered. The under copy serves as a record card and receives the same number; 12·5 cm. × 20 cm. is a suitable size. The name, ward, date, the nature of the specimen and the examination required are entered on the card and in the laboratory register. The request form remains in the office to await the report and the specimen with its card enters the laboratory. Cultures are made, the plates and tubes being labelled with the number of the specimen and the date, and a note is made on the back of card of the media used and the atmosphere in which the cultures were incubated. Results of the direct examination, macroscopic and microscopic, are entered on the front of the card to be included in the report. After overnight incubation the cultures are examined by the naked eye, and with a hand lens having a focal length of approximately three inches. Each different type of colony is noted, and working in order of predominance of growth, is labelled with small Roman letters *a*, *b*, *c*, etc. The degree of predominance is denoted by plus signs, $+ + + =$ profuse growth, $+ + =$ several hundred colonies, $+ =$ about a hundred colonies, $\pm = $ 10–50 colonies, and if there are fewer than ten the actual number is entered. The approximate diameter, measured in millimetres, of an isolated colony of each type is noted, followed by a sufficient description of the colony to enable anyone looking at the culture later to be able easily to distinguish it from the others present. Next, smears are made from single colonies of each type from all the plate cultures and

Gram-stained; the result of microscopic examination is entered on the card.

At this stage the record of a 24-hour aerobic blood agar (BA/O$_2$) culture from, say, a wound swab No. 493 might be :

No. 493 WOUND SWAB.
> BA/O$_2$
>> (*a*) +++ 2 mm. haemolytic yellow ; G + cocci in clusters.
>> (*b*) ++   0·2 mm. transparent non-haem. showing satellitism to *a* ; G − pleomorphic bacilli.
>> (*c*) ±   0·5 mm. haem. ; G + short chain strep.

The same procedure is followed for each plate culture, the Roman letters being continued for colonies on the anaerobic plate (BA/AnO$_2$) thus :

> BA/AnO$_2$
>> (*d*)   +++ 0·5 mm. haem. ; G + strep. (as *c*).
>> (*e*)   ++ 0·2 mm. non-haem. ; G + cocci in clusters (as *a*).
>> (*f*)   ++ 0·1 mm. transparent non-haem. ; G − bacilli (*as b*).

It is now possible to decide what further tests are necessary for final identification of these organisms. 493*a* is a staphylococcus and must be tested for coagulase production ; *b* is an haemophilus, in most cases it need not be tested further ; *c* is an haemolytic streptococcus and must be subcultured to glucose broth for Lancefield grouping. Single colonies are transferred to the appropriate media for these tests and a note is made, the results will be entered next day.

> (*a*) coagulase +
> (*c*) glucose broth Lancefield   A   C   G
>>> + − −

The final report will read as follows :—

> Direct examination : Numerous pus cells and Gram-positive cocci, a few Gram-negative bacilli.
> Cultures yield : *Strept. pyogenes*, *Staph. aureus* and an haemophilus.

The card is returned to the office, the report which has been entered on the front of it is typed on the request form and signed, and the card is filed. (See also page 28.)

Sometimes when a colony, which is present in the pool of inoculum on the primary plate, is spread out on another plate to obtain isolated colonies it proves to be mixed with another microbe not previously noted, or sometimes what was thought to be a single colony is actually a mixture. This often happens with anaerobes and leads to confusion unless a rule of notation is strictly followed. Let us take part of the record of another specimen to illustrate how this difficulty is overcome :

No. 832. PUS SWAB from post-partum uterus.

Day 1. BA/CO$_2$

      (*a*) ±   1 mm. non-haem. ; G + diphtheroid bacilli.

      BA/AnO$_2$

      (*b*) +++ 0·1 mm. non-haem. moist ; G + strep, a few
                            G − bacilli.

      (*c*) ±   0·5 mm. matt non-haem. ; G + diphtheroid as *a*.

    (*b*) subcultured to blood agar CO$_2$ and AnO$_2$

Day 2.

    (*b*) BA/CO$_2$ sterile 24 hours.

      BA/AnO$_2$

      (*ba*) +++ 0·1 mm. non-haem. moist ; G + strep.

      (*bb*) ++ minute non-haem. ; G − slender bacilli.

It occasionally happens, particularly in isolating clostridia, that a strain labelled (*bb*) is still mixed, and in that case the two strains are labelled (*bba*) and (*bbb*).

So far in the examples quoted it has been necessary to identify all the bacteria found because the specimens have come from sites normally sterile. When selective methods are used to recover known pathogens from sites with a normal flora, this method of recording cultures is modified. Examples of throat and rectal swab culture records will serve to illustrate the modification.

No. 262 THROAT SWAB—tonsillitis.

Day 1. BA/AnO$_2$

      (*a*) ++   0·5 mm. translucent β haem. ; G + strep.

          +++ 0·2 mm. α haem.

          +   0·5 mm. non-haem.

    Tellurite.[1]

      Sterile, reincubated.

    (*a*) to glucose broth for grouping.

Day 2. Tellurite.

      (*b*) ± 0·5 mm. black moist ; cocci.

                 A   C   G.

    (*a*) Lancefield group + − −

---

[1] Downie's blood tellurite agar.

Report.

Cultures yield *Strept. pyogenes*. No *C. diphtheriae* isolated.

No. 365 RECTAL SWAB—? dysentery in an adult patient.

Day 1.   Mac

$$+++ \text{ LF only}$$

DCA

$$(a) ++ \text{ 2 mm. NLF}$$
$$+ \text{ LF}$$

Selenite broth to DCA

(*a*) Slide agglutination Sh. flexneri (polyvalent) = − ve

Sh. sonnei   = + ve

To peptone water, urea slope and agar slope

Day 2.   DCA (from selenite broth)

$$++ \text{ LF}$$

(*b*) + 2 mm. NLF.   To peptone water and urea slope

Day 3.

	(*a*)	(*b*)
PW. indole	—	—
motility	—	+
Lactose	a_2	—
Glucose	A	AG
Mannite	A	—
Sucrose	—	—
Dulcite	—	—
Urea slope	—	+
Agar slope	NP	NP

(*a*) Tube agglutination *Sh. sonnei* serum $\frac{1}{20} \frac{1}{40} \frac{1}{80} \frac{1}{160} \frac{1}{320}$ c.

$$+ + + + \text{ tr. } -$$

Key		
Mac	=	MacConkey's bile salt agar
DCA	=	Desoxycholate citrate agar
PW	=	Peptone water
NLF	=	Non-lactose-fermenter
LF	=	Lactose-fermenter
A	=	Fermentation with acid formation
G	=	,, ,, gas ,,
ag	=	,, ,, small amount of acid and gas
a_2	=	small amount of acid on second day
NP	=	no pigment
c	=	saline control

Report.   Cultures yield *Sh. sonnei.*

(Note that 365*b* is *Proteus mirabilis*.   It is not reported because it is considered to be no more important than the other unidentified faecal flora.)

In these examples the description of colonies is limited to those which resemble the pathogens sought, for the rest a note of the total growth will suffice. The second sample shows the method of recording sugar fermentation. The presence of acid and gas is indicated by capital A and G, if a small amount only is seen a small Roman letter is used *a* and *g*; a number after the letter indicates the day of incubation on which the finding was noted. No number need be affixed to overnight results.

Sometimes when there are a large number of different types of bacteria to be identified the records overflow the space available on the card, they are then continued on another card which is stapled to the first and filed with it.

Reports

In order to avoid confusion there should be uniformity of reports sent from the laboratory. There is unfortunately a very wide choice of names for most of the common bacteria and, although it is a mistake to cling to the old-fashioned names solely because they are popular with physicians and surgeons who learned them many years ago, it is unnecessarily confusing if new names are continually being introduced in an effort to keep up to date. During the last twenty years advances have been made in classification which are of use to the clinician as well as to the bacteriologist ; members of the same genus behave in a similar way in the patient as in the test-tube and it is therefore important that the correct generic names should be generally adopted. If there is doubt about the correct name it can be sought in some standard text-book of bacteriology ; the names used in this text are the same as those in *Topley and Wilson's Principles of Bacteriology, Virology and Immunity* (sixth edition).

When reporting mixed cultures some indication should be given, whenever possible, of the most important pathogen. This is often done by stating the amount of growth, heavy, moderate or scanty, of each type of microbe hoping that from this the clinician will be able to gauge their importance. Unfortunately the unjustified assumption is often made that the proportion of each type of bacterium in the culture accurately reflects the proportion of each in the lesion. Many highly pathogenic bacteria which flourish in the tissues, as judged by their numbers in the stained film, and grow fairly well in pure culture may be outgrown in mixed culture ; moreover, even if the bacteria flourish both in tissues and on artificial media, since the oxygen tension in the wound is unknown, it is clearly impossible to judge which culture results reflect best the conditions in the wound when the proportion differs on the aerobic and anaerobic plate. The futility of assessing the proportion of different microbes in a wound from the culture of a single swab may be demonstrated by taking several swabs from different areas in the same wound on the same occasion, when, even if the wound is small, there is commonly a marked difference in the proportion of different types of bacteria found in the cultures.

It is clear then that there are many factors, some outside the bacteriologist's control, which influence the relative number of colonies of different types in mixed cultures. Therefore, the amount of growth of each type is not reported ; instead the bacteria are named in order of their importance as judged from the stained film, the plate culture and their reputation as pathogens. When, however, a series of cultures is made in an attempt to assess the results of treatment, for example in a streptococcal carrier, it may be useful to report the amount of growth of the pathogen, but in this case the specimens should be taken and the cultures made by the same person under the same conditions on each occasion.

Interpretation of results

The bacteriologist is often called upon to assess the significance of positive and negative results, particularly in specimens from carriers. Each case of this kind must be judged on its own merits, but the first thing to decide is how far the culture results represent conditions in the infected area. It is never possible to give a reliable opinion on results from a single specimen. If, when a series of, say, three throat swabs have been taken by an experienced sampler and have been seeded without delay, they have consistently yielded a few colonies only of the pathogen, it is reasonable to assume that there are only a few of them in the throat.

Negative results from faeces depend to a very great extent on the type of normal flora present. If they are mainly lactose-fermenters and grow poorly on the selective media used, it may be possible to isolate the pathogens when there are only 30 to 50 per ml. of faeces suspension cultured, but if there are many non-lactose fermenters including *Proteus vulgaris* and *Ps. aeruginosa* which flourish in these media it may be impossible to detect pathogens unless they are present in the order of 10^8 viable bacteria per ml. of suspension (Mohun and Stokes, 1942).

Although this difficulty does not arise in cultures of material which is normally sterile it may yet be difficult to assess the significance of positive and negative results. It is impossible to prove, in the logical sense, the absence of bacteria. By a negative result we may mean that routine aerobic and anaerobic cultures are negative after 48 hours' incubation. If the specimen has been efficiently taken from an untreated patient this excludes acute infection by the common pathogens, which may be all that is required ; but investigations lasting many weeks are necessary before it is possible to give an opinion that a lesion is not caused by bacterial infection, and this can never be certain until an alternative diagnosis is established. The significance of a positive result must be judged in relation to the patient's tissue response, and to the presence or absence of specific antibodies to the bacterium isolated. The presence of a bacterium in a normally sterile specimen does not necessarily imply infection by it ; the microbe may have entered the specimen during or after sampling. If the cultures were made in liquid medium it is impossible to judge the weight of original infection because heavy growth after overnight incubation may result from the presence of one viable contaminating organism. If, however, the same microbe is repeatedly found in pure culture in material from the same site when sampled with due care, it may be assumed to be present in the specimen before sampling and, if there are signs of inflammation, pus cells, swelling and fever, it is probable that the bacterium is the cause of the infection ; but even if all these signs are present it may still be only a secondary invader masking infection by a microbe such as *Myco. tuberculosis* or *Actinomyces bovis* which grow on artificial media with comparative difficulty. Suitably stained films may reveal the hidden pathogen, but in some cases it is only discovered later when elimination of the secondary invader has failed to cure the infection.

Computer aid for bacteriology records

Computer aid in diagnostic chemical pathology and haematology has become essential to deal with the work load in large laboratories. The need to use fully this costly equipment brings pressure to bear on the other main disciplines to join in. Most of the increased

work load in microbiology cannot be automated, nevertheless in
large departments a computer can be very helpful.

The need for preliminary reports and for time to check, for

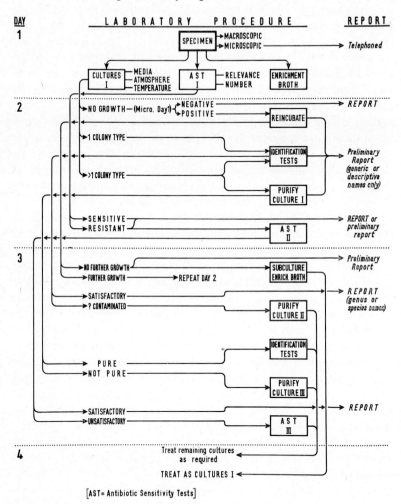

FIG. I.

example, antibiotic sensitivity tests before a permanent record is
made means that one cannot take over a system designed for chemical
pathology. The special needs of the totally different work procedure in microbiology which repeatedly involves human judgement

has to be understood by the computer staff. The report is not often numerical and flexibility is needed if it is to be as good as a manually produced report. This means including generic names and descriptive terms as well as species names. Figure 1 may help to explain the complexity of culture methods. When computer processing of microbiological data is introduced it is for the clinical microbiologist to insist that the computer serves the laboratory and not the laboratory staff the computer. For example, it is not necessary for the staff to memorize a numerical code for bacterial names, they can be converted by the computer, neither is it necessary for technicians to become punch type operators, an optical mark reader is appropriate for microbiology (Andrews and Vickers, 1974). Most laboratories in Britain have a fairly rapid turnover of staff and any system must take this into account. The aim should be for a technician to be able to learn the reporting system as quickly as he can learn to report in the conventional manner in an unfamiliar laboratory.

Additional print-outs

Once the computer is installed there is a temptation to do extra tests because the information from them can be processed. Before embarking on these one should be clear who is going to examine the print-outs and what purpose will be served by the work done.

Lists of antibiotic sensitivity tests performed on a particular species can be more easily scanned if the pattern of results is indicated by an arbitrary figure in the margin. A pattern of interest can then be picked out and any strain with an identical pattern will have the same margin number (Alexander, 1969).

The numbers of various species isolated over a period are remarkably constant and the computer should be used to highlight any significant change in the expected number so that it is noticed as quickly as possible. A significant change in less than a month is likely to be so dramatic that it will be noticed without computer aid therefore print-outs at less than monthly intervals are unlikely to be helpful.

In general, information on laboratory request forms is insufficiently reliable. The name and hospital number of the patient must be verified, other information required for programmes must also be checked before in-put.

A detailed description of a satisfactory records system is not of much value because in individual laboratories much will depend on the nature of the equipment available. Therefore problems common to all systems will be considered and possible ways of overcoming some of them will be indicated.

Advantages

The following advantages should be gained :

1. All specimens recently received can be listed according to the first letter of the patient's surname enabling staff to answer telephone enquiries easily. When a report is already on its way it will appear on the list and no search of the files is required, when it is being worked on the laboratory number and type of specimen will enable the secretary to locate the technician dealing with it so that he can comment. When no record is found on the list the specimen had not been dealt with before the list was made. No day book is required and lists can be produced twice daily.

2. The senior staff can easily check from the lists that reports have gone at the time expected and will investigate the cause of unusual delay.

3. Laboratory reports are produced automatically. Bacterial names are standardized and are clearly written, nevertheless, flexibility of reporting can be maintained.

4. Transcription between the pathologists or technicians report and the typed report is automatic and less liable to error.

5. Laboratory statistics including a weighting system for extra work done when cultures are positive can be calculated automatically.

6. Lists can be prepared of infections caused by potential hospital pathogens which helps the Infection Control Sister to investigate such infections earlier than might otherwise be possible.

7. Information required for research can be stored and produced when required. Accidental omission of such information is virtually excluded.

Essentials specific for microbiology

1. The clinical information on the request form must be constantly available to the person handling the specimen.

2. When specimens arrive late in the day the technical staff must be able to start processing them without waiting for the secretarial staff to type or punch particulars of patient and specimen. Use of no carbon required (NCR) paper for the request form which

gives a duplicate copy will answer this and the previous problem. Both copies receive the same laboratory number and work can then proceed.

3. Culture procedure depends on the number and kind of species isolated and human judgement is repeatedly necessary (see Fig. 1). Some form of work sheet or card for recording the tests made at each stage is therefore mandatory; records as described earlier in this chapter cannot be dispensed with. The report is automatically recorded but not the evidence which led to it.

Some systems attempt to include media inoculated and results of identification tests and, to a limited extent, this is reasonable but the need to have great flexibility if all culture specimens are to be included makes the system expensive; moreover, media and tests often change as improvements are discovered. Although rules are made for the basic procedures of primary culture nothing should be done to inhibit the laboratory worker from amplifying them in the light of clinical information or of the results of microscopic examination. Systems which include media to be inoculated and identification tests to be performed may succeed when known pathogens are to be excluded from samples such as faeces but will tend to be over-restrictive when an open mind is needed about bacteria isolated from specimens which are normally sterile.

4. Cumulative reports showing changes in normal values are one of the important advantages brought by computers. In microbiology, however, with the possible exception of reports in VD serology, updating is not essential. A list of microbiological results during an infective illness is a useful part of the summary of the notes when a patient is discharged and constant updating may help the clinician. In theory if the bacteriologist had before him a list of all previous results on say urine cultures or wound swabs from particular patients he might be led to ask for repeat antibiotic sensitivity tests or identification tests when these results had unexpectedly changed (they might of course also be wrong when they had not changed). In practice, however, such results are often confirmed and then one is left with an uneasy feeling that there might have been an error in the first result which can no longer be checked. There are many valid reasons for unexpected changes in bacterial flora. Regular checks by including positive quality control " specimens " is a better and more economical method of improving and gaining confidence in laboratory techniques.

5. A double check is needed, of the report before it is automatically recorded and of the print-out before it goes to the ward. Errors are more likely to be made and less easily recognized at the first stage. Moreover when, for example, a sensitivity test needs to be repeated this must be discovered as early as possible in the day. By the time the typed report returns from the computer the staff will be about to leave. Therefore a careful check should be made by the bacteriologist before the report forms leave for the computer. Final checking can be quickly done because faults in print-out are usually nonsensical and obvious.

Problems common to all disciplines

1. Provision must be made for processing specimens from patients before they have received their hospital identification number.

2. A system is needed which can be used without computer aid. For example, computer staff may not be available on Saturdays. Lack of written reports on Saturday means, in bacteriology, that no written report may be received until Tuesday from a specimen received in the laboratory the previous Thursday. Although obviously urgent reports can be telephoned on Saturday, as on any other day, reliance on telephoned reports for a day's work is unsatisfactory. In microbiology the work cannot be much reduced on Saturdays because cultures of specimens received on Friday will be ready for examination.

3. Any part of the mechanical process may break down and although every effort to restore the service will be made an alternative reporting system must be available.

Reporting without the computer

Reports can be quickly made without secretarial or computer help by using pre-printed self-adhesive stickers applied to the top copy of the duplicate no carbon required (NCR) request form, the bottom copy can be used as the working record card (see 3, page 27). Figure 2 illustrates a report with sticker applied and the key to abbreviations printed on the form. Stickers are preferred to printed forms because of the difficulty of getting ward staff to use the appropriate form and because the sticker applied can be varied according to the culture result. It is worth while to have three rolls of stickers, one combined microscopy and culture as in the figure and one each

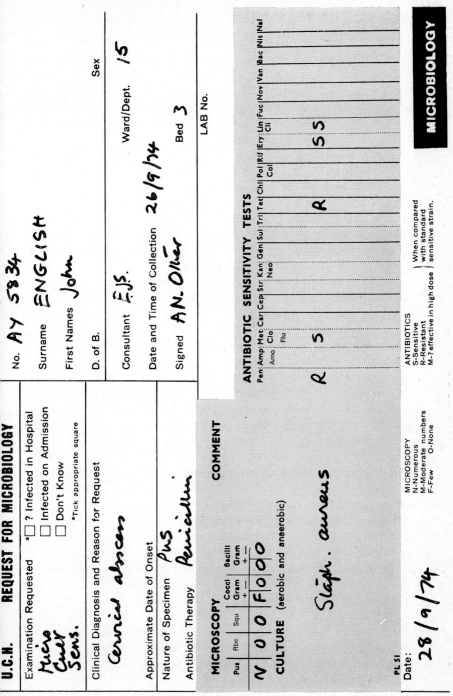

U.C.H.   **REQUEST FOR MICROBIOLOGY**

No. **AY 5834**

Surname **ENGLISH**

First Names **John**

Examination Requested   * ☐ ? Infected in Hospital
☐ Infected on Admission
☐ Don't Know
*Tick appropriate square

Micro
Cult
Sens.

D. of B.   Sex

Clinical Diagnosis and Reason for Request

Cervical abscess

Consultant **E.J.S.**   Ward/Dept. **15**

Approximate Date of Onset

Date and Time of Collection **26/9/74**

Nature of Specimen **Pus**

Antibiotic Therapy **Penicillin**

Signed **A.N. Other**   Bed **3**

LAB No.

COMMENT

MICROSCOPY

Pus	Rbc	Squ	Cocci Gram +	Cocci Gram −	Bacilli Gram +	Bacilli Gram −
N	0	0	F	0	0	0

CULTURE (aerobic and anaerobic)

Staph. aureus

ANTIBIOTIC SENSITIVITY TESTS

Peni:	Amp:	Met:	Car	Cep	Str	Kan:	Gen	Sul	Tri	Tet	Chl	Pol	Rif	Ery:	Lin:	Fuc	Nov	Van	Bac	Nit	Nal
	Amo	Clo				Neo						Col		Cli							
R		S								R				S S							
	Flu																				

MICROSCOPY
N-Numerous
M-Moderate numbers
F-Few   O-None

ANTIBIOTICS
S-Sensitive
R-Resistant
M-? effective in high dose

} When compared
with standard
sensitive strain.

PL51

Date: **28/9/74**

MICROBIOLOGY

PL40

Fig. 2.

of the microscopy and antibiotic sensitivity parts of the form separately. The separate stickers are used on the working cards as required and on the report when microscopy is reported and the culture is sterile or yields no significant growth. Rubber stamps of common negative reports also save time and improve legibility, they must of course be stamped on both top and bottom copy as they will not penetrate NCR paper.

Quality control

It is impossible for a clinical bacteriologist to know whether a high standard is maintained in his laboratory without frequent con-control tests. Failure to isolate a bacterial pathogen when clinicians expect one to be present can be put down either to a virus as the cause of infection or as evidence that the disease in non-infective; one or other of these explanations will in fact often be the case. When there is doubt about the result repeated attempts at isolation are often unrewarding because specific antibacterial treatment will interfere with the investigation.

A further problem is that important pathogens may be very rarely encountered. For example, in Britain a bacteriologist may routinely search for *C. diphtheriae* as a cause of sore throat but he is unlikely, however good his methods, to isolate more than one strain of this species in the whole of his working life and he may never encounter it. If he fails to isolate it when the need arises the patient and perhaps others may suffer.

The whole technique from the receipt of the specimen to the issuing of the report needs to be checked and this can most easily be done by examining simulated specimens. Individual laboratories can manufacture their own material or small groups of laboratories can organize a distribution between them, but the manufacture of test " specimens " is very time-consuming. Large-scale distribution has many advantages. Since there are so many possible variables in culture procedure including repeated human judgement, statistical methods are required for assessment and the larger the sample the more valuable the assessment will be. Specimen preparation on a large scale is more economical of special skills. It is worth while to use a computer to handle the results, especially of antibiotic sensitivity tests and other tests which are too numerous to evaluate quickly by any other means; moreover, no one is put in a position of authority to give the correct result because the majority result can usually be regarded as correct. Nevertheless, there is a place for evaluation by individuals generally regarded as knowledgeable in particular fields who can comment on possible reasons for errors and who can give an opinion on the rare occasions when a majority may be wrong. This is likely to happen when new techniques are introduced and the unreliability of those previously used has not been appreciated—for example, improved methods of testing staphylococci for sensitivity to penicillinase-resistant penicillins. Time for reporting should also be assessed because delay in reporting renders the best results of little practical use.

The term quality control is used here in preference to proficiency testing because it is the quality of the end product of the laboratory, the report, which is being tested. This includes the quality of materials used, the proficiency of the work done and the relevance of the tests to the clinical problem. No useful purpose is served by culturing, identifying and sensitivity-testing bacteria, however perfectly, when they are irrelevant to the infection and it is positively dangerous to test and report sensitivity to inappropriate drugs. Clinicians are overwhelmed by new drugs quite apart from antimicrobials. In this field, where the clinical bacteriologist must know about antimicrobial agents, he should advise his colleagues on their use and ensure that the report is not in any way misleading. For example, to report *Enterobacteriaceae* isolated from sites other than urine as sensitive to nitrofurantoin or nalidixic acid misleads, although the clinician should know that these drugs are inappropriate for infections outside the urinary tract. To report *Ps. aeruginosa* sensitive to streptomycin and gentamicin is misleading because streptomycin cannot be expected to succeed in treatment whereas gentamicin should do so and there is no indication of this in the report; the streptomycin result is irrelevant in most cases of Pseudomonas infection and is best omitted from reports on this organism.

In parts of the world where non-medical bacteriologists only are employed in diagnostic laboratories, proficiency in isolating, identifying and sensitivity testing is sufficient and the doctor receiving the report must judge for himself the relevance of the information on it; inevitably much irrelevant work will be done. In Britain where diagnostic laboratories are almost always in the charge of medically qualified bacteriologists the work done can be better aimed at the problem to be solved. In my view the contribution of the medical bacteriologist to the report should be included in quality control.

Anonymity of " specimen " and result

Laboratories which have not previously examined known positive specimens are usually appalled at the imperfection of their methods. Natural optimism has led them to believe that when, for example, an intestinal pathogen is present in faeces in fairly small numbers a " good " laboratory will always isolate it and that even improperly controlled antibiotic sensitivity tests are very seldom wrong. The shock of finding that this is not so is considerable. Distribution of large batches of the same material to many laboratories enables individuals to acquire a sense of proportion about their methods, by finding that all the laboratories make errors ; anonymity of the results will prevent embarrassment. The tendency to assume that when the result is wrong the simulated specimen is at fault is more easily overcome if the same strain is sent on several occasions, because failure to achieve the same result cannot be due to a fault in the specimen.

It is believed by some that anonymity of the specimen is very important and that care should be taken to ensure that those examining it should believe it is a normal specimen. This is very difficult to achieve in practice and, in my view, is unnecessary provided the time taken to report is assessed. The object of quality control is to provide material to enable laboratories to check their methods ; how they use the material should be left to them. If a laboratory is so keen to succeed that the most experienced technician always examines quality control specimens, much of the educational value of the service, which is its main purpose, is lost. Sufficient material should be provided to allow a duplicate set of cultures to be made from the same specimen. One can then be made and processed by an experienced worker and the other by one of the more junior staff who normally also handle such specimens. When the results differ it is usually possible within the laboratory to recognize and correct the error forthwith. When both results are wrong failure of human judgement is an unlikely cause. This very valuable educational exercise is one that can be undertaken even in a busy laboratory.

A scoring system will enable laboratories to appreciate individual and overall improvement but individual scores need not be taken too seriously. A laboratory reporting results achieved by all relevant staff which maintains a score in the top third of those reported in large-scale distributions is doing well.

Techniques and media vary greatly between laboratories and choice is often arbitrary from lack of evidence. Even when particular techniques prove excellent during special investigations it does not follow that they will prove equally satisfactory under field conditions. A study of the results reported from large-scale distributions of simulated specimens should provide conclusive evidence of the relative merits of the methods employed.

REFERENCES

ALEXANDER, M. K. (1969). Personal communication.

ANDREWS, H. J. and VICKERS, M. (1974). *J. clin. Path.*, **27**, 185.

FLEMING, A. (1929). *Brit. J. exp. Path.*, **10**, 266.

HAMILTON-MILLER, J. M. T. and BRUMFITT, W. (1974). *Brit. med. J.*, **iii**, 410.

MOHUN, A. F. and STOKES, E. J. (1942). Unpublished.

WILSON, G. S. and MILES, A. A. (1975). *Topley and Wilson's Principles of Bacteriology, Virology and Immunity*. Edward Arnold, London.

3

The culture of specimens normally sterile

Any bacterium recovered from a body fluid which is normally sterile should be regarded as a potential pathogen. The aim of the cultures is to recover any known human pathogen. If, however, a microbe previously considered to be harmless is found, it must not lightly be dismissed as a contaminant (see Chapter 1).

All strains isolated are identified at least as far as the genus and in most cases tests are carried further so that the species can be named. This is not unduly laborious because in many cases a single species only is found. Pleural and peritoneal fluids, however, commonly yield a mixed growth and anaerobes are often found in these sites. Mixed infection of the blood and cerebrospinal fluid does occasionally occur and it is important not to overlook it, since antibiotic therapy may fail if an unsuspected second invader is resistant to the drug chosen for treatment (Stokes, 1958).

BLOOD

When blood is cultured for the diagnosis of bacteraemia, the number of viable organisms which may be circulating in it is usually small, often about one or two per millilitre, sometimes less. Among the species which may be found are several which are delicate and grow with comparative difficulty in culture media, for example, *H. influenzae, N. meningitidis,* non-sporing anaerobes and the Brucella group. Almost all known pathogens, and some organisms usually considered harmless, have been recovered from the blood. It follows that a good technique is one in which a large volume is cultured under conditions which favour, as far as possible, the growth of all known pathogens. The blood must not be allowed to clot since the few bacteria present may form colonies in the interior of the clot making them inaccessible to examination. Finally the natural antibacterial power of the blood must be neutralized. Anticoagulants such as citrate and oxalate in the concentration necessary to prevent clotting are to some extent antibacterial; they may be

34

dispensed with by adding blood to large volumes of nutrient medium, so that clotting is prevented by dilution, or sodium polyanethol sulphonate (Liquoid) can be employed which is less inhibitory than other anticoagulants. Moreover, it neutralizes the bactericidal power of fresh blood and will withstand autoclaving ; it is used at a final concentration of 0·03–0·05 g per 100 ml. (von Haebler and Miles, 1938). See page 365.

Selective and enrichment broths are not recommended for blood culture because selective agents are likely to inhibit slightly the bacteria selected. Enrichment is not necessary because a good quality broth must be employed and it is enriched by the presence of the blood. Even when *Mycoplasma hominis*, which is demonstrably inhibited by antibody, is the cause of bacteraemia it can be isolated from Liquoid blood-broth without added enrichment (Stokes, 1955). The method to be described is aimed at isolating bacteria from the blood of patients in Britain. In countries where different bacteria are prevalent, contamination of the sample is more difficult to prevent and blood may have to travel long distances to the laboratory selective media might be advantageous.

METHOD: *Media (see Figure 3 and Chapter 10)*

Diphasic medium incubated in air

The two phases of this medium are 10 ml. Liquoid broth and a nutrient agar slope. The bottle is incubated in the upright position and it can be subcultured daily without opening by tipping the blood broth over the agar slope, being careful not to wet the cap. It is particularly useful when large numbers of cultures have to be subcultured. A disadvantage is that the broth must be added after the slope has been made and therefore contamination during preparation is possible. Moreover, colonies on the slope must be inspected through the bottle glass and when they are very small, or translucent and confluent, their presence may be missed. Colonies of rapidly growing organisms such as *Staphylococcus aureus* and *Salmonella typhi* can be easily seen and delay in reporting is minimal. Fungi, and bacteria which tend to grow in clumps may be more quickly recognized because all the blood broth is tipped over the agar and therefore the chance that a few clumps will adhere to the slope and form visible colonies is greater than when a loopful only is subcultured.

Liquoid broth (10 ml.) incubated in air plus 5–10% CO_2 (CO_2 bottle)

Incubation with added CO_2 is essential for the primary isolation of *Brucella abortus*. In our hands this medium has yielded growth of *Brucella abortus* unexpectedly from the blood of an elderly patient with fever and arthritis, from a patient already treated with tetracycline and from several febrile patients already diagnosed serologically ; the addition of liver broth when brucellosis is suspected is not necessary. The cap of the bottle is loosened and when no CO_2 incubator is available it is incubated in a jar which is evacuated to 10 cm. mercury and then 7·5 cm. of this vacuum is filled with CO_2 from a bladder which has been filled from a cylinder. Failing this a candle jar can be used. Additional CO_2 is also required for the isolation of some streptococci causing endocarditis and rarely for other CO_2 dependant species. Its presence is not known to inhibit medically important aerobes.

Thioglycollate broth 100 ml. (anaerobic bottle)

Anaerobes are present in at least 10 per cent of positive blood cultures from patients in general hospitals (Stokes, 1958). The only strict aerobes likely to be present are *Brucella species* and *Pseudomonas aeruginosa*. *Pseudomonas* will usually grow on the surface of thioglycollate medium but sometimes fails to do so (Washington, 1972). The medium will support the growth of small inocula of sporing and non-sporing anaerobes some of which are inhibited by Liquoid. Clotting is prevented by dilution. In spite of the finding by Roome and Tozer (1968) that large volumes of medium are required to inhibit the bactericidal power of fresh blood in aerobic cultures, Shanson (1974) found no advantage in increasing the volume of this medium for the recovery of small inocula of anaerobes ; it was the best of the anaerobic methods investigated by him being superior to cooked meat medium either with or without Liquoid and to other varieties of thioglycollate broth. These differ in ability to support the growth of small inocula of non-sporing anaerobes and in some the organism, having grown, quickly dies and may therefore be missed when there is an interval of several days between subcultures. The formula quoted, page 365, has been found to be satisfactory (Shanson, 1974).[1] The volume and depth of the medium are also important and should not be reduced.

[1] See also Shanson and Barnicoat (1975).

Temperature of the medium

It is probably wise to warm the media before inoculation by leaving a few bottles ready for use in the incubator. Model and Peel (1973) have cast doubt on the need for this provided the cultures can be incubated directly after venepuncture, but further evidence is needed before the practice can confidently be abandoned.

Time of sampling

The number of bacteria found in the blood varies from time to time in the same patient and it is clearly desirable to take the sample when many are present. When a patient suffers from occasional rigors and is afebrile in the intervening periods, sampling as soon as possible when he feels a rigor coming on is likely to succeed whereas cultures attempted at other times may fail. The time of maximum opportunity cannot often be predicted. Other febrile patients are not likely to have such extreme variation in bacterial population of their blood and cultures may be taken at any time. When treatment is urgent, two cultures (6 bottles) can be taken within an hour so that treatment may proceed.

Obtaining the sample

Except in the case of small children and infants (see below) a member of the laboratory staff familiar with sterile culture technique should, when possible, take the blood and inoculate appropriate media at the bedside. This has the additional advantage that a first-hand account of the patient's condition and any antimicrobial treatment given can be obtained. The operator washes and thoroughly dries his hands. It is not necessary for him to wear a surgical gown or mask. He inspects the site of venepuncture and cleans the patient's skin over it thoroughly. Isopropyl alcohol followed by acetone or ether are convenient because they remove superficial dirt and grease and leave the skin dry. Water-based antiseptics are contraindicated because those which are sufficiently benign to apply to the skin may become contaminated with bacterial growth ; the aim is to leave the skin clean and dry.

Venepuncture is performed in the usual way, 20 ml. blood being withdrawn into a sterile syringe, the needle is then removed and the blood is delivered into the three media bottles. Provided the syringe is not allowed to touch the bottle neck, flaming the rim with a spirit lamp is unnecessary, but exposure of the medium must be minimal. Meticulous aseptic technique is required. Venepuncture practised for other purposes does not need to take account of the introduction of small numbers of harmless bacteria and technique although nominally " aseptic " often falls short of that required for taking satisfactory samples for culture. When blood is regularly obtained by staff unskilled in culture technique, contamination may be more easily avoided if the culture media are capped with perforated metal caps with penetrable liners, so that the cap does not have to be removed. The cap must then be covered so that the penetrable part is sterile and does not have to be " sterilized " immediately before penetration. The needle should be changed before inoculation to avoid the risk of introducing skin contaminants with the blood.

Using strict aseptic precautions 5 ml. is inoculated into each bottle, more than 5 ml. per bottle is liable to clot. The remainder should be transferred to a plain sterile bottle and the serum from it is then separated and kept frozen for serological tests when required.

When a small volume of blood only is obtained it is better to inoculate fewer bottles than to distribute very small quantities among all three. When there is enough for two, inoculate the thioglycollate medium and the diphasic medium which should then be incubated in air plus 5–10 per cent CO_2 with the cap loosened. When there is only enough for one bottle from an adult, thioglycollate is usually the best choice. *Brucella* and *Pseudomonas spp.* are the only strict aerobes likely to be encountered in Britain.

Small children and infants are best venepunctured by paediatric staff and less than 20 ml. is usually obtained. The best technique will then depend on circumstances in individual cases but in general all the blood should be cultured, none being saved for serology. When less than 4 ml. is available it should be inoculated into the diphasic bottle only which is then incubated in air plus CO_2. The addition of CO_2 is important because some strains of *Strept. pneumoniae, N. meningitidis* and *H. influenzae* are CO_2 dependent. Volumes between 4 ml. and 10 ml. should be divided between the diphasic medium incubated as above, and thioglycollate broth. When more than this is obtained either some can be saved for serology or it can be divided between all 3 bottles.

Number of samples

The idea that the more samples taken from each patient the greater the likelihood of a positive result is erroneous. A laboratory having to subculture innumerable blood broths, most of which are sterile, cannot be expected to give proper attention to those which really matter. It has been shown that when the first two samples are negative the chance of a positive result from further samples is remote (Crowley, 1970). More than one sample should, however, be taken before treatment is given because of the risk of contamination which may ruin a single sample. Two samples (6 bottles) are sufficient from patients suspected of acute generalized infection in whom treatment is urgent. Three samples (9 bottles) are recommended in other patients. It may sometimes be worthwhile to take another three samples when these prove to be negative but the bacteriologist should be consulted before doing so.

Recognition of growth

The diphasic bottle should be inspected and tipped daily. The CO_2 and anaerobic bottles should be inspected and subcultured twice weekly. They are gently mixed without wetting the caps and a loopful is subcultured to blood agar incubated in the appropriate atmospheres, i.e. with added CO_2 from the CO_2 bottle and anaerobically plus 10 per cent CO_2 (see below) from the anaerobic bottle. Staphylococcal and coliform growth is easily seen ; streptococci often form visible colonies on the surface of sedimented red cells when these are not lysed by streptococcal haemolysins. Discrete colonies should be transferred by pipette to a sterile tube taking with them as small a volume of fluid as possible so that Gram-stain, subculture and antibiotic sensitivity tests may be more easily and reliably performed. It should not be assumed that when no growth is seen the blood broth is sterile. It is possible to obtain confluent growth when a loopful is subcultured from such bottles. Moreover, bacteria such as non-sporing anaerobes which form small translucent colonies on blood agar may never produce visible opacity in blood broth. Growth in bottles is also very difficult to recognize when the red cells have degenerated and there is no longer clear supernatant fluid above them.

The diphasic bottle should be subcultured to blood agar incubated in air for at least 2 days before it is finally discarded. This is to ensure that growth which cannot be seen through the bottle glass or which will not grow on the agar slope is not missed.

Subculturing can be done economically by spreading a loopful from each bottle over a sector of blood agar, ⅛th of an 8·5 cm. Petri dish. When growth is seen on the plate culture, a second subculture from the bottle should be spread to obtain single colonies to check that only one species is present. This will also ensure that the growth seen on the first subculture is not a plate contaminant accidentally spread.

All CO_2 subcultures should be inspected next day and reincubated for at least one more day.

The anaerobic subcultures should be incubated in a jar plus about 10 per cent CO_2 which aids the growth of non-sporing anaerobes (Watt, 1973), and when no growth has been observed in the bottle they should be inspected after 2 days' and 4 days' incubation. Delicate anaerobes are more likely to grow when they are not frequently exposed to air which is lethal to them. Inspection of plates and their return to the jar should be as rapid as possible. A hand lens should be used to inspect all plate cultures ; colonies of non-sporing anaerobes are often difficult to see. When growth on the anaerobic subculture is seen, subculture immediately to another blood agar plate from both bottle and plate and incubate in a jar without delay ; subculture also to blood agar incubated in air plus 10 per cent CO_2, to check that the organism is an anaerobe. The addition of CO_2 for this purpose will prevent CO_2 dependent organisms, which sometimes grow in liquid anaerobic culture, from being misidentified as strict anaerobes.

When apparently significant growth is first seen in the bottles appropriate procedures are indicated in Figure 4. The significance is doubtful when the Gram-films from the bottles are not identical. Growth from each must be identified and when any isolate proves to be an unlikely contaminant repeated blood cultures will usually be necessary to decide its significance.

Antibiotic sensitivity should be tested from each bottle when either *Staph. aureus* or *Staph. albus* is isolated because a major difference between them indicates a difference between strains and casts doubt on the significance of the growth. With species less commonly present on the skin this precaution can be omitted.

Figure 5 shows the procedure when growth is seen in only one bottle. This may be because growth in one bottle is visible first or

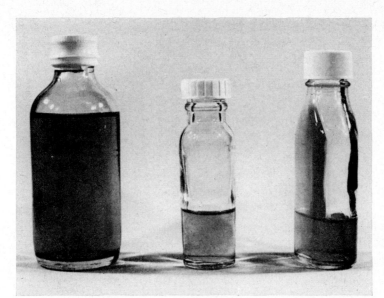

FIG. 3.

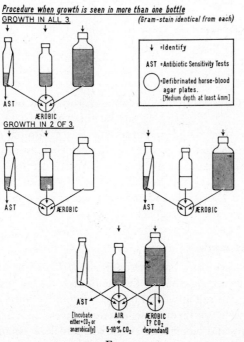

Procedure when growth is seen in more than one bottle

GROWTH IN ALL 3.   (Gram-stain identical from each)

↓ = Identify

AST = Antibiotic Sensitivity Tests

◯ = Defibrinated horse-blood agar plates.
[Medium depth at least 4mm]

AST

ÆROBIC

GROWTH IN 2 OF 3.

AST   ÆROBIC

AST   ÆROBIC

AST   AIR   ÆROBIC
[Incubate   +   [? CO₂
either + CO₂ or   5-10% CO₂   dependant]
anaerobically]

FIG. 4.

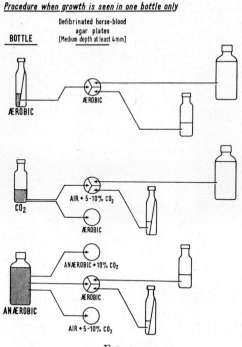

FIG. 5.

because the organism is only capable of growth in one of them and this must be tested. Growth in the diphasic or CO_2 bottles can sometimes be recognized on Gram-stain as a familiar contaminant and further investigation of other bottles can then await identification and the normal subculturing routine. Growth deep in the anaerobic bottle, however, should be investigated without delay because Gram-stains of Clostridia and anaerobic cocci can easily be mistaken for *Bacillus spp.* and micrococci.

Period of incubation

Blood cultures from patients clinically believed to have subacute bacterial endocarditis should be incubated for at least 3 weeks. When growth has not appeared by this time and 3 cultures (9 bottles) have been made in favourable circumstances, i.e. before antimicrobial therapy and when the patient was febrile, a bacterial cause of the infection is unlikely. Other possible causes should be investigated, e.g. *Coxiella burneti* infection, and non-infective conditions should be considered.

Blood from patients suspected of brucellosis should be sub-cultured 2–3 times weekly for at least 4 and preferably 8 weeks. Delayed growth is common and may appear especially late when the patient has already received treatment. Both aerobic bottles should be incubated in air plus 5–10 per cent CO_2. The additional CO_2 will not inhibit growth of aerobes in the diphasic bottle. For other clinical categories prolonged incubation may be worth while, especially when antimicrobial treatment has been given before sampling, but in many cases an alternative diagnosis will have been reached within a few days and much laboratory time can be saved when this is discovered and the cultures can be discarded.

Prevention of laboratory contamination

The CO_2 bottles are most likely to be contaminated because their caps have to be loose during incubation. The screw cap can be exchanged for a sterile plastic or metal cap which covers the neck of the bottle and makes handling easier. Cotton wool plugs are not recommended because the rim of the bottle-neck is exposed. Sub-culture is best carried out in a laminar-flow cabinet but this is not essential. The loop wire should be sufficiently long to sample the bottles without introducing the handle of the loop into them. When flaming, the bottle should be carried sideways to the flame, not scooped towards it collecting airborne contaminants *en route*.

When bottles are incubated in jars with plates they must be on top of the plates and either in racks made for the purpose or supported with clean paper to prevent them from falling over. Do not use cotton wool for this purpose, it is often contaminated with spores.

Penicillinase and other antimicrobial inhibitors

Penicillin or ampicillin are the most likely antibiotics to have been given before blood culture is requested, and penicillinase may be routinely incorporated in culture media but there are disadvantages in this. Since penicillinase is unstable it must be introduced after autoclaving which may lead to contamination, moreover, it does not survive indefinitely so that large amounts have to be introduced if a sufficient quantity is to be guaranteed to neutralize penicillins at the time of culture. Since it is a biological product and is not normally tested for its effect on the growth of small numbers of a wide variety of pathogens, one cannot be sure that it does not also have slight antibacterial properties.　For these reasons and because the culture of blood in the presence of antibiotics is to be discouraged, it is better to introduce an appropriate dose of penicillinase when it is required.　This must be done by skilled laboratory staff because of the danger of contamination,　There are a variety of betalactamases some of which inactivate penicillins but have little effect on cephalosporins and others in which the position is reversed.　A preparation which inactivates both groups of drugs can be obtained commercially and is best for routine use in blood cultures (Waterworth, 1973), see page 16.

Para-aminobenzoic acid is an inhibitor of sulphonamides, it is also a growth factor for some species and will withstand autoclaving, it should be added routinely to blood culture broth, see page 364.

Pour plates are of limited value though highly recommended by some.　They are best made by transporting molten agar to the bedside and making the plates directly.　Alternatively 3 ml., i.e. 1 ml. blood, can be withdrawn from one of the aerobic culture bottles to make a pour plate on return to the laboratory.　There is no doubt that in some cases growth may be recognized earlier and, moreover, the number of colonies seen when *Staphylococcus albus* is isolated may help to distinguish between true infection and contamination. However, laboratory contamination of these plates, especially when they are made by unskilled staff at night, makes interpretation difficult and diagnosis is usually rapid using the diphasic medium. When *Staphylococcus albus* bacteraemia is suspected several blood cultures have usually been taken and although a pour plate in addition would be helpful, this happens so rarely that it is not thought to be a sufficient reason for including it as a routine.

Glucose.　The addition of glucose is not recommended although it may increase the rapidity of growth of some species.　The bacteria having grown tend to die rapidly because of the changes in pH brought about by fermentation and it may prove impossible to subculture them successfully (Waterworth, 1972).　The small amount in the thioglycollate broth recommended is not apparently sufficient to have this effect since strains which died rapidly in other media survived well in it (Shanson, 1974).　Glucose is also used as a means of reducing Eh with the hope of culturing strict anaerobes but is totally inadequate for this purpose.

INTERPRETATION

Recognition of contamination

Almost any bacterial species capable of growth in blood broth at 37° C. has at some time caused bacteraemia. Therefore the presence in blood cultures of a species not normally considered to be pathogenic is not evidence of contamination, positive evidence is required as follows :

1. *Growth in one bottle only of a species capable of growth in all three*

Note that this can also be due to small numbers of organisms in the blood at the time of venepuncture when the few bacteria in the sample happen to be delivered into one bottle. When this is the case subsequent blood cultures can be expected to yield the same species.

2. *Isolation of different species from different bottles especially when they are common culture contaminants*

Note that bacteraemia of more than one species may be seen, particularly after heroic abdominal surgery when a mixture of aerobes and anaerobes can be isolated.

The commonest contaminants are *Staphylococcus albus*, micrococci, diphtheroids and aerobic sporebearers which enter the culture from the patients or the operator's skin or from the air of ward or laboratory. When this happens in the anaerobic bottle incubation should continue and subcultures should thereafter be made on kanamycin blood agar which will inhibit many aerobes but will allow almost all anaerobes to grow. When such a suspected contaminant grows in one of the aerobic bottles repeated subculture is not helpful but the strain should be kept so that comparison with it can be made should similar growth appear later in other bottles, however unlikely a potential pathogen it at first appears to be.

Cultures during antimicrobial treatment

If the laboratory is to give the best possible service it must be understood that to attempt culture in the presence of treatment should be done only in exceptional circumstances. Negative results for bacterial growth on such specimens are generally worthless. The idea that because the patient is not responding to treatment the pathogen will grow is erroneous. An example is the *in vitro* sensitivity of *Salmonella typhi* to tetracyclines which are useless in the treatment of typhoid fever. Blood cultures from undiagnosed typhoid patients treated with tetracycline will usually, therefore, be negative or growth will be delayed until the drug has degenerated and the bacilli have recovered sufficiently to grow.

It is usually possible when treatment has been given, to stop it completely for 36 hours and then take cultures. A quicker result can be expected by stopping the drugs and sampling when the blood is free of them. A policy of culturing all blood sent regardless of the circumstances leads to an overwhelming number of cultures none of which can be properly investigated through lack of laboratory time. The bacteriologist should keep in close touch with clinicians to advise culture early, when appropriate, and to ensure that culture during antibiotic treatment is not undertaken unnecessarily.

When fever develops during treatment of bacteraemia the possibility of superadded infection due to resistant bacteria or fungi must be borne in mind especially when antibiotics have been given intravenously. A more likely explanation is that the fever is a drug reaction. There are various ways in which this can be tested. Consultation between physician, pharmacologist and bacteriologist is likely to solve the problem better than sending additional blood cultures which will usually be sterile.

When a sensitive organism is isolated and the patient fails to respond to appropriate treatment the possibility of a second unrecognized organism in the blood should be considered. This is most likely to happen when faecal organisms are found, the second invaders being non-sporing anaerobes which grow more slowly. Continued subculture from the anaerobic bottle on kanamycin blood agar will usually reveal the additional pathogen.

Isolation of viruses from blood is worth attempting when there are strong reasons for suspecting serious acute generalized infection due to *Herpes virus, Coxiella burneti* (Q fever), Rickettsia or Chlamydia. A sequestrine sample should be sent *unfrozen*, with minimum delay. Early in acute infection the attempt is worth making, later and in chronic infection, for example, Coxiella endocarditis, the diagnosis is made serologically.

Bacteria found in febrile disease

Acute	Subacute
Staph. aureus	Strept. viridans
Strept. pyogenes	Strept. faecalis

Strept. pneumoniae   *Strept. spp.*
Enteric group   Anaerobic cocci
Cl. welchi   *Staph. albus*
Anaerobic streptococci   *Haemophilus spp.*
Bacteroides spp.   Corynebacteria
Mycoplasma hominis   Lactobacilli
N. meningitidis
N. gonorrhoeae
Listeria monocytogenes
Vibrio fetus

Note : Any of the bacteria listed under Acute, with the possible exception of *Strept. pyogenes* will sometimes cause subacute disease. Rarely other bacterial species and fungi are isolated particularly in immunosuppressed patients. Significant isolation of a species which will not grow under the conditions recommended has not been reported.

It is possible to culture organisms, particularly of the salmonella group, from samples of clotted blood taken for agglutination tests, if the patient is not available for another blood sample or when antibiotics have been given after taking the clotted sample. The clot is broken by shaking with sterile beads in a screw-capped bottle and is then distributed into two tubes, one containing broth, the other bile broth or tetrathionate broth. Growth can be expected only if the blood was heavily infected ; negative results are of little significance.

MARROW

In typhoid fever positive cultures are sometimes obtained from the marrow when blood culture is negative. In any undiagnosed general infection when the blood is sterile, the possibility of a positive result from marrow culture should be considered; if the marrow is to be examined microscopically it should also be cultured. The method is the same as for blood, but since the quantity of marrow withdrawn for culture is unlikely to exceed 2·5 ml. it is best to place the whole of it in the diphasic bottle incubated in air plus 5–10 per cent CO_2. When a very small amount only is available, nutrient broth is added to the culture so that the proportion of marrow plus broth to Liquoid broth is approximately 1 : 2 ; otherwise the concentration of Liquoid may be too high and may inhibit the growth of delicate bacteria.

CEREBROSPINAL FLUID

The fluid should arrive still warm and either be examined immediately or placed in the incubator for examination within an hour.

MACROSCOPIC EXAMINATION

Note the colour and the presence of turbidity, deposit, or clot. When there is no clot count the number of white cells per cubic millilitre of fluid using a white blood cell counting chamber. Then transfer about 2 ml. of the fluid to a sterile centrifuge tube and centrifuge at about 3,000 r.p.m. for five minutes. Transfer the supernatant fluid to a separate clean container for chemical examination. Seed the deposit on two blood agar plates and into cooked meat broth and make smears on three clean glass slides. If the specimen is clotted transfer three small pieces of clot on to clean glass slides, tease them out and allow them to dry. Cells and bacteria, if present, are most likely to be found in the clot. Add some sterile beads to the remaining fluid and clot and shake thoroughly to break it up before the fluid is centrifuged.

MICROSCOPIC EXAMINATION

Stain one film with Gram's stain for bacteria, one with Leishman's stain for a differential cell count and leave the third for auramine or Ziehl-Neelsen stain for acid-fast bacilli, if required.

It is usually possible from the total and differential cell counts to discover whether or not the patient has bacterial meningitis. When the fluid is turbid with white cells, mostly polymorphs, the causal organism is likely to be *N. meningitidis*, *Strept. pneumoniae*, or *H. influenzae*. The Gram-stain may show organisms resembling one of these. A positive result is a very valuable guide to treatment which is urgent. No effort should therefore be spared in searching for bacteria. It is particularly important in tuberculosis because positive cultures cannot be observed quickly. Several drops of deposit should be placed on one spot on the slide, each being allowed to dry before the next is added. In this way bacteria can be concentrated in a small area for examination. A preliminary report may now be sent, but on no account should the investigation be allowed to stop at this stage ; morphological appearances may be deceptive and a satisfactory diagnosis has not been made until the causal organism has been cultured and identified by appropriate tests. When the cell count shows a few lymphocytes only and when no abnormality is seen in fluid from a patient with meningism an attempt to isolate virus is worth while. The fluid should either be dealt with by the virus laboratory forthwith or be frozen at a temperature not exceeding $-60°$ C. until it can be received there. A sample of clotted blood, a throat swab in virus transport medium and a stool sample (not a rectal swab) should also be sent.

CULTURE

The deposit is seeded on two blood agar plates, incubated in air plus 5–10 per cent CO_2 and anaerobically, and into cooked meat broth. (Some strains of meningococci, pneumococci and haemophilus fail to grow on first isolation unless CO_2 is added to the atmosphere.) The remainder of the fluid which has not been centrifuged should be incubated overnight. When the primary cultures are sterile it is sometimes possible to recover the pathogen by repeating the cultures from the incubated fluid next day, or from subcultures from the cooked meat broth.

In most cases of meningitis if the fluid has been efficiently sampled before treatment, a pure growth of the pathogen will be obtained.

BACTERIA FOUND

> *N. meningitidis*
> *Strept. pneumoniae*
> *H. influenzae*
> *Myco. tuberculosis*
> *List. monocytogenes*

The species listed are most commonly found in meningitis, the last two on the list are normally associated with lymphocytes rather than polymorphs in the infected fluid. Listeria resemble diphtheroids in the deposit and must not be disregarded. Almost all medically important bacteria have been isolated from the cerebrospinal fluid, moreover, if saprophytes find their way into the subdural space either as a result of injury or when a needle is introduced, they may cause inflammation of the meninges, which are more vulnerable to attack than other tissues (Smith and Smith, 1941).

SEROUS FLUIDS

(pleural, pericardial, ascitic, synovial, bursa, hydrocele)

Macroscopic and microscopic examination are made in the manner described for cerebrospinal fluid (page 31) omitting the total white cell count. When the fluid is purulent it is treated as pus (page 63).

CULTURE

The deposit is seeded on two blood agar plates and into cooked meat broth. One plate is incubated in air plus 5–10 per cent CO_2, the other anaerobically plus about 10 per cent CO_2 (Watt, 1973).

When the plate cultures are sterile they are reincubated. The cooked meat culture is examined ; if there is evidence of growth it is well mixed and seeded on two blood agar plates for incubation as above. If there is no evidence of growth it is reincubated and subcultured on the following day.

When growth is present on the original blood agar plates the cooked meat broth may be discarded. It is used in the first place to encourage the growth of both anaerobes and aerobes when there are very few of them in the specimen. Even if large numbers of viable organisms were originally present they may have diminished very considerably by the time the fluid is cultured, particularly if it contains natural or therapeutic bactericides. The chance of re-covering pathogens is increased by inoculation of a fluid medium with the whole of the deposit from a large volume of centrifuged fluid. The significance of growth from a primary liquid culture is always difficult to assess, since one viable contaminating organism may yield abundant growth after overnight incubation. When growth is satisfactory on the original plates and all bacteria seen in the original Gram-stained smear have been accounted for, there is no need for the cooked meat broth culture which may be discarded.

Fluid from a closed serous space, if sampled with due care, usually yields a single pathogen within 48 hours ; but if anaerobes are to be excluded the plates and broth should be incubated for at least a week because non-sporing anaerobic bacilli and anaerobic streptococci often need prolonged incubation on first isolation. They are commonly found in symbiosis but either may be found alone or in the company of aerobes. Fluids containing them are usually, but not always, foul smelling and there may be signs of gas at the infection site.

PROCEDURE

Examine the plate cultures, using a hand lens, and make a note of the amount of growth and the colonial morphology. Then pick a single colony of each type, Gram-stain and inoculate into suitable medium for identification. Colonial differences may be found to be due to variants of one species but such colonies must be assumed different until proved the same. Fluids from cavities which have opened either spontaneously, e.g. bronchopleural fistula, or at operation, frequently become infected with several species; each must be identified.

BACTERIA FOUND

> *Strept. pneumoniae*
> *Strept. pyogenes*
> *Staph. aureus*
> *Haemophilus* spp.
> Non-sporing anaerobic bacilli
> Anaerobic streptococci
> *Myco. tuberculosis*
> *Actinomyces*

Secondary invaders

> *Strept. viridans*
> *Strept. faecalis*
> *Klebsiella* spp.
> *Esch. coli*
> *Ps. aeruginosa*
> *Proteus* spp.
> Other coliforms

URINE

Urine is the only specimen normally sterile which need not routinely be cultured anaerobically. This seems illogical but can be justified on the grounds that anaerobic infection of the urinary tract is rare and that anaerobes in significant numbers will not be missed provided that the Gram-stained smear of the deposit is examined in all samples showing pus cells. When bacteria are seen and the aerobic culture is sterile, the sample must be cultivated on solid medium in an anaerobic jar. Liquid culture of urine or urine deposits is unsuitable unless the sample has been obtained direct from the bladder by aspiration at operation, or through the abdominal wall. The normal human urethra is not sterile and even a carefully taken catheter specimen will usually contain a few microbes which will flourish in liquid culture and give totally misleading results.

In order to avoid the risk of infection due to catheterization, clean mid-stream specimens are sent for culture. Careful sampling and either cultivation within an hour or storage at 4° C. immediately after voiding are essential. No laboratory technique will give a reliable result on a carelessly taken specimen left for many hours at room temperature. For suitable collecting methods in antenatal clinics, from bedridden women and young children, see " Any Questions " (1966). Commercially prepared spoons or dip slides can be bought which when dipped into the freshly voided specimen will yield growth of significant numbers of urinary pathogens after incubation. These are invaluable for use by general practitioners when refrigeration of the specimen is impracticable and delay cannot be avoided, but they are fairly expensive and there is no advantage to be gained by using them in hospital (Shrestha and Richardson, 1973).

A simple chemical test which detects absence of the very small amount of glucose normally present in urine, due to its metabolism by infecting microbes, is useful in differentiating between true bacilluria and heavy contamination. Difficulty in collecting a clean sample does not interfere with the test provided it is made a few hours after voiding and the urine is kept in a cool place meanwhile. The disadvantage is that the specimen must be taken after fasting overnight and cannot be used in diabetes (Emmerson, 1972).

Since the common urinary pathogens are also likely contaminants from vagina, faeces or perineal skin, a culture method which distinguishes contamination from infection is required.

In most hospital laboratories large numbers of urine samples are sent from antenatal patients, and from patients with a variety of symptoms, to exclude occult urinary infection. It is too costly in time and materials to investigate all these specimens fully. A screening method for bacilluria is therefore employed, simple enough to be carried out by unskilled staff in clinics. All urines are screened, but in order to report as quickly as possible culture and sensitivity test results on specimens from patients with acute infection, some method is needed to distinguish them from the large number of non-infected urines also received. A number of methods have been published that recorded here is modified from Mansell and Peacock (1973). It is simple and has the advantage that cells are also enumerated without centrifugation or the use of expensive counting chambers. The method correlates well with the counting chamber method. Urines showing 3 or more pus cells per field are cultured in full, see **(b)** below.

Casts are often insufficiently numerous to be seen without centrifugation. When examination for them is requested 15 ml. urine should be centrifuged and the deposit examined in a wet preparation using an ordinary slide and coverslip.

MACROSCOPIC AND MICROSCOPIC EXAMINATION

Note colour and nature of sediment, if any. Mix the sample thoroughly and using a pipette fill the well of a well-slide which takes about 3 drops ($\equiv$ o·1 ml.). Place a coverslip[1] on the filled well and examine under a ×40 objective and ×10 eyepiece. Scan the well in its centre, choose a representative field and count the cells. Discard the slide into 5 per cent hypochlorite solution for thorough washing, rinsing and drying next day. Slides containing several wells can be obtained.[2]

In order to assess the significance of cultures Gram-stained films of centrifuged deposits will be needed for all purulent specimens and all showing positive screening cultures. These can be prepared after overnight refrigeration. Refrigerate all specimens.

CULTURE METHODS

(a) Screening Culture (Leigh and Williams, 1964)

The urine is well mixed and a standard strip of sterile blotting paper[3] 6 mm. wide with a bent " foot " (6 × 12 mm.) is dipped into it. Alternatively a straight strip marked 12 mm. from the end can be dipped and applied to the medium as far as the mark. Excess fluid is removed against the side of the jar and the foot is then touched on the surface of a well-dried cystine-lactose-electrolyte-deficient

---

[1] No. 1 22 × 22 mm. 0·14–0·16 mm. thick from Chance Bros., 29 St. James St., London, England.
[2] 76 × 25 mm. × 1·25 mm. ground edges with polished depressions 15 mm. diameter, approximately 1 mm. deep, from Baird and Tatlock, Chadwell Heath, Essex, England.
[3] Postlip Mill fibre-free fluffless from T. B. Ford Ltd., London, E.C.4, or Bacteruritest, Mast Laboratories Ltd., Liverpool, England.

medium (CLED) plate. Two strips are cultured from each sample ; six urines can be cultured in duplicate on each plate. Figure 2 shows results of a screening test when the numbers of bacteria in the fluid are known. Note that the duplicate results are similar and that growth from a specimen with significant numbers of bacteria is not likely to be ignored.

TABLE 1

Examples of screening culture results

Results		Report
Culture	Gram-stained Smear	
Sterile	Not examined	Screening culture negative.
Two or more colonial types	Squamous cells and various organisms	Specimen contaminated. Please repeat if required.
Apparently pure but scanty	Squamous cells and various organisms	Screening culture negative.
Apparently pure moderate or heavy growth	One type of organism only and few or no squamous cells	Screening culture under investigation.

After overnight incubation the majority of cultures will be sterile or show at most one or two colonies. Those which yield a heavy growth will be fully cultured from the refrigerated uncentrifuged urine unless there is evidence of contamination, i.e. several colonial types on the CLED plate or various types of organisms and squamous cells in the Gram-stained smear. CLED medium is well suited to this purpose because it prevents swarming as effectively as MacConkey agar but is less inhibitory allowing contaminants to reveal their presence. Examples of reports on screening cultures are seen in Table 1.

All screened urines are refrigerated overnight because, if the screening culture proves positive and the patient has meanwhile received antibiotic, it will be impossible to obtain a further satisfactory sample for full culture.

(b) Full culture

The refrigerated specimen is well mixed and a 5 mm. loopful is seeded on to each of 3 media ; a blood agar plate incubated either in air or anaerobically, a CLED plate and a lysed blood agar plate for antibiotic sensitivity tests (see page 212). Inoculum on the blood agar and CLED plates are spread out to obtain single colonies; half plates may be used for economy, but a whole plate is required for the sensitivity test.

Although anaerobic incubation of the blood agar culture is not essential, it has two advantages. First there is no danger of missing an anaerobic infection. (These are rare, less than 1 per cent of all positive cultures, and usually occur in patients with carcinoma or

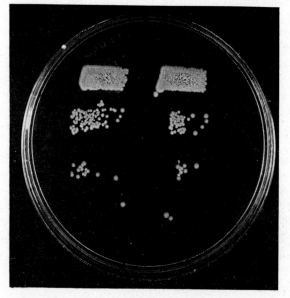

FIG. 6.

Screening test method (test made on dilutions of counted *Esch. coli* culture).

Dilution 1. Equivalent to most infected specimens, 2 million/ml.
Dilution 2. 200,000 bacilli/ml., borderline significance.
Dilution 3. 20,000 bacilli/ml.⎫
Dilution 4. 200 bacilli/ml.⎭insignificant.

papilloma of the bladder.) Second, anaerobic incubation reveals contamination from the vagina. It is not unusual to see an apparently pure growth of a coliform on the aerobic sensitivity plate when the anaerobic culture reveals several colonial types. Gram-stained films of deposits from such specimens closely resemble smears of vaginal swabs. Anaerobes and microaerophilic bacteria exist normally in large numbers in the vagina.

It has been shown by parallel plate counts that approximately 50 colonies per 5 mm. loopful of uncentrifuged urine is equivalent to 100,000 organisms per ml. of urine (Guttman and Stokes, 1963). Plate counts are a more accurate method of estimating numbers than the average of 3 standard loopfuls, but there seems little point in greater accuracy unless the sampling technique can be relied on High counts due to contamination are not uncommon in urine sent from clinics and wards, especially in warm weather.

It is very important that the significance of growth should be carefully assessed in the laboratory and that the results of antibiotic sensitivity tests should not be reported unless the bacteriologist is satisfied that the organisms tested are likely to be causing damage. *Esch. coli* and other common pathogens of the urinary tract are as likely as other harmless bacteria to contaminate specimens. Lack of judgement in the laboratory may lead to unnecessary treatment with antibiotics which are potentially harmful to the kidneys.

Usually infections are caused by a single species, but mixed infections are not unknown. When two species are constantly present in significant numbers in two or more specimens carefully taken, it is probable that both of them are infecting the urinary tract, but this is uncommon except in patients with a cystostomy when mixed infection is the rule.

PROCEDURE

Examine the colonies, record the colonial morphology and Gram-stain. Subculture colonies for identification tests.

When microbes seen in the Gram-stained smear have not appeared, and the blood agar culture was incubated aerobically, request another specimen for anaerobic culture. If in the meantime the patient has received antibiotic treatment it may yet be possible to cultivate anaerobes by reincubating the original blood agar plate anaerobically and by cultivating the refrigerated urine as described for pus.

When the specimen is purulent but yields no growth and no microbes are seen in the deposit, *Mycoplasma* may be the cause. Reincubate the blood agar plate anaerobically. Genital strains of *Mycoplasma* grow equally well thus and the medium will retain moisture which is essential for them ; minute colonies should be visible within 5 days provided the medium is of sufficiently good quality to support their growth.

Although *Mycoplasma* have been recovered from the urine of normal people, this is true of all urinary pathogens and when a profuse growth is obtained from uncentrifuged purulent urine and no other cause is found, it is reasonable to assume they are the infecting agent. Only highly nutrient defibrinated blood agar is likely to yield growth (Stokes, 1968) and special media may have to be employed (see below).

" **Abacterial pyuria.** " In this rare condition pus cells are found in the urine of a patient with symptoms of urinary infection, but routine cultures including special methods for *Myco. tuberculosis* fail to reveal a causative organism. The routine method is then extended as follows. First check that the patient is not excreting antibacterial drugs which may affect the cultures, then obtain a fresh warm specimen preferably by suprapubic aspiration. Examine the deposit under dark ground illumination or with a phase contrast microscope, and stain films of it with Gram, Ziehl-Neelsen and Giemsa stains. Examine the Giemsa film for inclusion bodies of Chlamydia. Seed the deposit on two blood agar plates on Nagler's medium [1] and in 50 per cent human serum broth (2 tubes). Incubate the Nagler plate, one blood agar plate and one serum broth culture in air plus CO_2 and the remaining cultures anaerobically plus CO_2. Examine them daily for seven days. If there is macroscopic evidence of growth in the liquid culture examine it microscopically and subculture on blood agar. If all cultures remain apparently sterile, seed the liquid cultures on to blood agar plates on the seventh day and incubate these secondary plates in air and anaerobically for seven days before discarding them. Liquid cultures are worth doing only on aspirated specimens because minor contamination of transurethral specimens, even taken by catheter, is inevitable.

BACTERIA FOUND

Esch. coli	*Strept. viridans*	Anaerobic cocci
Klebsiella	*Salmonella* species	*Bacteroides* spp.
Proteus	(*Shigella* species)	
Ps. aeruginosa	*Brucella* species	

[1] Nagler's medium, which is usually employed for the rapid identification of *Cl. welchi*, is highly nutrient especially for *Mycoplasma*. For preparation see Chapter 10.

Other coliforms
Strept. faecalis   Leptospira species
Staph. albus   Mycoplasma species
Staph. aureus   (Haemophilus species)

Typhoid and salmonella carriers. When there are large numbers of these organisms in the urine the routine method will reveal them, but if a patient is an intermittent urinary carrier the organisms may be scanty and special methods will be necessary to recover them.

The whole of the deposit from about 30 ml. of fresh clean urine is seeded into 10 ml. of one of the following media: tetrathionate broth, selenite F, or bile broth and subcultured on MacConkey's medium after overnight incubation.

Leptospira icterohaemorrhagiae. Leptospira may be isolated from the urine in the second and subsequent weeks of Weil's disease. The urine should be made alkaline, when necessary, by giving the patient potassium citrate because the spirochaetes die rapidly in acid solutions.

Method. Centrifuge about 40 ml. of fresh *warm* catheter urine and immediately inoculate about 1 ml. of deposit intraperitoneally into a guinea-pig. Make cultures in Fletcher's rabbit serum medium and examine the remainder under dark-ground illumination or by phase-contrast microscopy. The urine must be carefully handled since it may be highly infective. It is usually difficult to demonstrate the spirochaetes directly but they should be easily seen in the guinea-pig blood or peritoneal fluid, if this is examined microscopically several times from the fourth to the tenth day after inoculation. The guinea-pig usually dies about the tenth to the fourteenth day with typical post mortem findings, i.e. haemorrhages and jaundice. Samples of blood are obtained during life by heart puncture. Peritoneal fluid is most easily obtained by puncturing the abdomen with a sharp fine Pasteur pipette ; the small quantity of fluid thus obtained is easier to handle than fluid taken with a syringe and needle. Peritoneal puncture is the method of choice.

Pathogenic viruses can be isolated from urine in various generalized infections but as far as is known they do not cause primary infection of the urinary tract in Britain. Most virus diseases are diagnosed more easily by other means but in cytomegalovirus infection in children, and in immunosuppressed adults, examination of the urine is worth while. It may be possible to see inclusions in cells in the centrifuged deposit stained by Giemsa. Urine for culture should be sent in an equal volume of virus transport medium frozen at −60° C. A throat swab or nasopharyngeal washings should also be sent preserved in the same manner.

TISSUE (biopsy and autopsy material)

Cultures of tissue removed at operation or post mortem may sometimes reveal pathogens which have previously escaped isolation, either because the lesion was deep-seated, beyond the reach of usual sampling methods, or because discharges from it were heavily infected with contaminating organisms. In the latter case cultures of tissue may prove successful because, unlike pus, it can first be washed free from contaminating organisms and then ground up so that the infecting bacteria within it are released into the surrounding medium.

The specimens are treated according to the nature of the investigation. Sometimes it may only be necessary to search for a known pathogen to confirm a diagnosis already made ; for example, the culture of heart valves from a patient with endocarditis when the pathogen was isolated from the blood during life, or the culture of autopsy material from experimental animals infected with known strains. The method with uncontaminated specimens is to break up the tissue in a mechanical stomacher or to grind it in a sterile Griffith's tube[1] (see Fig. 7) or pestle and mortar, adding a little peptone water if necessary. The fluid from the grindings is seeded on solid and liquid media incubated in an atmosphere known to be suitable for the growth of the pathogen. Solid medium is used to enable identification to be made when a few organisms have been allowed to contaminate the tissue during manipulation. Liquid medium usually gives a better chance of recovery if the numbers of the pathogen in the specimen is small, but a few contaminants will ruin the result.

Sometimes tissue, usually a lymph gland, is cultured when all other investigations have failed to establish the diagnosis in a patient with prolonged pyrexia, when the infective nature of the pathological process is itself in doubt. In such a case it is necessary to provide conditions suitable for the growth of all known bacterial pathogens before a confident report can be made that there is no evidence of bacterial infection. This involves much labour and an extravagant

[1] Obtainable from Baird and Tatlock (London) Ltd.

use of medium, but as the result may be of vital importance, no effort should be spared. The following procedure is recommended :

The specimen is cut in half, one half for histological examination is placed in formol-saline, the other in a dry sterile screw-capped container or in a sterile Petri dish is sent to the laboratory for culture without delay. On arrival, a small piece is cut off and its inner surface is smeared on two clean glass slides for Leishmann and Giemsa stains to show the presence of protozoa and inclusion bodies ; the rest is ground as described above. Two smears are made for

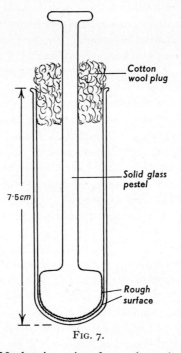

Cotton wool plug

Solid glass pestel

7·5 cm

Rough surface

FIG. 7.

Gram and Ziehl-Neelsen's stains from the grindings and a little peptone water is added if necessary, to make the volume of the remaining fluid sufficient to inoculate the following media :

Three blood agar plates and three glucose (0·5 per cent) serum (10 per cent) agar plates, for incubation in air, air plus 5–10 per cent CO_2 and anaerobically plus CO_2 :

two 4 per cent glycerol blood agar slopes and 4 per cent glycerol blood broth ; to be incubated in air plus 5–10 per cent CO_2 for the diagnosis of glanders :

two tubes of 30 per cent human serum broth, two of serum (10 per cent), glucose (0·5 per cent), broth; to be incubated, one of each, in air plus CO_2 the other anaerobically :

for aerobic incubation only, one tube of Robertson's cooked meat broth, and media for Mycobacteria, see Chapter 6. The remainder of the fluid is inoculated intramuscularly into a guinea-pig and intraperitoneally into a guinea-pig and three mice.

The plate cultures and glycerol blood agar slopes are incubated for 7 days and examined after overnight incubation and then on alternate days. The liquid cultures are incubated for 4 weeks and subcultured every fourth day on to solid medium of similar composition incubated in the same atmosphere. These secondary plates should be incubated for at least 4 days. They are seeded whether or not there is visible growth in the liquid cultures because occasionally growth, which is insufficient to cause turbidity, may be demonstrated by subculture on solid medium.

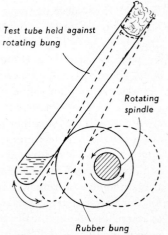

Test tube held against rotating bung

Rotating spindle

Rubber bung

FIG. 8.—Mechanical shaker.

In recording the work done on this type of specimen it is helpful to label the original liquid cultures with large Roman letters in addition to the specimen number so that subcultures can be easily distinguished. Thus the aerobic 30 per cent serum broth culture of specimen No. 635 is 635A and all subcultures from it will be so labelled.

All the animals should be inspected daily. If the mice show signs of sickness one of them is killed and an autopsy is performed with cultures of heart blood, spleen and peritoneal fluid (see page 184). The others are left to see if the infection will prove fatal. If they die autopsies are made. Should the guinea-pig which was inoculated intraperitoneally show signs of ill health, it is wise to take samples of its heart blood and peritoneal fluid for culture and microscopy, because if the animal survives there may be less chance later of recovering the pathogen. The fluids are examined by dark-ground illumination or phase-contrast microscopy and smears of them are stained by Gram and Ziehl-Neelsen's methods. If all the animals are alive and well after six weeks they are killed and examined for evidence of chronic infection; lesions found are cultured and examined histologically.

Tissue which is likely to be contaminated, i.e. all autopsy specimens collected more than four hours after death and tissue from superficial or " open " lesions, must be washed free of contaminating bacteria before they are prepared for culture. A small piece of tissue showing macroscopic signs of recent infection is placed in a tube containing about 2 ml. of peptone water and is then agitated thoroughly on a mechanical shaker. (One can be very simply made by placing a rubber bung eccentrically on the hub of any wheel which can be rotated either mechanically or by hand. Fig. 8. The tube to be shaken is held gently against the rapidly rotating bung.) The piece of tissue is then transferred with the aid of fine sterile forceps or a loop to another tube of peptone water and is shaken again. This is repeated until six tubes of peptone water have been used in washing. The piece of tissue is then ground with about 0·5 ml. of sterile peptone water. It is better to use peptone water than saline or distilled water for washing because it can be relied on not to kill delicate organisms. The grindings are then treated in the manner described for uncontaminated specimens (page 59). The technique needs experience because if a rather friable specimen receives too much washing all the bacteria may be removed ; on the other hand if the washing is insufficient the surface contaminants will remain and spoil the cultures. The risk of losing all the bacteria can be avoided by seeding the washings on blood agar incubated aerobically and anaerobically plus CO_2. Contaminating organisms will be most plentiful in cultures from the first washings and will gradually diminish in numbers in the subsequent cultures. The last tube from which growth is obtained, and cultures of the ground tissue itself, should yield an almost pure growth of the pathogen.

BACTERIA FOUND

In subacute and chronic infections
Anaerobic streptococci
Anaerobic non-sporing bacilli
Erysipelothrix
Listeria
Streptococci, staphylococci, coliforms, *Pasteurella* and other bacteria which more commonly give rise to acute infections.

Myco. tuberculosis, Actinomyces israeli and fungi often cause chronic lesions. Methods for their isolation will be considered later (Chapter 6). *Pseudomonas mallei, Myco. leprae* and *Leishmania* are rare causes. Active syphilis and yaws can be excluded if serological tests are negative.

The somewhat laborious method described is intended for use on the rare occasions when it is necessary to exclude infection by any known pathogenic microbe, other than viruses. In an individual case the clinical findings may indicate infection by a member of a particular species, in which case selective media may be used *in addition* to those already recommended. If part of the routine is omitted the report is correspondingly modified. It is not reasonable to report " cultures sterile " or " no evidence of infection " until the whole investigation has been completed. As this takes eight weeks (including cultures for *Myco. tuberculosis*) the microscopic findings are reported separately and a preliminary report on the cultures is sent at the end of the first week.

PUS

Although pus is usually sampled with a swab, it is more satisfactory whenever possible to take a liquid sample with a syringe or pipette. When the lesion is deep-seated and opens through a sinus, the whole of the inner dressing should be sent in a sterile jar. Swabs should be placed in Stuart's transport medium for the preservation of anaerobes and microbes sensitive to drying. There must be no delay in submitting swabs in transport medium because some species will grow in it at room temperature and may obliterate the true pathogens. When liquid pus is sent a swab of the pus in transport medium should also accompany it.

MACROSCOPIC AND MICROSCOPIC EXAMINATION

Note the colour, consistency, odour and any peculiarity such as the presence of granules. The discharge may be coloured by bacterial pigments, e.g. pyocanin or by the breakdown of haemoglobin.

In all cases make a smear for Gram-stain ; if the lesion is subacute or of long duration make another for Ziehl-Neelsen or auramine stain. The microscopic examination is very important and should never be omitted (see page 12). When a swab is examined the smears should be made first on sterile slides. One can then be sure that the

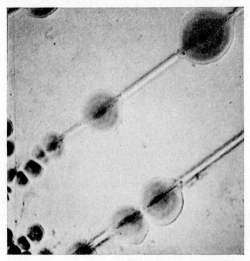

FIG. 9.

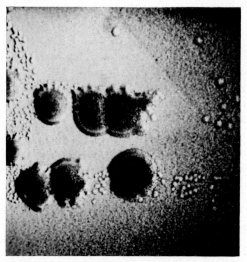

FIG. 10.

FIGS. 9 and 10.—Primary blood agar cultures of abdominal pus after 48 hours' incubation.

FIG. 9.—Aerobic culture. *Esch. coli* colonies only.

FIG. 10.—Anaerobic culture. Numerous colonies of a *Bacteroides* spp. are also seen (they were not apparent after overnight incubation).

bacteria seen have come from the lesion and cannot have been accidentally picked up from a minute contaminating colony on one of the culture plates.

CULTURE

Seed the specimen on two blood agar plates (one incubated aerobically, plus 5–10 per cent CO_2, the other anaerobically plus 10 per cent CO_2), on MacConkey agar and in cooked meat broth. Refrigerate the remainder of liquid specimens. If there is any indication from the clinical condition of the patient or, when preliminary examination of the specimen incriminates a particular species, additional cultures may be made on selective media. For example, kanamycin blood agar should be included for the selection of anaerobes in pus from abdominal and chest wounds and from all deep-seated abscesses. In some cases it may not be necessary to seed the pus on MacConkey agar which is used mainly for rapid identification of bacteria in mixed cultures, but the routine of two blood agar plates and a cooked meat broth is followed in all cases.

When antibiotic sensitivity tests are required seed the swab heavily and evenly on a separate sensitivity test blood agar plate (page 212) before immersing it in cooked meat broth.

PROCEDURE

Examine the plate cultures with a hand lens in a good light. When they are sterile after overnight incubation reincubate them. When growth is seen, describe each type of colony on all the plates, pick off single colonies of each type, Gram-stain them and compare the appearance of the bacteria in the smears with those seen on the stained film of the specimen. There may be considerable discrepancy; if so reincubate the cultures in the hope that those microbes seen in the original film which have failed to appear will grow after further incubation.

When the aerobic plate cultures reveal a pure growth of an organism resembling a common facultative anaerobe (e.g. *Staph. aureus*) and the anaerobic culture also shows only one type of colony, and further, if the Gram-stain of both is identical, then it is reasonable to assume that they are the same and to continue identification from one plate culture only, disregarding corresponding growth on the other. If however there are several types of colony on each plate culture, it is impossible at this stage to be sure that they correspond, and even after Gram-staining it is not often possible to disregard one of the plates.   They may apparently yield growth of the same species when an anaerobe is present in the following circumstances :

(*a*) When a strict aerobe on the aerobic culture is similar microscopically to the anaerobe.

(*b*) When a facultative aerobe with a colony variant is seen on the aerobic plate which fails to show the variant under anaerobic conditions.   An anaerobe may then be mistaken for the variant.

It is necessary therefore to subculture all types of colony from

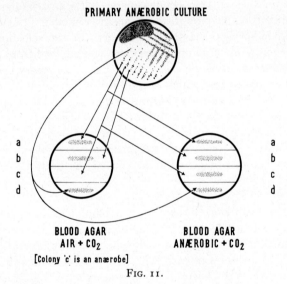

PRIMARY ANÆROBIC CULTURE

a
b
c
d

a
b
c
d

BLOOD AGAR
AIR + CO$_2$

BLOOD AGAR
ANÆROBIC + CO$_2$

[Colony 'c' is an anærobe]

FIG. 11.

a mixed anaerobic culture to an aerobic blood agar plate.   If they grow they can then be compared with the aerobes already under investigation and those which correspond can be discarded, see Fig. 11.

Further investigation for anaerobes.   Less than 1 per cent of anaerobes encountered in pus in civilian practice are *Clostridia* and if the much commoner non-sporing anaerobes, which are found, for example, in about 30 per cent of samples of abdominal pus, are not to be missed it is necessary to reincubate the plates for a further day (Stokes, 1958).   Both plates should be reincubated, since when small colonies appear on the anaerobic plate on the second day they cannot be assumed to be anaerobes unless they have failed to appear after 48 hours' incubation aerobically. Routine reincubation of primary plates need not delay the final report because when growth is present on the first day identification almost always takes a further day, and when it is absent reincubation is needed in any case. Colonies are described and Gram-stained and when necessary retested for ability to grow in air plus CO_2. Unless the original sensitivity plate was cultured anaerobically anaerobes must now be tested. Identification is desirable but our knowledge of the pathogenic role of various species of non-sporing anaerobes is at present rudimentary and full identification will not affect treatment, therefore they can be reported forthwith as *Bacteroides* spp., see also Chapter 5, or anaerobic cocci. *Clostridia* usually appear after overnight incubation and must be identified (see page 160). Sometimes colonies seen only on the anaerobic plate prove to be aerobes not found on the original aerobic culture, where they have been swamped by the growth of bacteria which are partly or completely inhibited by anaerobiosis. Haemolytic streptococci are sometimes isolated in this way when they would have been missed but for the anaerobic culture.

When growth is present on the original plates further procedure with the original cooked meat broth depends on the original Gram-film. When bacteria seen microscopically have not grown the cooked meat culture must be investigated. After 2–3 days' incubation it is subcultured on kanamycin blood agar incubated anaerobically plus CO_2. A parallel plate incubated in air plus CO_2 is also needed so that anaerobes can be recognized without further tests. Anaerobes will often be discovered in pus from deep wounds if the cooked meat broth is subcultured routinely, but isolation from the liquid culture only is hard to interpret. Inclusion of a selective anaerobic plate originally for such material is preferred; numbers of colonies grown then indicate numbers in the specimen not growth in the enrichment broth.

When numerous bacteria of apparently one kind are seen in the smear but growth is scanty, say about 50 colonies per plate, further organisms should be sought because such a sample should yield a heavy growth. When the organisms seen are Gram-negative bacilli and only a scanty growth of coliforms has appeared overnight, *Bacteroides* may be present in much larger numbers (see Fig. 10.). Similarly, anaerobic cocci may be accompanied by staphyloccoci from the skin which are of secondary importance.

When the original plates are sterile the cooked meat broth may bring to light the pathogen, aerobe or anaerobe, because it can be inoculated with a larger volume of pus and therefore the chance of a positive result is greater.   This often happens when the majority of organisms have been killed by an antibiotic.   The significance of growth from liquid cultures however is often hard to assess (see page 50).   Cooked meat cultures can be discarded when all bacteria seen have been isolated on the original plates.

INTERPRETATION OF RESULTS

Some bacteria are so often associated with pus that they are termed pyogenic, but this, like pathogenic, is not a precise term and organisms not usually called pyogenic are found in pus.   If cultures yield a pure growth of an organism and no other type was seen in the stained smear of the specimen, it is usually, and often correctly, assumed to be the only invading microbe ; but a Gram-negative coliform bacillus in the smear may be any member of the coliform group including *Proteus*, *Pseudomonas*, *Salmonella* and *Shigella*, it may be a parvobacterium or an anaerobe.   The presence of an unsuspected second infecting organism may be revealed when the first is inhibited by antibiotics, either in the patient's tissues after therapy, or during laboratory tests.

When pyogenic organisms are found, they are usually considered to be the primary cause of the infection and the rest to be secondary invaders or saprophytes.   Sometimes disease caused by secondary invaders is more severe than the original infection.   For example gas-gangrene caused by secondary invasion by *Cl. welchi* of a deep wound already infected with pyogenic cocci, and wound diphtheria in which the original infection may have been very mild.

BACTERIA FOUND

(*a*) Open wounds and burns

(i).	(ii).
Staph. aureus	*Proteus*
Staph. albus	*Ps. aeruginosa*
Micrococcus species	Enterococci
Strept. pyogenes	*Strept. haemolyticus*
Esch. coli	(other than Group A)
Diphtheroids	*Strept. viridans*
Aerobic spore bearers	Anaerobic streptococci
	Bacteroides species
	Cl. welchi

(iii). *Haemophilus*	*Cl. tetani*
Past. septica	*Cl. oedematiens*
Strept. pneumoniae	*Cl. septicum*
Actinomyces species	*Cl. histolyticum*
C. diphtheriae	

This is a general frequency list which will vary with the age of the wound, its site and history. Pus from war wounds yields a higher proportion of anaerobes because of the extensive tissue damage and delay in surgical treatment which favours their growth.

(*b*) Abscesses and other closed lesions

The skin and environmental contaminants, *Staph. albus*, *Micrococcus* Diphtheroids and spore bearers, are less frequent. All the rest listed above are found, and in addition:

Myco. tuberculosis	*Ps. mallei*
Brucella species	*Erysipelothrix* species
Salmonella species	*B. anthracis*
N. gonorrhoeae	*Mycoplasma* species
N. meningitidis	

Occasionally fungi and protozoa are encountered.

Of these, the bacteria usually called pyogenic are *Staph. aureus*, *Strept. pyogenes*, *Strept. pneumoniae*, *N. gonorrhoeae*, *N. meningitidis* and coliforms.

REFERENCES

Any Questions (1966). *Brit. med. J.*, **ii**, 1439.
CROWLEY, N. (1970). *J. clin. Path.*, **23**, 166.
EMMERSON, A. M. (1972). *J. Obst. and Gynaec.*, **79**, 828.
GUTTMANN, D. and STOKES, E. J. (1963). *Brit. med. J.*, **i**, 1384.
LEIGH, D. A. and WILLIAMS, J. D. (1964). *J. clin. Path.*, **17**, 498.
MANSELL, P. E. and PEACOCK, A. M. (1973). *J. clin. Path.*, **26**, 724.
MODEL, D. G. and PEEL, R. N. (1973). *J. clin. Path.*, **26**, 529.
ROOME, A. P. C. H. and TOZER, R. A. (1968). *J. clin. Path.*, **21**, 719.
SHANSON, D. C. (1974). *J. clin. Path.*, **27**, 273.
SHANSON, D. C. and BARNICOAT, M. (1975). *J. clin. Path.*, **28**, 407.
SHRESTHA, T. L. and RICHARDSON, N. E. G. (1973). *J. clin. Path.*, **26**, 819.
SMITH, W. and SMITH, M. M. (1941). *Lancet*, **ii**, 783.
STOKES, E. J. (1955). *Lancet*, **i**, 276.
STOKES, E. J. (1958). *Lancet*, **i**, 668.
— (1968). *Proc. roy. Soc. Med.*, **61**, 457.
von HAEBLER, T. and MILES, A. A. (1938). *J. Path. Bact.*, **46**, 245.

WASHINGTON, J. A. II (1972). *Microbiol., Appl.* **23,** 956.
WATERWORTH, P. M. (1972). *J. clin. Path.,* **25,** 227.
— (1973). *J. clin. Path.,* **26,** 596.
WATT, B. (1973). *J. med. Microbiol.,* **6,** 307.

4

Specimens from sites with a normal flora

The examination of these specimens presents the bacteriologist with a dilemma. On the one hand the clinician wants as quickly as possible, and within a few days at the latest, a detailed list of the bacteria in the specimen, on the other hand the bacteriologist knows that it is impossible in most cases to satisfy this demand in so short a time. Two courses are open to him. He may sacrifice the scientific integrity of his report by identifying perhaps one or two easily tested strains, guessing the rest by morphological appearances and his knowledge of the normal flora of the site from which the specimen was taken ; or he may limit his investigation with a view to excluding known pathogens, making no attempt to identify the rest of the bacteria found. The first method is unsatisfactory for the following reasons. The report is misleading because the clinician will assume that all the bacteria have been equally well investigated and that their identity is certain, whereas in fact only one or two of the species mentioned have been identified and the chance that the guess-work applied to the rest is correct varies with the experience of the bacteriologist but at best falls short of 100 per cent. Probably the individual patient will not suffer, unless the bacteriologist or his assistants become overworked and over-confident and carry the guess-work a stage further so that the majority of bacteria are identified by colonial and microscopic morphology, or even colonial morphology alone. This departure from the scientific method is the result of trying to solve the dilemma by appeasement. A more reasonable way of overcoming the difficulty is to explain it to the clinician and reach an agreement with him on the policy to be adopted. A convenient routine which reveals what he needs to know, quickly in most cases, without violating scientific principles, is to seek for known pathogens using appropriate selective media, and to report their presence or absence, taking no account of other bacteria which happen to grow on these media. This routine can be extended on the demand of the clinician if his patient is suffering

71

from an infection showing unusual symptoms or signs. He will then personally contact the bacteriologist, giving details of the case, and they will arrange a suitable series of tests and will keep full clinical and laboratory records which can be referred to later should similar cases be reported.

When the bacteriologist examines the plate cultures he will note any change in the normal flora as judged by the appearance of the colonies, as well as searching for those resembling known pathogens. If further tests show that although no known pathogens are present the flora is unusual for the site from which the specimen was taken, suspicion of pathogenicity may be cast on bacteria previously thought to be harmless. In this case the demand for a full investigation will come from the bacteriologist.

The amount of routine work which is done over and above the exclusion of pathogens will depend on the laboratory facilities. If the staff is hard pressed it may be just a note of the total amount of growth on the plate cultures, followed by a record of identification or exclusion of microbes resembling pathogens. If more time be available each type of colony can be described, Gram-stained and fully identified. In this case the work done will be recorded in the laboratory but will not be reported unless the bacteriologist thinks it has some special significance. It is an excellent exercise for bacteriologists in training, often brings to light interesting bacterial variants and may occasionally lead to the diagnosis of a rare infection which would otherwise have escaped notice.

It is better to use the laboratory for detailed investigation and full records of unusual cases, than to squander time and effort on half-hearted attempts to identify all the bacteria in each culture. Hospital laboratories are seldom equipped to investigate satisfactorily the normal human flora as a part of the routine work. In many cases—for example vaginal swab cultures—the colonies that appear overnight on blood agar give only the vaguest reflection of the total number and variety of species in the specimen. If it is seeded on several different enriched media, incubated in two or more different atmospheres and the cultures are examined frequently over a long period of incubation, a surprisingly large variety of bacteria can be isolated. Logically these are of as much importance as the ones of unproven pathogenicity which grow readily on blood agar, but it is clearly impossible to investigate them in every case.

To summarize : There are four types of investigtaion which can usefully be made in a hospital laboratory on specimens from sites with a normal flora.

1. Exclusion of known pathogens.
2. Exclusion of known pathogens plus partial investigation of normal flora.
3. Full bacteriological investigation of clinically rare infections.
4. Full investigation of all the bacteria found in all specimens.

The first of these must be done as thoroughly as possible in all cases because on its reliability the well-being of the patient and all those exposed to the risk of infection from him depend. In well-equipped laboratories it should be replaced by the second type. The extra work need not be reported unless it accompanies a request from the bacteriologist to combine forces with the clinician in an effort to establish the cause of a rare infection.

The fourth type is not part of the routine work but is research which may suitably be undertaken in a hospital laboratory. The results are for scientific record and perhaps publication and may have little to do with the treatment of the individual patients from whom the specimens were taken.

UPPER RESPIRATORY TRACT

PATHOGENS TO BE EXCLUDED

Group I   *Strept. pyogenes*   *C. diphtheriae*
           Strept. haemolyticus
           (other than Group A)

Group II *Strept. pneumoniae*   *N. meningitidis*
           Haemophilus species   Vincent's fusiform bacillus
           Klebsiella species   and spirilla
           Staph. aureus   *Candida albicans*

OTHER BACTERIA FOUND

           Strept. viridans   Non-haemolytic
                           streptococci
           N. pharyngis   *Strept. faecalis*
           N. catarrhalis   *Esch. coli* and
                           other coliforms
           Staph. albus   Yeasts and fungi
           Micrococcus species   Other undifferenti-
                           ated saprophytes

Diphtheroids
Aerobic and anaerobic spore bearers
Non-sporing anaerobic bacilli
Anaerobic streptococci

Throat swabs

Throat swabs from cases of acute tonsillitis are always examined for haemolytic streptococci and *C. diphtheriae*. In Britain by far the most common bacterial cause of upper respiratory infection is *Strept. pyogenes*. Although diphtheria is becoming increasingly rare the consequences of overlooking the infection may be so serious, both to the patient and to those in contact with him, that it is necessary to search for the organism in every case of febrile sore throat.

The pathogens in Group II (page 73) are relatively benign and provided the specimen has been properly taken small numbers of them can be disregarded because they are often found in the upper respiratory tracts of healthy people. When one of them forms the major part of the growth it is generally, and probably correctly, assumed to be the cause of the infection because cultures from healthy people seldom yield these organisms in profusion.

(a) Microscopic examination

In cases clinically suggestive of Vincent's angina and in all cases of tonsillitis with membrane or exudate, a film is made on a sterile glass slide, stained with strong carbol fuchsin for three minutes without heating and examined for spirochaetes and fusiform bacilli. If a satisfactory specimen has been taken from a Vincent's infection they will be seen in large numbers. A few are often found in smears from normal mouths ; there is no need therefore to search diligently for them as large numbers only are significant. Pseudo-membranes are also seen in thrush caused by *C. albicans* which is a primary infection in infants but in adults is more often seen after prolonged antibiotic therapy. Masses of yeasts will be seen in the exudate. Stained films are useless for the diagnosis of infections other than these.

The morphology of the normal throat flora is extremly varied, streptococci can almost always be found and it is impossible to differentiate haemolytic streptococci from other streptococci by their microscopic morphology. Diphtheroids are very common and often resemble *C. diphtheriae*. Moreover, films from diphtheria membrane sometimes show nothing resembling a diphtheria bacillus. The stained film may therefore be omitted when it is not necessary to exclude Vincent's angina or thrush.

(*b*) Exclusion of *Strept. pyogenes*

The swab is seeded on a blood agar plate and incubated in an anaerobic jar plus 10 per cent CO_2.[1] The anaerobic atmosphere selects haemolytic streptococci from the rest of the throat flora. All strains grow well under these conditions, many yield larger colonies than they do on plates incubated in air. Moreover, larger zones of haemolysis are seen and it is not at all uncommon to find strains of *Strept. pyogenes* (*Strept. haemolyticus* Lancefields Group A) which show very feeble haemolysis when incubated in air but clear zones on anaerobic plate cultures. Further, many throat commensals which flourish when grown in air yield minute colonies in this atmosphere. There is no doubt that in some cases anaerobic culture reveals the presence of *Strept. pyogenes* which would otherwise be missed (Cook and Jebb, 1952).

Anaerobiosis affects the other potentially pathogenic species listed as follows. Haemolytic streptococci (including varieties other than group A) and *Strept. pneumoniae* grow well, many strains better than in air. *Haemophilus* species is not greatly affected, most strains show slightly smaller colonies under anaerobic conditions but the presence of CO_2 helps their growth. This is true only of culture on a highly nutrient defibrinated horse blood agar poured at least 3 mm. deep; on some blood agar *Haemophilus* species grows poorly. The other aerobic species listed are inhibited in varying degree. The non-sporing anaerobes which are commonly found in the throat do not overcrowd the culture because they grow slowly and show, at most, very minute colonies after overnight incubation.

This method of selection is preferred for the following reasons. Streptococcal growth is often actually stimulated and haemolysis improved; no special medium is required; all satisfactory swabs from the upper respiratory tract yield facultative anaerobes so that no further test of a satisfactory specimen by seeding on non-inhibitory medium is needed.

[1] For the encouragement of *Haemophilus*, see this page.

PROCEDURE

1st Day.   Examine the plate culture with a hand lens by reflected light.   If the specimen was satisfactory there should be confluent or almost confluent growth in the pool of inoculum.   Now examine it by transmitted light and note the presence or absence of beta-haemolytic colonies.   If there are none the report can be sent " No *Strept. haemolyticus* isolated."   If beta-haemolytic colonies are present Gram-stain a single colony.   If it is a streptococcus inoculate 0·2 per cent glucose broth with an exactly similar colony, or the remains of the same one, for Lancefield grouping.   When it is impossible to pick single colonies choose the least contaminated one and subculture on blood agar.   If the colony seeded on blood agar came from a very crowded area it is best to incubate this second plate also anaerobically, but if it was in contact with only one other colony aerobic culture may yield isolated colonies.   Remember that a minute contaminating colony on an anaerobic culture may grow ten times as large in air and obscure the streptococcal growth.

Bacitracin sensitivity is a valuable aid to rapid identification provided its limitations are realized.   It has the advantage that the result is not affected by slight contamination (page 121).

(c) Exclusion of *C. diphtheriae*

The swab is seeded on blood agar and blood tellurite agar. Tellurite selects *Corynebacteria* from other throat flora very efficiently.   It inhibits all growth to some extent, but the *Corynebacteria* are much less affected than others.   Even so, the plates should be incubated for 48 hours before discarding them as negative, because some strains fail to appear earlier.   It also has the advantage of helping to differentiate typical strains of the three types of *C. diphtheriae*.[1]   This however is of secondary importance in hospital work, because information about the type of infecting bacillus makes no difference to the treatment.   The rapid identification of *C. diphtheriae* is all important ; differentiation into gravis, mitis or intermedius strains follows later because it may have epidemiological significance.   The blood agar culture is made for two reasons.

---

[1] Of the numerous tellurite media available, Downie's blood tellurite agar has been found easy to prepare and yields a good growth of most strains of *C. diphtheriae* after overnight incubation.

First, because in all cases of sore throat suspected of diphtheria haemolytic streptococci must also be sought : the two infections are often difficult to distinguish clinically and if the cultures are negative for diphtheria, *Strept. pyogenes* is the most likely pathogen. Second, because it is necessary to check that the specimen was satisfactory. Sterile cultures on tellurite medium may result either from the lack of organisms in the throat capable of growth on it, or because the swab was unsatisfactory due to contamination with antiseptic or to some other fault in sampling, or to antibiotic therapy.

When penicillin or other antibiotic has already been given, the bacteriologist should warn the clinician that diphtheria cannot be excluded. *C. diphtheriae* is sensitive to all the common antibiotics but antitoxin is also required for treatment. At one time penicillinase was spread on half the plate culture to antagonize penicillin in the specimen but proved ineffective, growth being equivalent on the two halves. There is no way in which infection by highly sensitive microbes can be excluded when swabs are sent from antibiotic-treated patients.

PROCEDURE

Incubate the plates overnight, the tellurite culture in air, the blood agar anaerobically plus CO_2.

1st Day. Examine the blood agar plate and make a note of the amount of growth. When it is very scanty the plates should be discarded and another specimen requested. If it is likely to be difficult to obtain another specimen continue the investigation, but make a note that a negative result is of no value. When haemolytic streptococci are found do not discard the tellurite culture, streptococcal carriers do sometimes become infected with diphtheria.

Make smears from isolated black colonies on the tellurite medium and stain with methylene blue. Although the bacilli do not show their classical morphology when grown on tellurite medium, they are recognizable as diphtheroids and strains of *C. diphtheriae* are usually highly pleomorphic compared with other species. As the colonies are picked, note their consistency. All three types emulsify easily; gravis colonies are friable and have the consistency of margarine, mitis colonies are butyrous. Although many strains of cocci are inhibited by Downie's medium, some will grow and produce black colonies which may be mistaken for diphtheroids. They are easily distinguished when stained with methylene blue, but care must be taken not to mistake barred diphtheria bacilli for chains of cocci. Some strains of *Proteus* also yield black colonies on this medium but they are moist and brownish black and not likely to be mistaken for diphtheroids.

Table 2 lists descriptions of the various colonies commonly encountered on Downie's blood tellurite medium.

TABLE 2

	Corynebacterium diphtheriae			*Diphtheroids*	*Cocci*
	Gravis	Mitis	Intermedius	Hofmann and Xerosis Group	Micrococci and Neisseria
18 to 24 hrs.	0·5 to 0·8 mm. grey black matt ± haemolytic	0·5 to 0·8 mm. grey black matt haemolytic	Minute grey black matt non-haemolytic	0·1 to 0·8 mm. grey matt non-haemolytic	Minute deep black shiny or rugose surface
36 to 48 hrs.	1·5 to 2 mm. grey black matt friable gently sloping edges with central eminence occasionally typical "daisyhead".	1·5 to 2 mm. grey black dull or faintly glistening surface butyrous shape of colony similar to gravis strain but no daisyhead.	0·8 to 1 mm. grey black matt colonies very flat resemble small drops of indian ink.	0·5 to 2 mm. grey black often with pale peripheral zone sometimes shiny. Pool of inoculum often dense black with separate colonies paler.	0·5 to 1·5 mm. deep black shiny or rugose surface often difficult to emulsify.

Single colonies are as dark or darker in colour than those in crowded areas.

Corynebacterium ulcerans is a rare cause of diphtheria. The growth resembles *C. diphtheriae mitis* but biochemical reactions may be mistaken for *C. diphtheriae gravis*, see Chapter 5. Haemolysis of horse blood agar is very rarely caused by harmless Corynebacteria. This is a sign worth looking for by subculture from likely colonies on the tellurite plate to blood agar incubated in air. Sucrose fermenting strains which show any degree of haemolysis should be tested for toxigenicity, see Chapter 5.

When an epidemic strain has been identified by biochemical tests and its behaviour on the tellurite medium is well known, it is reasonable when examining subsequent specimens from the same patients

and from contacts, to rely on the colonial morphology for a preliminary diagnosis and to test the strains later at leisure. But
in routine hospital practice when single cases are investigated
fermentation reactions are essential for the identification of diphtheroids. Morphology is not sufficiently reliable for diagnosis and
atypical strains are not infrequently encountered. An exception
can perhaps safely be made when the appearance of the colonies
suggests a diphtheroid and the stained film shows absolutely
regular short diphtheroid bacilli with no long forms. Stains for
metachromatic granules are not of much diagnostic value because
young virulent gravis strains may show no granules, and diphtheroids,
particularly of the *C. xerosis* group, frequently show a large number
of them.

Having established from colonial and microscopic appearance
that diphtheroids are present, pick single colonies to a Loeffler slope
and Hiss' serum water for fermentation and virulence tests (see
Chapter 5) and also for further examination of microscopic morphology. Reincubate the tellurite culture.

Preliminary reports. When at this stage a culture from an
infected throat shows a heavy growth of slightly haemolytic pleomorphic diphtheroids and no other obvious pathogen, a preliminary
positive report should be sent to initiate antitoxin therapy and
isolation of the patient if these precautions have not already been
taken. A heavy growth of diphtheroids is so common from nasal
swabs that it is wiser to await the confirmation of the fermentation
tests before giving an opinion unless the morphology is typical in
every respect. It is usually impossible to give a preliminary report
on swabs from carriers.

2nd Day. Examine the reincubated tellurite culture. If it was
negative on the first day but black colonies have now appeared,
examine them as previously described. If diphtheroids were found
on the first day they will now show greater differentiation of colonial
morphology and when experience of the medium has been gained
it is possible to guess fairly accurately whether they are *C. diphtheriae.*
Stain a film from the Loeffler slope to see if the microscopic morphology bears out the diagnosis, and seed appropriate sugar media
for final identification (page 134).

(*d*) **Other throat pathogens** (Group II, page 73).

These organisms are commonly found in moderate numbers in healthy throats so that heavy predominant growth of them only is significant.

1. *Strept. haemolyticus* other than Group A will be differentiated by bacitracin sensitivity and Lancefield grouping (pages 120–125).

2. *Strept. pneumoniae* grows well anaerobically. It can therefore easily be identified if it is present on the anaerobic blood agar culture made for the recovery of *Strept. pyogenes*. Atypical strains closely resemble *Strept. viridans*; it is therefore necessary to subculture suspected colonies to blood agar for " Optochin " sensitivity, or to 5 per cent serum broth for a bile solubility test (page 126).

3. *Haemophilus.* Haemolytic strains of *H. parainfluenzae* sometimes cause sore throats and their colonies are easily mistaken for haemolytic streptococci. Gram-stain reveals pleomorphic Gram-negative bacilli. Satellitism to other colonies may be seen on the original plate, but if there is no evidence of it a satellitism test is required for identification (see page 154). Most strains are slightly inhibited by anaerobiosis. Although the colonies are small, a heavy growth will not easily be overlooked on an anaerobic culture. Some strains of *Esch. coli* and *Bacillus spp.* are haemolytic when incubated anaerobically. They may be found in small numbers in throat swab cultures.

4. *Staph. aureus.* This is not a common throat pathogen and seldom causes simple sore throat. It sometimes causes quinsy and can be cultured from the pus when the abscess bursts. It is a facultative anaerobe and although under strictly anaerobic conditions the colony size may be reduced tenfold, a predominant growth can be easily recognized after overnight incubation if the plate is carefully examined. Strains showing haemolysis aerobically are non-haemolytic on anaerobic plate cultures.

It is clear that if throat swabs are seeded on blood agar to be incubated anaerobically and on tellurite blood agar, good results can be expected for the recovery of haemolytic streptococci, *C. diphtheriae* and *Strept. pneumoniae*, but that these two cultures alone are not ideal for the recovery of less common pathogens, some of which may need aerobic incubation. If numerous throat swabs have to be examined, routine seeding on three plates becomes very extravagant. This difficulty may be overcome by reincubating the anaerobic culture in air whenever no likely pathogen is found on the first day. It will not usually delay the result because in such cases the tellurite culture needs reincubation before a negative report can be sent. The method of choice for throat swabs depends on the prevalence of the various infections in the population investigated. In Britain *Strept. pyogenes* is much the most important cause of tonsillitis. Diphtheria is now so rare that many general hospital laboratories encounter less than one case in a decade. The proportion of sore throats which can reasonably be attributed to the other pathogens listed is in our experience of the order of one for every twenty cases of streptococcal infection. Under these circumstances it is important to use a good selective method for *Strept. pyogenes* and routine anaerobic culture is well worth while. If unlimited resources are available, aerobic blood agar cultures should also be made.

Sore throats associated with malaise and either low-grade fever or none are usually viral in origin. A throat swab in virus transport medium frozen at $-60°$ C. and a sample of clotted blood should be sent during the first 2 days of symptoms. The sooner the sample is taken the better the chance of isolation. A further sample of blood will be needed later to show a rise of antibody titre if virus is isolated.

Nasal swabs

In all cases of rhinitis with a purulent nasal discharge three cultures are made. The swab is seeded on blood agar incubated in air plus 5 per cent CO_2 for *N. meningitidis*, haemophilus and *Staph. aureus*; on blood agar incubated anaerobically for the selection of haemolytic streptococci and *Strept. pneumoniae* and on blood tellurite medium for *C. diphtheriae*.[1]

It is difficult to assess the significance of *Staph. aureus* isolated from nasal swabs. A heavy growth of this organism is common from apparently healthy noses but sometimes the formation of crusts on the nasal mucosa and nasal discharge may be caused by it.

Carriers. When nose and throat swabs are taken in a search for carriers, the full investigation can be modified. If *Strept. pyogenes* be sought, the swab is seeded on blood agar only which is

---

[1] For preliminary examination of *Strept. pyogenes*, *C. diphtheriae*, and *Strept. pneumoniae*, see under throat and sputum cultures. For *N. meningitidis*, see under genital cultures.

incubated anaerobically.   Swabs from potential carriers of diphtheria are seeded first on to a small area of blood agar to test that the specimen is satisfactory (there is no need to spread the inoculum) and then on to tellurite blood agar.

Nasopharyngeal swabs and aspirates

Strept. pneumoniae, *N. meningitidis* and members of the haemophilus group are more commonly found in the nasopharynx than in the nose or throat.   Nasopharyngeal swabs are essential for the isolation of *N. meningitidis* from suspected carriers and for the recovery of *Bord. pertussis* from cases of whooping cough.

Meningococcal carriers are swabbed through the mouth, using a curved protected swab (West's swab) to prevent accidental contamination, alternatively a laryngeal swab may be employed (see page 179).   The swab is seeded immediately on warm blood agar or chocolated blood agar and incubated in air plus 10 per cent CO_2. In children a pernasal swab may be preferred which must be seeded on selective medium for gonococci and meningococci (see page 366), to suppress the growth of nasal flora.   The organism can be found in healthy people ; carrier rates sometimes rise to 60 per cent in an apparently healthy population.   The significance of a positive finding is, therefore, doubtful and swabbing of all contacts of patients with meningitis is not recommended.   Spread within the family can be prevented by giving all members prophylactic treatment for one week.   Three days after the end of treatment members of a family with children should be checked for clearance ; nasopharyngeal swabs should be taken in the laboratory and cultured directly.

Examination of aspirated muco-pus from the nasopharynx of patients with pneumonia, especially children, who have no sputum sometimes enables a bacterial diagnosis to be achieved ; the material obtained is comparatively free from contamination.   A fine plastic catheter with a small collecting bottle attached can be obtained for the purpose.   The catheter is gently introduced through the nose and suction is applied with a syringe attached at the distal end ; the whole is sent in a plastic bag to the laboratory.   Even a very small sample in the tubing only may give a useful result, it is removed by cutting the tubing.   The specimen is treated as sputum, it is also suitable for virology and some of it can be extracted from the tubing and bottle and sent in an equal volume of virus transport medium frozen at $-60°$ C.   Smears of specimens from young children should also be sent for immunofluorescence of respiratory syncytial virus.

Bordetella pertussis

When there is no obstruction to the nasal passages, pernasal swabs are more satisfactory than nasopharyngeal swabs for the investigation of whooping cough. They are very small swabs made by wrapping cotton wool round thick loop wire instead of round the usual rigid metal or wooden applicator ; they can be passed through the nose to the nasopharynx. Plates exposed a few inches from the mouth during coughing often yield a heavy growth of the pathogen in typical whooping cough and have the advantage that they do not distress the patient. They are not, however, as satisfactory as per-nasal swabs for the investigation of early and atypical cases.[1]

Special medium is essential for the isolation of *Bord. pertussis*. The classical Bordet-Gengou medium, screened with penicillin gives good results when the commensals of the upper respiratory tract are penicillin sensitive ; but, when pertussis has to be isolated from a swab contaminated with penicillin resistant organisms, some additional selector is required. The most efficient medium for general use is Lacey's medium which contains a sulphonamide, highly selective for *Bord. pertussis*, in addition to penicillin (Lacey, 1951), but this is tedious to prepare and not commercially obtainable. The selective sulphonamide, Mand B 738 incorporated in charcoal blood agar instead of in Lacey's medium is almost as efficient and can be bought, see page 367. *Acinetobacter parapertussis* is inhibited by the selective medium. For routine investigation two cultures must therefore be made, one on selcetive medium for *Bord. pertussis* and one on blood agar screened with penicillin and streptomycin for *Acinetobacter parapertussis* (Lacey, 1954).

Plates are seeded at the bedside, without spreading the inoculum, and are incubated in air in a moist atmosphere. They are examined on the third and fourth days for typical pearly colonies (see page 155).

Laryngeal swabs

The larynx can be sampled under indirect vision using a laryngeal mirror, but the swabs are more often inserted " blind." The swab

---

[1] If more than one site is sampled in each patient, e.g. pernasal swab plus either nasopharyngeal or supralaryngeal swabs, the recovery rate can be increased by 40 per cent (Lacey, 1953).

is made on a metal applicator which is bent to an angle of 120 degrees two inches from its end, and is enclosed in a tube (18 cm. × 3 cm.) for sterilization. It is passed through the mouth over the back of the tongue, and when the long straight part of the applicator lies along the tongue the swab should be in the larynx. The patient will be stimulated to cough and material from the trachea as well as from the larynx itself will be deposited on the swab. For a more detailed account of the method, see page 179.

It is essential to swab the larynx in cases of suspected laryngeal diphtheria because tonsillar swabs may be negative when the patient has an extensive laryngeal membrane. This is a rare condition and the larynx is more often swabbed for the diagnosis of pulmonary tuberculosis in patients with no sputum (see page 179). Laryngeal swabs are also of use for the diagnosis of acute lung infections in children when a nasopharyngeal aspirate is unobtainable. The specimen is cultured on media employed for sputum.

Epiglottis

Haemophilus influenzae type b causes acute epiglottitis in young children (Addy *et al.*, 1972) and may be recovered from the blood and from the upper respiratory tract in this dangerous condition. A clinical diagnosis must be made and treatment given immediately. Results of culture will confirm the diagnosis but by the time this is known the child will either be dead or well on the way to recovery.

Mouth swabs

Ulcers of the mouth are sometimes caused by bacteria. The commonest infection is Vincent's angina, which is diagnosed by finding numerous spirochaetes and fusiform bacilli in a stained film of the exudate (see throat swab). Ulceration due to *Strept. pyogenes* and *Staph. aureus* is uncommon but may be severe and lead to cancrum oris. If swabs are taken carefully from dental abscesses they should be relatively uncontaminated with mouth flora and are treated as pus (see page 63). Thrush, caused by the yeast *Candida albicans*, is common in children. Stained films of the exudate show numerous yeasts which can be identified by culture (page 195). When the stained film reveals no yeasts or Vincent's organisms, the swab is seeded on two blood agar plates for aerobic and anaerobic incubation, and examined next day for haemolytic streptococci and *Staph. aureus* or for a heavy growth of any other organism not commonly found in profusion in the mouth. Aphthous ulcers are not, as far as is known, caused by bacterial infection ; it may occasionally be necessary to take cultures from them to exclude the infections previously mentioned. Syphilis and tuberculosis must also be borne in mind as possible causes of mouth ulcers.

Herpes simplex virus sometimes causes painful ulcers inside the mouth particularly in immunosuppressed patients (Aston *et al.*, 1972). Mouth lesions in herpangina or hand, foot and mouth disease caused by members of the Coxsackie A group of viruses may also be seen. A swab in virus transport medium frozen at $-60°$ C. should be sent with a sample of clotted blood. Isolation and a rise in antibody titre to the virus isolated will confirm the diagnosis.

Antrum wash-out fluid

Pus aspirated from the antrum is treated as described in Chapter 3 but it is worth while in addition to set up a primary " Optochin " sensitivity test for the rapid identification or exclusion of *Strept. pneumoniae*. Primary tests of sputum can be made on the same plate (method, see page 127). Common pathogens found are *Strept. pneumoniae, Strept. pyogenes, Staph. aureus* and *H. influenzae.* Non-sporing anaerobes, particularly cocci, may also be isolated in almost pure growth. Saline washings from patients with chronic antrum infections are always contaminated with the upper respiratory tract commensals and detailed investigation of them is of doubtful value. Moreover the saline may kill delicate organisms within an hour or two, so that predominance of growth of different species in cultures is more than usually unreliable as an indication of their predominance in the material sampled. Usually the main purpose of the investigation is to find which, if any, of the available antibiotics or chemotherapeutic agents is likely to be successful in treatment.

Procedure

As soon as possible after the fluid has been collected, make a stained film and seed on three blood agar plates for aerobic and anaerobic incubation and for antibiotic sensitivity tests by the paper disc method described in Chapter 7. Common pathogens encountered are *Strept. pneumoniae* and *Haemophilus* (for their identification, see Chapter 5). Often a very scanty growth of several different species is found, which do not resemble known pathogens. This may be reported forthwith " Cultures yield no significant growth."

SPUTUM

Examination for *Myco. tuberculosis* and fungous infections will be considered in Chapter 6. Other primary pathogens which must be excluded in respiratory infections are : *Strept. pneumoniae, Strept. pyogenes, Staph. aureus, H. influenzae* and very rarely Friedlander's bacillus.

General considerations [1]

In acute bacterial pneumonia the pathogen is seen in smears and recovered almost pure from all the cultures. With the exception of tubercle bacilli, all the important lung pathogens can be recovered from the upper respiratory tract of apparently healthy people. Their recovery in small numbers from sputum is not therefore sufficient evidence that they are the cause of a particular infection. Often when sputum is examined from an acutely ill patient the cultures yield a scanty mixed growth of mouth and upper respiratory tract commensals such as *Strept. viridans*, Neisseria of the catarrhalis and pharyngis groups, *Staph. albus*, Micrococci and nonhaemolytic streptococci. A few colonies of *Strept. pneumoniae, H. influenzae*, diphtheroids, yeasts, coliforms and haemolytic streptococci (often Lancefield's group C) may also be found. The proportion of each type of colony on the different cultures often varies and this cannot be accounted for by the different atmospheres in which the plates were incubated. Moreover if different portions of purulent material are cultured from the same specimen different organisms may be found. The cause of many acute respiratory infections is still largely a matter for speculation ; if they are fully investigated a proportion prove to be virus or mycoplasma pneumonias or rickettsial infections but in a large number of cases the patient recovers and no satisfactory laboratory diagnosis is made. Some of these are not true infections but are due to aspiration of secretions followed by blockage of small bronchi and collapse of the lung distally. Physiotherapy will do more to help the patient than antibiotic treatment aimed

[1] See also page 71.

at the commensals isolated. Most of the remainder are probably caused by viruses but proof must await more effective diagnostic virological methods. From the physician's point of view two questions are asked ; first, " Is the patient suffering from an acute bacterial respiratory infection ? " Second, " Is treatment with specific chemotherapeutic agents likely to be beneficial ? " Routine examination of sputum is therefore aimed at answering them.

Interpretation of sputum cultures is facilitated by liquefaction and dilution of the specimen (Dixon and Miller, 1965). Pneumococci and Haemophilus normally present in the nasopharynx which may contaminate the specimen will not be sufficiently numerous to appear in the cultures. Moreover, untreated sputum is very difficult to sample for culture and liquefaction avoids the risk of accidentally choosing a non-purulent part of the specimen. A true pathogen will be present in sufficient numbers to appear in almost pure culture from the diluted specimen.

Liquefaction must be rapid and preferably carried out at room temperature to avoid overgrowth in the specimen of contaminating staphylococci and *Enterobacteriacae*. A solution of *N*-acetyl-L-cysteine will liquefy sputum in 30 minutes at room temperature (18° C), if it is constantly shaken (Mead and Woodhams, 1964). The solution must be freshly made, for convenience bottles containing each of the three reagents are prepared ready for use. The procedure should be carried out in a safety cabinet and the bottles should be shaken upright to avoid wetting the caps.

PROCEDURE

Choose a purulent part of the specimen, tease two small pieces of it on two slides to make films for Gram and acid-fast stains. Streak the remainder across a blood agar plate for direct Optochin sensitivity. This is most easily done with a short sterile swab as used for antibiotic sensitivity tests see Chapter 7 ; at least 6 samples of sputum can be tested on one plate (page 127). This often saves time because it enables pneumococci to be differentiated from *Strept. viridans* after overnight incubation. Samples of antral pus can be tested on the same plate. An additional advantage is that this pre-liquefaction culture checks that the specimen was satisfactory and that lack of growth on the dilution culture means few organisms were present not that the sputum pot contained disinfectant.

DILUTION METHOD (Dixon and Miller (1956) modified)[1]

Materials

 N-acetyl-L-cysteine 2 g. in sterile screw-capped bottles.

 80 ml. phosphate buffer (sterile) containing 0·15 universal indicator, the solution should be pale green, pH 7.

 13 ml. sterile 4 per cent NaOH.

 10 ml. 1/4 strength Ringer solution (sterile) in a sterile screw-capped bottle.

METHOD

Add the cysteine to the buffer and mix well, the solution will turn pink. Add the sodium hydroxide and mix well. Add an equal volume of this fresh green solution to sputum and shake for 30 minutes at room temperature being careful not to wet the cap of the sputum pot.

Using an automatic pipette transfer 0·1 ml. homogenate to the Ringer solution and mix well. Culture one 5 mm. loopful of the dilution to blood agar and incubate in air plus 5–10 per cent CO_2 because some strains of *Strept. pneumoniae* and *H. influenzae* fail to grow on first isolation without it. Chocolate blood agar can be used but good quality defibrinated horse-blood agar is satisfactory and there is no need to use a special medium.

(*a*) Sputum from acute respiratory infection

Microscopy. A clinical diagnosis of acute bacterial pneumonia can often be strengthened by examining a Gram-stained film of a muco-purulent part of the sputum when the pathogen will be seen in areas full of pus cells, uncontaminated by other bacteria. Because the relation of the bacteria to pus cells is important in assessing their significance the film must be made before liquefaction.

Culture. Examine the cultures, using a hand lens, and record a description of the colonies seen : Gram-stain them. If any resemble the important lung pathogens inoculate single colonies into suitable media for identification. Colonies of *Strept. pneumoniae* can easily be mistaken for *Strept. viridans*, and occasional *Strept. viridans* have typical "draughtsman" moist colonies. Bile solubility or " Optochin " sensitivity tests, described on page 127, are always required if cultures yield a moderate or heavy growth of small alpha-haemolytic colonies which are Gram-positive cocci.

---

[1] Dilution tenfold less than described is satisfactory for sputum from hospital patients.

Significance of Culture Results : Reports. When *Strept. pneumoniae* is recovered in almost pure culture from sputum, it is considered to be the primary pathogen. Staphylococcal pneumonia may occur as part of a generalized septicaemia but *Staph. aureus* is also found as a secondary invader in acute influenza-virus pneumonia. When it is found in pure culture in sputum from acute primary pneumonia and there is no evidence of recent staphylococcal infection elsewhere, a virus infection should be suspected. In such a case the infection may fail to respond to rigorous antibiotic therapy. *H. influenzae* is well known as a secondary invader in influenzal pneumonia and is probably incapable of causing acute infection of a healthy normal lung, although it may do so post-operatively and in the elderly.

When the predominant organism has been identified the report may read thus:

" Smear : Numerous pus cells and Gram-positive-capsulated diplococci; no acid-fast bacilli seen.

" Cultures yield a mixed growth mainly *Strept. pneumoniae* which is sensitive to penicillin."

When cultures yield a scanty mixed growth, or when an organism, not one of the known pathogens, is predominant on one plate culture and absent on the others, no attempt is made to identify all the species found. The report might then read :

" Smear : Numerous pus cells, small numbers of various organisms ; no acid-fast bacilli seen.

" Cultures yield a scanty mixed growth of doubtful significance."

The result of antibiotic sensitivity tests should not be reported unless the bacteriologist is satisfied that the organisms tested are likely pathogens. It is foolish to expect that treatment with a drug which is active against commensals will necessarily benefit the patient. In treated patients such reports may be a positive danger. Resistant commensals appear within a day or two of the onset of treatment and physicians have been known to change the antibiotic in spite of a good clinical response because the laboratory reported that organisms in the sputum were resistant.

This is particularly likely when patients have received ampicillin and cultures yield antibiotic resistant *Enterobacteriacae* or *Pseudomonas*. Since their colonies are large the quantity of growth, and hence its significance, tends to be overestimated. The dilution procedure recommended avoids this pitfall because when they are contaminants they will not appear on the dilution culture. When an almost pure growth of them is seen it is important to check that a fresh specimen has been received and that large numbers cannot be due to growth in the sample after collection. Serious and even fatal infection is sometimes caused by them especially as a complication of immunosuppression. The difficulty of assessment is similar to the problem of contaminated urine samples.

(*b*) Chronic bronchitis, emphysema, asthma

It is doubtful whether routine blood agar cultures of sputum from such cases are of any value as an aid to diagnosis or treatment.

Acute exacerbations in chronically infected patients are usually caused by pneumococci or haemophilus. As these species are at present almost invariably sensitive to antibiotics, the former to penicillin and ampicillin, the latter to ampicillin, it is reasonable and quicker to make a clinical therapeutic trial, particularly as there can be no guarantee that the pathogen will be isolated from the first sample of sputum examined (May, 1952, 1953). Cultures should be made when an exacerbation is sufficiently severe for admission to hospital, as guidance may be needed if treatment fails and the chance of a useful result is greater in these circumstances.

(*c*) Bronchiectasis

In this condition the bronchi are abnormally susceptible to acute infection. During an attack the sputum is investigated as already described for acute pneumonia. Cultures may be requested of sputum from a patient with localized chronic disease, to test the antibiotic sensitivity of the organisms before operation. Numular sputum from bronchiectatic cavities yields a very varied growth of any or all of the organisms previously mentioned. An extra blood agar culture incubated anaerobically plus 10 per cent CO_2 should be made from a purulent part of the untreated specimen because non-sporing anaerobes may predominate; they are unlikely to survive the prolonged exposure to air which is inevitable during liquefaction.

(*d*) Lung abscess and bronchoscopy specimens

These specimens are treated as pus (see page 63).

VOMIT

It is sometimes necessary to examine vomit for the presence of food-poisoning organisms. The most likely pathogens in Britain are members of the salmonella and dysentery groups, enterotoxin-producing strains of *Staph. aureus* and *Cl. welchi*. Other microbes are probably capable of causing food poisoning if they are allowed to multiply in the food before it is eaten. A heavy growth of any organism, particularly any known human pathogen, obtained from most of the specimens of vomit examined in an outbreak of food poisoning should lead to its further investigation as the possible cause of the outbreak (see Chapter 9).

PROCEDURE

Seed the specimen on two blood agar plates for aerobic and anaerobic incubation and on to selective media for the salmonella and dysentery groups (see method for faeces).

RESULTS

Negative cultures do not exclude bacterial food poisoning. Staphylococci are often killed by cooking, but this makes no difference to the symptoms which are caused by the relatively heat-stable toxin. When *Staph. aureus* is isolated, it is not usually possible in a routine hospital laboratory to find whether it is an enterotoxin producing strain. Bacteriophage typing and toxicity tests may be made at a reference laboratory to establish it as the cause of the outbreak. In *Salmonella* food poisoning, faeces culture is likely to give more positive results than examination of vomit.

Botulism. In many cases the toxin only is ingested, the organisms having multiplied, formed their toxin and died before the food was eaten. *Cl. botulinum* is unlikely to grow in human tissues and if it is still viable it will be more easily isolated from the remains of food than from the patient's excreta. The vomit may be tested for toxin by filtering it through a membrane after centrifugation and inoculating it into mice, controls being protected with the antitoxins.

FAECES

Most faecal specimens examined in a routine laboratory come from patients with intestinal symptoms suspected of *Salmonella* infection or dysentery. Numerous selective media have been

described for faeces culture; all have their advantages and limitations. It is important to realize that nearly all the selective media
recommended act by inhibiting not only intestinal commensals
but, to a lesser extent, the pathogens themselves. The success of
a selective medium for any particular specimen depends on the
absence in the faecal flora of harmless organisms which grow well
in it. *Proteus, Ps. aeruginosa* and other non-lactose-fermenters are
commonly found in loose stools from any cause and may outgrow
the pathogen. It can be shown experimentally that a method which
will give positive results from inoculated samples of faeces when
there are only 50 to 100 viable *Salm. paratyphi B* per ml. of faeces
suspension, at other times will fail unless there are at least 10^8 per
ml. (Mohun and Stokes, 1942). Negative results are often due to
overgrowth of the pathogens by *Proteus mirabilis.* No selective
medium is known which inhibits *Proteus* and *Ps. aeruginosa* and
can yet be guaranteed to yield a good growth of the common intestinal
pathogens after overnight incubation.[1] Specimens from patients in
the acute stage of infection are so heavily populated with the pathogens that they can easily be found, but the variable nature of the
faecal flora accounts for the difficulty of recovering them from some
specimens late in the infection and from carriers.

MACROSCOPIC AND MICROSCOPIC EXAMINATION

Note colour, consistency and any abnormality such as pus, blood
and excess mucus.

Select, if possible, part of the stool which contains pus and mucus,
put a small loopful of it in a drop of saline on a glass slide, mix well
and cover it with a cover slip. Examine under 16 mm. and 4 mm.
objective. Note pus cells, parasites, cysts and ova (see page 97).

CULTURE

When the stool is liquid it can be seeded without preparation.
When it is solid take a piece about the size of a small pea on the end
of a swab or stout wire loop and make a thick emulsion in peptone
water. Seed the specimen or emulsion on MacConkey agar and
desoxycholate citrate agar and inoculate about 1 ml. of it into 2
bottles of selenite broth;[2] incubate the plates and one broth culture

---

[1] Novobiocin improves tetrathionate broth in this respect (Jeffries, 1959).
[2] Faeces from a suspected typhoid carrier should, in addition, be cultivated on Wilson and Blair's bismuth sulphite medium.

at 37° C and the second broth culture at 42° C. which increases its enrichment for Salmonella (Harvey and Price, 1968).

1st Day. (*a*) When lactose-fermenters only are found on the MacConkey agar, make a note of the growth. (It should be very heavy if the specimen is satisfactory, unless the patient has received antibiotics by mouth). When the sample comes from a patient 3 years old or more, discard the plate. Cultures from infants and young children must be examined for the serological types of *Esch. coli* known to cause infantile enteritis (see page 138). This may also be required occasionally in outbreaks of diarrhoea in adults since particular types of *Esch. coli* foreign to the country of origin of the patient may cause travellers' diarrhoea (Rowe *et al.*, 1970). Typing by a reference laboratory will be required when an apparently similar strain is isolated in abundance from stools of the sufferers and no other cause is found. The desoxycholate citrate culture usually shows much less growth of lactose-fermenters and may be sterile. If no non-lactose-fermenting colonies are found reincubate it. Subculture a large loopful of each of the selenite broth cultures on to desoxycholate citrate agar[1] incubated at 37° C.

(*b*) When non-lactose-fermenting colonies are found on the plate cultures pick one of each kind from each plate into peptone water and on to a urea slope and incubate. Gram-stain the remains of the colony. It is essential that *single* colonies should be picked from the plates. If there is the slightest chance that one of them may be contaminated, because it lies in a crowded area of the plate, seed first on MacConkey agar and pick a single colony next day. It is false economy in time and materials to risk contaminating a culture which is to be used for fermentation tests. Moreover the contamination may not be recognized and may lead to delay in identification or a wrong diagnosis.

Non-lactose-fermenting colonies are easily overlooked, particularly on desoxycholate citrate agar, because they are translucent and the same colour as the medium. *Sh. sonnei* may be mistaken for a poorly fermenting strain of *Esch. coli* because even after overnight culture the colonies sometimes develop a pale pink colour. If the MacConkey agar plates have been poured rather deeper than usual or are heavily coloured with neutral red, non-lactose-fermenting colonies may appear pink because the colour is transmitted through

[1] XLD agar may prove to be superior as it differentiates non-lactose-fermenting colonies better (Taylor, 1965).

them from the medium, translucency is a better guide than colour. All translucent colonies should be regarded as non-lactose-fermenters until proved otherwise.

After two to four hours incubation the urea slope may appear pink. This is due to the growth of a urease producer and is evidence that the colony seeded on it is not a member of the Salmonella or Shigella groups; the peptone water culture may therefore be discarded. When there is no evidence of urease production, make a hanging drop preparation from each and examine it under the 4 mm. objective for motility. If the bacillus is actively motile it is not a member of the dysentery group. Inoculate each peptone water culture into a series of media for biochemical tests.

When larger numbers of stools and rectal swabs are examined a short range of tests including sucrose and Kligler's composite medium saves time and materials (Kligler, 1918 ; Bailey and Lacey, 1927). In laboratories where most Enterobacteriacae come from urine or wound cultures and therefore need full identification it may be economical to put non-lactose fermenters which fail to split urea through the full range or tests rather than to prepare an extra composite medium.

2nd Day. Examine the subcultures from the selenite broths and proceed with them in the manner described for the original plate cultures. If the primary desoxycholate citrate agar was reincubated, examine it again for non-lactose-fermenting colonies. If it is still negative it can be discarded. When small colonies are seen the possibility of *Yersinia* infection should be borne in mind, see Chapter 5.

Examine the tubes inoculated from the peptone water cultures of non-lactose-fermenters. The media inoculated for biochemical tests are chosen for their usefulness in excluding as rapidly as possible non-lactose-fermenters which are not members of the *Salmonella* or *Shigella* groups. No attempt is made to identify these organisms and they are not mentioned in the report. The following is a list of media for first-line tests on non-lactose-fermenters, with notes on the value of each of them.

(*a*) *Peptone water* : Examined after 2 to 4 hours' growth for motility. If it is actively motile *Shigella* is excluded. After overnight incubation test for indole. If it is present the organism is not a *Salmonella* (see Chapter 5).

(b) *Fermentation of carbohydrates* (see also *Note*, page 148).

> *Glucose and mannite :* All members of the *Salmonella* group ferment these sugars with the formation of acid and gas except *Salm. typhi* which ferments them without gas production. Many members of the dysentery group ferment them both forming acid but no gas.

Sucrose : Not fermented by *Salmonella*. Salicin can also be employed for the exclusion of *Salmonella* and *Shigella* which do not ferment it.

Lactose : Confirms that the bacterium under investigation is a non-lactose-fermenter. Note that *Sh. sonnei* ferments lactose late.

Dulcite : Fermented by almost all *Salmonella* except *Salm. typhi* which may not ferment it. Not fermented by *Shigella*.

(c) *Urea slope : Proteus* and other organisms produce urease which changes urea to ammonium carbonate and ammonia. The intestinal pathogens do not give this reaction.

(d) *Agar slope :* Pigmented growth or colour diffusing into the agar excludes *Salmonella* and *Shigella*. The culture will be used for agglutination tests when necessary.

After overnight incubation refer to Tables 10 and 12 (pages 141, 147), which indicate the possible identity of strains. Cultures giving positive reactions alien to the *Salmonella* and *Shigella* groups may be discarded. The intestinal pathogens normally give typical fermentation reactions in glucose and mannite after overnight incubation, but dulcite fermentation and indole formation may be delayed. When in doubt check the Gram-stain, and test for purity by plating on blood agar and MacConkey agar, then set up the second line tests in the table and reincubate the original rack of primary tests.

When Kligler's medium and sucrose only have been inoculated strains giving reactions seen in Table 3 after overnight incubation need further investigation, others can be discarded.

Growth on the Kligler slope can be used for slide agglutination. When this confirms the probable identification the physician can be telephoned. Weak positive slide agglutination should be ignored and the strain fully identified before a report is sent. Final identification must await a full set of biochemical reactions and further agglutination tests see Chapter 5.

Although cholera has not been endemic in Britain for more than

TABLE 3

Probable Identity	Motility	Indol	Sucrose*	Kligler's			
				Butt	Slope	Gas	Blackening
Salm. typhi	+(−)	−	−	Yellow	Pink	−	+
Salmonella species	+(−)	−	−	Yellow	Pink	+	+
Shigella species	−	+−	−	Yellow	Pink	−	−

* *Sh. sonnei* may show very weak fermentation after overnight incubation.

a century it may be necessary to exclude it in patients with diarrhoea who have visited endemic areas within the previous 2 weeks ; special medium will be required.

PROCEDURE (from Furniss and Donovan, 1974)

Inoculate 20 ml. alkaline peptone water, pH8, with 2 ml. liquid faeces or the same quality of emulsified faeces from a suspected carrier.

Seed heavily on thiosulphate-citrate-bile salt-sucrose-agar (TCBS) medium.

After 5 hours' incubation, inoculate another TCBS plate from the top of the peptone water culture and also another alkaline peptone water from this first tube.   Incubate for a further 5 hours or overnight and inoculate a third TCBS plate from the second alkaline peptone water.

Vibrio cholerae forms 2 mm. yellow colonies on TCBS.   It is Gram-negative but not necessarily curved and vibrio-like.   Slide agglutination with polyvalent serum will confirm a tentative diagnosis.   When it is negative it should be repeated from growth on nutrient agar.   When it is positive the nearest reference laboratory should be telephoned forthwith so that final identification can be made and the appropriate public health measures instituted without delay.

Patients arriving by air from the Far East may suffer from infection by *Vibrio parahaemolyticus* which can also be isolated on TCBS.   It does not ferment sucrose and the colonies will therefore appear green.   This organism is not a public health hazard in Britain.

Non-bacterial enteritis is common in children in the summer and is most often associated with ECHO virus. Recently orbivirus particles have been seen in the faeces of children with acute diarrhoea in Australia (Bishop *et al.*, 1974) and in Britain (Flewett *et al.*, 1974) who named them rotaviruses. Faeces unfrozen without transport medium should be sent as early as possible in the disease with a sample of clotted blood. When virus has been isolated a second sample of blood will be necessary to demonstrate a rise in titre. Enteroviruses can often be isolated from the faeces of healthy people and isolation alone is not therefore sufficient evidence of infection.

Examination for ova, cysts and parasites

If *E. histolytica* is to be demonstrated in its active state, a fresh stool passed into a warm bedpan, or a fresh sigmoidoscopy specimen must be sent for immediate examination.

PROCEDURE

Select a part with pus and mucus, emulsify it in a drop of warm saline and examine it immediately using the 16 mm. or 4 mm. objective. *E. histolytica* will be seen actively motile with ingested red cells. For cysts and ova emulsify a little faeces in a drop of 1 per cent eosin in saline on a glass slide. Cover with a cover slip and examine. The cysts and ova appear as clear bodies in contrast to vegetable matter which is stained. (N.B. Drops of oil may be mistaken for parasites ; if they are present ask for another specimen.) Repeat the examination using Lugol's iodine instead of eosin and the internal structure of the ova and cysts will be seen.

FORMOL ETHER METHOD OF RITCHIE (Allen and Ridley, 1970)

MATERIALS : 10 per cent formalin (1 vol. 40 per cent formaldehyde + 9 vol. distilled water)
  ethyl ether
  conical centrifuge tubes 15 ml. capacity
  swab sticks to reach bottom of tubes
  copper wire gauze, hole aperture 400–450 μm
  evaporating basin.

METHOD : Emulsify a pea-sized sample of faeces, which must not contain paraffin or barium, in 7 ml. formalin in a centrifuge tube. Sieve by pouring through gauze into an evaporating basin. Wash out the tube. Return fluid from the basin to the tube. Add 3 ml.

ether and shake vigorously for 30 seconds (or mix with an electric mixer for 15 seconds in a wide bore tube). Centrifuge at 3,000 r.p.m. for exactly 60 seconds or at 2,000 r.p.m. for 90 seconds.

A layer of debris will have accumulated beneath the ether layer ; loosen it gently from the side of the tube with a swab stick. Pour the tube contents into discard, one or two drops of supernatant and a small deposit will remain. Mix well, transfer to a slide with a pipette (to avoid debris on the wall of the tube) and examine under a coverslip. This modification gives a good yield of Ascaris, Taenia and Schistosoma ova in addition to other parasites. The concentration varies from 15 to 50 times depending on the parasite and to some extent on the consistency of the specimen.

GENITAL SPECIMENS

The bacteriological diagnosis of genital infections is comparatively straightforward in acute venereal disease, but many non-specific acute infections are encountered in which even prolonged and detailed investigation fails to reveal sufficient evidence to incriminate a particular species as pathogen. Examination of these specimens from women is particularly difficult because they are often contaminated by the normal vaginal flora which is largely composed of slow-growing organisms many of which are microaerophilic or anaerobic and which show very poor colonial differentiation for the first two or three days of incubation. The lactobacilli, diphtheroids, anaerobic non-sporing bacilli, anaerobic Gram-positive cocci, *Listeria* and *Haemophilus* isolated from the vagina have been studied by many bacteriologists, but so far these groups of organisms are not satisfactorily classified, and although there is no doubt that some of them can under suitable conditions assume a pathogenic role, in individual cases it is often impossible to assign it to a particular species. Unless the suspected organism is a well established pathogen, its presence in large numbers in the cultures is not by itself sufficient evidence of pathogenicity. (For evidence required to establish a strain as pathogen see Chapter 1.)

It is convenient to consider specimens according to the clinical state of the patient from whom they are taken.

Acute Infections

(a) Puerperal sepsis and septic abortion

Bacteria found

Recognized pathogens	Other bacteria
Strept. pyogenes	*Corynebacteria* species
Cl. welchi	*Lactobacillus* species
Anaerobic streptococci	*Staphylococcus* species
Staph. aureus	*Streptococcus* species
Strept. Group B	Anaerobic Gram-positive cocci
Strept. faecalis	Anaerobic Gram-negative cocci
Esch. coli	
Listeria monocytogenes	
Mycoplasma hominis	
Bacteroides species	

Any of the bacteria, particularly anaerobes, which invade wounds (see Chapter 3).

Since puerperal fever may be rapidly fatal if untreated, the usual practice when pyrexia is discovered is to swab the cervix and begin treatment without waiting for the results of culture. Fortunately many common uterine pathogens are penicillin sensitive and the patient is often convalescent by the time the report reaches the ward. Nevertheless these infections should be thoroughly investigated. When *Strept. pyogenes* or antibiotic-resistant 'hospital' *Staph. aureus* are found in puerperal infection it is a sign that the aseptic surgical technique may have failed, and it is important to discover how this has occurred so that future accidents may be avoided. Infection by the other pathogens listed above indicates that bacteria from the patient's own vagina or perineum have invaded the uterus and epidemic spread of the infection is extremely unlikely. If the cultures yield a heavy growth of an organism, such as a non-sporing anaerobic bacillus, which is not an established pathogen, every effort should be made to find if it is the pathogen in this case. Such an investigation may include blood cultures, tests for antibodies in the patient's serum, repeated swabs during the illness and convalescence, and animal pathogenicity tests.

PROCEDURE

Stuart's transport medium is recommended for high vaginal swabs because anaerobes survive well in it and microbes such as *Mycoplasma* which will not stand drying are preserved.

Make a smear on a sterile glass slide for Gram-stain. Seed the swab on two blood agar plates, for incubation in air plus CO_2 and anaerobically, and on to MacConkey's bile salt agar ; also on to sensitivity test blood agar suitable for primary sensitivity tests, which is best incubated anaerobically since anaerobes are so commonly encountered.

1st Day. Examine the cultures carefully for colonies resembling the species in the first list above. When any of them are present in profusion they are considered to be the cause of the infection because they have been proved pathogenic on numerous previous occasions. *Strept. pyogenes* is so rarely found in a healthy vagina that it is probably the cause of the infection even if only a few colonies are found on the plate cultures. When however a satisfactory specimen yields only a few colonies of *Cl. welchi, Esch. coli, Strept. faecalis, Staph. pyogenes* or anaerobic streptococci, it may well be that they are harmless and the pyrexia may be caused by infection with a slow-growing species which has not yet appeared, or by infection elsewhere, perhaps in the urinary tract or chest. Single colonies of all types resembling bacteria in the first list above are Gram-stained and subcultured into suitable medium for identification tests. At this stage a preliminary report may be sent. Reincubate blood agar and sensitivity test cultures and note further growth, usually of non-sporing anaerobes. Anaerobic streptoccoci often need 48 hours' incubation before colonies are visible and they may not appear until the third day.

When in the Gram-stained smear of the specimen pus cells are few, or absent, and on the second day none of the recognized pathogens listed above has appeared, a report can be sent worded to show that complete identification of all species has not been attempted (see page 6). When the smear shows evidence of infection as judged by numerous pus cells, few squamous cells and lack of the normal varied vaginal flora, the specimen is treated as pus. Prolonged incubation may lead to the implication of non-sporing anaerobes, *Mycoplasma, Listeria monocytogenes* or some other slow-growing microbe as the cause of a particular infection.

(*b*) Acute cervicitis in non-puerperal women, Acute urethritis, Vulvo-vaginitis in children

BACTERIA FOUND

N. gonorrhoeae, Strept. pyogenes.
Any of the bacteria listed under (*a*).

Specially prepared swabs and Stuart's transport medium [1] are required for gonococci; they are also satisfactory for other bacteria. A smear is best made at the bedside. Best results for the isolation of *N. gonorrhoeae* are achieved when patients attend a clinic where trained staff culture directly to warm selective medium. A small, sterile disposable loop introduced into cervix or urethra collects a much better sample for direct culture than a swab, which is usually too large to reach the site of infection. Culture plates are incubated immediately in the clinic and at the end of the session are taken without delay for incubation in the laboratory in air plus 5–10 per cent CO_2.

PROCEDURE [2]

Gram-stain the smear. Seed the swab from transport medium on to two blood agar plates and selective medium (see page 366). Incubate one blood agar and the selective medium in air plus 5–10 per cent CO_2 and the other blood agar anaerobically plus 10 per cent CO_2. Occasional strains of *N. gonorrhoeae* are inhibited by the selective medium, therefore for maximum success plain blood agar or chocolate agar should also be included, this is very extravagant when a large clinic is served. An alternative is to use the selective medium alone which will succeed for almost all actively infected patients and repeat cultures on two media when a negative result is obtained from purulent material.

1st Day. Gram-stain any colonies resembling *N. gonorrhoeae*. If none are found, reincubate the culture because many strains need 48 hours' incubation before growth is visible on first isolation. Do not allow the culture to stand on the bench until it is cold. Examine it quickly and reincubate it.

2nd Day. After 48 hours' incubation colonies of *N. gonorrhoeae* are typically 0·5 to 1·5 mm. in diameter, moist, domed, with entire edge, translucent and easily emulsifiable. They are never matt,

---

[1] For preparation, see page 362.
[2] This is equally applicable to the examination of discharges for *N. meningitidis*, the colonies closely resemble *N. gonorrhoeae*.

rugose, heavily pigmented or absolutely opaque.   All colonies which resemble them are Gram-stained.   Any which prove to be Gram-negative cocci are identified by staining with fluorescent antibody (page 133).   Alternatively they may be subcultured to blood agar for further tests, or, if growth is sufficient and the colonies are well isolated, to hydrocele fluid sugar slopes (see page 132).

Fluorescence microscopy. Fluorescent antibody techniques can be applied to smears of pus from suspected gonorrhoea.   In positive cases the cocci fluoresce but occasionally non-specific positives are seen.   Therefore culture followed by rapid identification of the Gram-negative cocci by fluorescence microscopy (see page 105) is recommended.   A reliable report 1 to 2 days later is preferable to risking an early false positive report.   *N. gonorrhoeae* and *N. meningitidis* have antigens in common and the immunofluorescent test will be positive with either.   Fermentation tests will confirm the presumptive diagnosis of gonococcal infection, see page 132.

(c) Acute vaginitis

Bacterial infection of the vagina in adults is rare in Britain. Occasionally it is caused by haemolytic streptococci but the common pathogens are a protozoon, *Trichomonas vaginalis*, and a fungus, *Candida albicans*.

PROCEDURE

A wet specimen is required for *Trichomonas*.   It may be taken either with a swab, if a microscope be available near at hand for its examination, or with a wide-mouthed blunt-ended pipette placed in a plugged test-tube if the examination is likely to be delayed for a few hours.   Make a wet preparation by rubbing the swab in a drop of saline on a glass slide or by delivering a drop of discharge from the pipette on to the slide (it may be necessary to dilute a thick purulent discharge with saline).   Cover the drop with a clover slip and examine it under the 16 mm. and 4 mm. objective.   In Trichomonas infection, examination of the discharge shows numerous pus cells and motile flagellated protozoa.   Swabs may also be placed directly in an enrichment medium containing chloramphenicol which can be obtained commercially and this is the most convenient method for individual patients when direct microscopic examination is unavailable, an additional swab in Stuart's transport medium should also be sent.   Positive cultures alone may mislead because a few

Trichomonas may be found in normal people and these will grow in the liquid medium. A preparation in normal saline from the swab in transport medium may show numerous motile Trichomonas even after several hours. This is always a significant finding but infection cannot be excluded when none are seen.

The swab for culture should be taken from an inflamed area of the vaginal wall. It is examined as a swab of pus. When *Candida albicans* is the cause it is usually easily found in the Gram-stained films from the swab and, since it grows on blood agar, it will appear in the cultures made for bacteria. Sabouraud's agar screened with antibiotic selects Candida from vaginal bacteria and the swab should also be seeded on it. Small numbers of Candida, including *C. albicans* are often present in the absence of vaginitis, therefore isolation does not necessarily imply infection.

Gonorrhoea in adults is a disease of cervix and urethra not of the vagina. Vaginal swabs are therefore not suitable for the diagnosis of gonorrhoea except in childhood. Gonorrhoea of the rectum and throat is occasionally seen and swabs should be cultured on selective medium, see page 366.

Chronic Infections

(a) Leucorrhoea

A large number of gynaecological patients complain of chronic vaginal discharge. The causes of this symptom are manifold and in only relatively few cases can leucorrhoea be attributed primarily to infection. In Britain there are two important pathogens which need to be excluded. They are, *Trichomonas vaginalis* and *Candida albicans*. There are almost certainly other infective causes of chronic vaginitis, but in most hospital laboratories it is impossible to make a full investigation of specimens from every case of vaginal discharge. The cultures yield a profuse growth of microaerophilic and anaerobic bacteria which need many days incubation before pure cultures can be obtained for identification. Moreover, to establish any one of them as pathogen would need prolonged investigation and a search for more patients similarly infected. A compromise must therefore be made and the following routine has been found satisfactory both to clinician and bacteriologist.

Procedure

Examine microscopically a wet preparation and Gram-stained film of all specimens. When numerous pus cells, yeasts or Gram-negative diplococci are seen, the preliminary diagnosis must be confirmed by culture. When there are very few pus cells the condition is most unlikely to be infective and cultures are not made. When chronic gonorrhoea is suspected take *cervical* and urethral swabs in transport medium and use the method already described for acute cervicitis. When a patient shows any unusual sign, such as ulceration of the vaginal wall, and cultures may be expected to yield valuable information, a special note is sent with the swab explaining the need for culture in the first instance and the specimen is treated as pus.

(b) Syphilis

Clean around the site of the suspected chancre or gumma with sterile gauze and a little sterile saline if necessary. Allow some of the tissue fluid from the lesion to run into a glass capillary tube, seal the ends either in a flame or with plasticine, and take it to the laboratory for examination under dark-ground illumination or phase contrast microscopy. Stained films of the fluid are much less satisfactory. Spirochaetes other than *Trep. pallidum* are often found in fluid from chronic ulcers. A fluorescence test employing absorbed syphilitic serum will confirm the diagnosis (see page 288).

SKIN

BACTERIA FOUND

Staph. aureus, other Staphylococci and micrococci
Strept. pyogenes, other haemolytic streptococci
Strept. viridans and non-haemolytic streptococci
Enterobacteriacae and *Pseudomonas*
Diphtheroids (some anaerobic or micro-aerophilic)
C. diphtheriae
Myco. tuberculosis and other mycobacteria
Anaerobic streptococci
Erysipelothrix species
Actinomyces species
Pseudomonas mallei and *B. anthracis* very rarely.

Staph. aureus is the only pathogen which commonly inhabits the normal skin. Carrier rates vary from 10 to 20 per cent in apparently healthy adults, and the organism may remain in the skin for months or even years. Skin carriers are almost always nasal carriers and the skin population is probably maintained by reinoculation from the nose. *Staph. albus*, micrococci and diphtheroids are often found on normal skin but they are usually harmless. The other microbes listed are present either as casual superficial contaminants, which are quickly removed by the antibacterial action of the skin secretions and by washing, or they are found in discharges from skin lesions.

(a) Acute infections

Examine swabs from inflamed areas, impetigo, superficial wounds, burns, etc., by the method described for pus. Bacteria found are *Strept. pyogenes* and *Staph. aureus* very commonly, *C. diphtheriae* rarely. The skin of the buttocks, lower abdomen and thighs is often contaminated with coliforms and *Strept. faecalis*; these may be found in other areas, especially after antibiotic therapy when they replace the antibiotic-sensitive pyogenic cocci.

Fluid from blisters is difficult to sample without accidental contamination. The following procedure is usually successful.

Select an intact bulla with as firm a layer of skin over it as possible. Clean the skin with liquid soap, spirit and ether. It is very important that it should be dry before sampling. Pierce the skin of the bulla with the broken end of a sterile Pasteur pipette and suck out the fluid. Place drops of it immediately on to two blood agar plates and into cooked meat broth. Incubate the plates in air and anaerobically. The fluid is often found to be sterile when broken blisters are heavily infected with pyogenic cocci.

Virus in vesicle fluid can be seen by electron microscopy. The fluid should be sent in a capillary tube sealed at each end. When herpes simplex is suspected and there is no possibility of smallpox a small drop of fluid can be sent dried on a glass slide but this will not be suitable for culture and is potentially infective. It should be covered with another clean slide to avoid contamination of the container. A swab in virus transport medium frozen at −60° C. should be sent for culture in all cases when examination for virus cannot immediately be performed.

(b) **Chronic ulcers** [1]

It is usually impossible to determine the cause of a chronic spreading ulcer by taking swabs from the weeping surface except when acid-fast bacilli are seen (see page 109).　Secondary invaders, e.g. *Proteus, Ps. aeruginosa* and the pyogenic cocci, may sometimes be avoided by sampling in the following manner.　Clean the healthy skin just beyond the edge of the ulcer with liquid soap, spirit and ether, then pierce the skin with a sterile hypodermic needle so that its point travels near the edge of the ulcer but below the surface ; gently press the indurated tissue around the needle, withdraw it and using a sterile syringe squirt any tissue fluid which may be in it on to two glass slides, two blood agar plates, into cooked meat broth and 0·2 per cent glucose broth for *Erysipelothrix*.　If the quantity of fluid be insufficient, deliver one small drop on to blood agar and another on to a glass slide.　The plate culture is best incubated anaerobically, as obligate aerobes are much less common than anaerobes.

When a biopsy is performed take a small piece of the tissue in a sterile tube, wash it free of contaminants, grind it in a sterile tube or stomacher and make cultures as described for culture of tissue (page 59).

Bacteria found

> *Strept. pyogenes*
> *Streptococcus* species
> Anaerobic streptococci
> *Staph. aureus*
> *C. diphtheriae* (rarely)
> *Mycobacterium* species
> *Actinomyces* species
> *Pseudomonas mallei*
> *Erysipelothrix* species

Bear in mind also the possibility of fungous infections other than actinomycosis and when necessary, stain smears for Leishman-Donovan bodies.

---

[1] It is assumed that before these investigations are begun syphilis has been excluded by negative serological tests.

(c) Conjunctival Swabs

When the conjunctiva is acutely inflamed, the discharge is treated by the method described for pus.

The discharge is best cultured at the bedside, (outpatients can attend the laboratory), because lysozyme in tears will kill some bacteria during transit. When direct culture cannot be made a swab in Stuart's transport medium should be sent. The anaerobic culture sometimes enables *Strept. pneumoniae* or *Strept. pyogenes* to be more easily recognized but anaerobic infection of this superficial site is very unlikely. Carbon dioxide, 5–10 per cent, should be added for aerobic culture for the isolation of CO_2 dependant strains of Haemophilus, Neisseria and *Strept. pneumoniae*. Additional primary culture on a Loefflers serum slope will aid recognition of Moraxella.

Conjunctivitis is also caused by *Adenovirus* and *Chlamydia*. A swab in virus transport medium frozen at $-60°$ C. should be sent when *Adenovirus* culture is required. *Chlamydia* inclusions may be seen in a smear of the discharge stained by Geimsa, but infection in Britain is not often sufficiently florid to be able to diagnose the condition thus. A swab in chlamydia culture medium should be sent direct to the virus laboratory. When delay is unavoidable it should be kept at $+4°$ C. Culture may be positive even after a few days held at this temperature in the medium.

BACTERIA FOUND

Adults and Children	Infants
Haemophilus species	*Staph. aureus*
Strept. pneumoniae	*Staph. albus*
Staph. aureus	*Enterobacteriacae*
Haemolytic streptococci	*Pseudomonas*
Neisseria	*Neisseria*
Moraxella	Rarely other bacteria
Enterobacteriacae	listed under adults
Pseudomonas	
Corynebacteria species, *Staph. albus* (as commensals)	

REFERENCES

ADDY, M. G., ELLIS, P. D. M. and TURK, D. C. (1972). *Brit. med. J.*, **i**, 40.
ALLEN, A. V. H. and RIDLEY, D. S. (1970). *J. clin. Path.*, **23**, 545.
ASTON, D. L., COHEN, A. and SPINDLER, M. A. (1972). *Brit. med. J.*, **iv**, 462.
BAILEY, S. F. and LACEY, G. R. (1927). *J. Bact.*, **13**, 182.
BISHOP, R. F., DAVIDSON, G. P., HOLMES, I. H. and RUCK, B. J. (1974). *Lancet*, **i**, 149.
COOK, J. T. and JEBB, W. H. H. (1952). *Mth. Bull. Minist. Hlth Lab. Serv.* January **11**.
DIXON, J. M. S. and MILLER, D. C. (1965). *Lancet*, **ii**, 1046.
FLEWETT, T. H., BRYDEN, A. S., and DAVIES, H. (1974). *J. clin. Path.*, **27**, 603.
FURNISS, A. L. and DONOVAN, T. J. (1974). *J. clin. Path.*, **27**, 764.
HARVEY, R. W. S. and PRICE, T. H. (1968). *J. Hyg.*, **66**, 377.
JEFFRIES, L. R. (1959). *Ibid.*, **12**, 568.
KLIGLER, I. J. (1918). *J. exper. Med.*, **28**, 319.
LACEY, B. W. (1951). *J. gen. Microbiol.*, **5**, vi.
—— (1953). Personal communication.
—— (1954). *J. Hyg. Camb.* **52**, 273.
MAY, J. R. (1952). *Lancet*, **ii**, 1206.
—— (1953). *Ibid.*, **ii**, 534.
MEAD, G. R. and WOODHAMS, A. W. (1964). *Tubercle*, **45**, 370.
MOHUN, A. F. and STOKES, E. J. (1942) Unpublished.
ROWE, B., TAYLOR, J. and BETTELHEIM, K. A. (1970). *Lancet*, **i**, 1.
TAYLOR, W. I. (1965). *Amer. J. clin. Path.*, **44**, 471.

5

Identification of bacteria

Introduction

Given cultures of infected material, a text-book of systematic
bacteriology and a knowledge of bacteriological technique, it should
be possible in theory to identify any bacteria found; in practice,
however, the road to identification is full of pitfalls. This chapter is
a guide to the inexperienced traveller. It is not an abbreviated
system of bacteriology but deals with the special problem of rapid
identification in a hospital laboratory. It should be used in con-
junction with, not as a substitute for, the standard text-books.

It is often possible after examination of the primary overnight
plate cultures to gain a fairly accurate idea of the nature of the
organisms present. It is, however, extremely important to approach
identification with an open mind and not to be too much influenced
by colonial appearance, by the clinical symptoms and by a knowledge
of the bacteria commonly found at the site of sampling. Previous
experience indicates in which order tests are to be made but none
must be omitted on the grounds that the presence of a particular
species is improbable (see Chapter 1).

In every case the first step is to examine isolated colonies carefully
with a hand lens by reflected and transmitted light, then pick a single
colony of each type, Gram-stain smears of them and examine them
under the microscope.

The early bacteriologists, many of whom were familiar with histological techniques, spent much time staining and examining bacteria
microscopically. Now it is known that bacterial cells have no
constant morphology comparable with tissue cells. Their appearance can be greatly altered by slight changes in the composition
of the medium in which they grow. Bacteria seen in smears of
infected discharges often look quite different after culture on blood
agar and different again after growth on selective media. Moreover,
the presence of sub-lethal concentrations of antibiotics in tissue
fluids, or in cultures has a profound effect on the morphology of
organisms which are sensitive to them.

Some variations are so commonly met with that they become helpful in diagnosis, for example the extreme pleomorphism of *C. diphtheriae* grown on tellurite medium ; in general, however, the presence of atypical forms makes identification more complex.

Examination of a Gram-stained film can only be expected to indicate to which family the microbe belongs and which biochemical and serological tests are most likely to disclose its identity. Stains to show granules, capsules and flagella and special spore stains are seldom used because the structures which they reveal are not of diagnostic importance. In all cases the next step after Gram-stain is to seed the remains of the colony, or another exactly like it, into a medium which is likely to yield prolific growth. This pure culture is then used for biochemical and serological tests.

Essential tests for the diagnosis of each medically important species will now be described, starting with the colonial appearance after overnight incubation on the media recommended in Chapters 3 and 4 (see Table 4). When in a particular investigation the results are equivocal, the reader is advised to turn to a text-book of systematic bacteriology where he will find descriptions of further tests appropriate to the problem.

Note. Colonial morphology is so variable that description is useful only when the preparation and constituents of the medium are defined as precisely as possible. The species of blood in blood agar will obviously make a difference to the appearance of haemolytic organisms but colonies will also vary with the presence of anticoagulant and its nature, with the method of extraction of the meat broth base, the peptone, the thickness of the medium in the plates and the concentration of agar. Some medium sold as MacConkey agar contains bile salt which commonly inhibits species which are normally described as capable of growth on it ; Difco MacConkey agar yields colonies as described here. The description of colonies on blood agar applies to medium prepared in the hospital department, method page 363, Oxoid No. 2 nutrient agar base is a satisfactory alternative to that described.

Procedure when tests fail to identify

When a bacterium gives reactions which fail to fit the description of any known species it may in fact be a new species, but one of the following possibilities is more likely.

(*a*) The culture may be contaminated.

TABLE 4

Table to indicate to which genus a bacterium is likely to belong when all that is known about it is the appearance of its growth on the original plate cultures, i.e. blood agar (aerobic) blood agar (anaerobic) and MacConkey's bile salt agar.[1]

Blood agar (aerobic) after overnight incubation

1. *Colonies* 1 *mm. or more in diameter.*

Gram-positive cocci	Staphylococci
	Micrococci
	(Streptococci)
Gram-negative cocci	*Neisseria*
Gram-positive bacilli	*Corynebacteria*
	Bacillus
Gram-negative bacilli	(*a*) Growth on MacConkey's medium *Enterobacteriacae*, *Pseudomonas* (*Brucella*, *Pasteurella*, *Yersinia*)
	(*b*) No growth on MacConkey's medium (*Pasteurella*)

2. *Colonies less than* 1 *mm. in diameter.*

Gram-positive cocci in pairs or short chains

    (*a*) Growth on MacConkey's medium
        Strept. faecalis (group D)
        Streptococci of groups B, C, G

    (*b*) No growth on MacConkey's medium
        Strept. pyogenes (group A)
        Streptococci of groups other than group D
        Strept. pneumoniae
        Strept. viridans

Gram-positive cocci in clumps	(Staphylococci)
	Micrococci
Gram-negative cocci	*Neisseria*

[1] See Note, page 110.

(Table 4 cont.)

Gram-positive bacilli	*Corynebacteria*
	Lactobacilli
	Actinomyces
	Listeria, Erysipelothrix
Gram-negative bacilli	Parvobacteria

Blood agar (anaerobic + 10% CO_2) after overnight incubation, no equivalent growth on aerobic culture

1. *Colonies about 1 mm. or spreading.*

   Clostridia

2. *Colonies minute.*[1]

Gram-positive cocci	Anaerobic streptococci
	Anaerobic cocci
Gram-negative cocci	*Veillonella*
Gram-positive bacilli	*Lactobacillus* species
	Actinomyces
	Corynebacterium species
Gram-negative bacilli	*Bacteroides, Fusiformis* (a few are Gram +).

To reveal contamination sub-culture on a blood agar plate so that numerous well isolated colonies are obtained. Even if no contaminants are visible after overnight incubation, pick a single colony and repeat the subculture in the same manner twice consecutively. A single colony from the third blood agar plate is assumed to be pure and a culture from it is re-tested. Identification tests may be repeated from a single, well isolated, colony on the first subculture to save time, but it is useless to repeat the tests without any attempt at purification.

(*b*) The strain may have lost its true staining properties or colonial morphology with the result that the wrong tests have been made.

To reveal lost staining properties and colonial variation, subculture to a Loeffler's serum slope because growth on this medium usually yields typical microscopic morphology. After overnight

---

[1] These small colonies may fail to appear for several days. Some may prove to be microaerophilic, not true anaerobes.

incubation Gram-stain very carefully timing each stage. When there is no change in morphology subculture from the Loeffler slope to a blood agar plate and examine the colonies carefully next day for any sign of reversion to a different colonial form. (Common variants are described with typical members of their species later.)

(*c*) It may be a microbe which lies outside the field of medical bacteriology but would be familiar to veterinary or plant pathologists.

Make three comparable cultures, on blood agar, and nutrient agar incubated at 37° C. and on nutrient agar incubated at room temperature in the dark. Also, since growth may be inhibited by broth, test in peptone water at 37° C. and at room temperature, and in broth and peptone water at 50°–56° C. Medically important bacteria grow at least as well on blood agar as on nutrient agar and usually better at 37° C. than at other temperatures.

Appropriate tests for the identification of members of each medically important genus will be found under separate headings in this chapter.

STAPHYLOCOCCI

Gram-stained smear of infected discharge. The cocci appear rather large, mainly in pairs, a few single. They may be intra- or extra-cellular. Dead cocci stain poorly and can easily be mistaken for *Neisseria*.

Culture : 18 to 24 hours

(*a*) *Blood agar aerobic.* Smooth, slightly domed colonies with entire edge 1 to 2 mm. diameter; sometimes beta-haemolytic, opaque or translucent, pigmentation variable but usually not marked in a young culture.

(*b*) *Blood agar anaerobic* (plus about 10 per cent CO_2). Opaque domed colonies 0·1 to 0·5 mm. diameter. Strains which are beta-haemolytic on the aerobic culture show no haemolysis. No pigment formed.

(*c*) *MacConkey's bile salt agar.* Colonies 0·1 to 0·5 mm. opaque. Pigment often marked which may mask the pink colour of lactose fermenting strains. See Note, page 110.

(*d*) *Microscopic examination.* Gram-positive cocci in clusters regular in size, some strains stain poorly and show a mixture of positive and negative cocci.

Staph. aureus is the most important pathogen. It can be identified by its power to form human plasma-coagulating substances, coagulases which are possessed by no other Gram-positive coccus.

Coagulase Test

1. *Tube culture method.* Inoculate 5 ml. of broth with a single colony of the staphylococcus. Before incubation add ten drops of sterile human plasma, obtained from a discarded bottle of blood from the hospital blood bank, and incubate for 18 hours. (The proportion of plasma to broth is important.) Control positive and negative tests should be set up each time because although some samples of plasma kept at about 4° C. will give positive results for several months, others fail to do so after a few weeks. Heparin is added to the plasma to prevent clotting due to the removal of citrate either by organisms which metabolize it, or by calcium in the medium which will combine with it.

False positives. Provided the test is made as described, these do not occur. Heavy granular growth may be mistaken for clot but it will disintegrate on shaking, whereas a clot will contract but will remain solid.

False negatives. If the culture is incubated longer than 20 hours the clot may lyse. Contaminated cultures will give negative results when fibrinolysin-producing organisms, such as streptococci are present.

2. *Slide coagulase test.*[1] Place two large loopfuls of saline separately on a glass slide. Emulsify part of a colony in each of them. Add a small loopful of plasma to one drop of emulsion, the other is the control. When coagulase is present the bacteria will clump together as in a slide agglutination test. The mechanism is entirely different from specific agglutination ; it is due to the formation of an envelope of fibrin round each coccus which causes them to clump. When positive, the test is usually satisfactory, but negative results must be checked with the tube culture or standard test. Results are easily misinterpreted and the tube culture method is therefore preferred, it should always be performed on cultures from serious infection, e.g. blood culture in endocarditis.

---

[1] This test demonstrates bound coagulase or clumping factor which is different from free coagulase, the enzyme responsible for positive tube tests.

3. *Standard method.* Pick a single staphylococcal colony and seed into nutrient broth. Incubate overnight. Sterile plasma is obtained from human blood which has been prevented from clotting by the addition of one-eighth of a volume of 2 per cent trisodium citrate. Take 1 ml. of a 1 in 10 dilution in saline of this plasma, add five drops of the overnight culture and incubate in a 37° C. water bath for three hours; then stand on the bench. A clot is usually seen immediately after incubation but its formation is occasionally delayed.

Of these three methods the tube culture test is the most generally useful. It is much quicker than the standard test, and has proved reliable when freshly isolated strains have been tested in parallel by both methods.

The tube culture method is used routinely, the slide coagulase test when a positive result will enable the report to be sent without delay and the standard test occasionally to check equivocal tube culture results.

The demonstration of DNase production is useful for identification when coagulase tests are equivocal or when plasma is not available. Colonies are spot-inoculated on agar medium containing deoxyribonucleic acid ; a positive control, *Staph. aureus*, and a negative control, *Esch. coli*, should be included. After overnight incubation the plate is flooded with normal hydrochloric acid which causes precipitation with the nucleic acid, making the medium cloudy. Spots of heavy growth of DNase producing bacteria are surrounded by a clear zone where the nucleic acid has been hydrolysed by the enzyme they have formed, (Blair *et al.*, 1967).

Pigmentation

The name " aureus " is misleading because pigmentation is variable and cannot be relied on for identification. If after overnight incubation the culture is allowed to stand on the bench for 24 hours, pigmentation may become more marked and colonies, which were apparently identical immediately after incubation, may now show a difference in colour. As a result *Staph. aureus* may be recognized when it is mixed with *Staph. albus*, which would not otherwise be possible except by coagulase testing a very large number of colonies, or by employing a selective medium (Barber and Kuper, 1951). Difference in pigmentation is often well marked on MacConkey's bile salt agar which, if it is included in the primary culture media, may show two varieties of staphylococcus when the colonies on blood agar are apparently all of one kind.   It must be stressed that there are many harmless yellow pigmented Gram-positive cocci and that the differentiation is only of value in demonstrating two types of colony *both* needing a coagulase test.   It may well be that the pigmented coccus is harmless and the pale one a strain of *Staph. aureus* incapable of forming much pigment.   Different strains of *Staph. aureus* may appear golden, buff coloured, yellow, creamy, white or even greyish in colour.

Variants

Other variants are also encountered.   One is very common ;   a pure growth of *Staph. aureus* shows on the aerobic blood agar plate two types of colony, one large and typical the other small, about 0·5 mm. in diameter.   Both are found to be coagulase positive, the large one on subculture usually gives rise to large colonies only, the small one to a mixture of the two types.   A less common variant is one which on the primary aerobic blood agar culture shows a pure growth of very minute colonies.   They are so small as to be barely visible to the naked-eye after overnight incubation.   The variant needs CO_2 for maximum growth and if the culture is incubated in air plus 5 to 10 per cent CO_2 all the colonies become large and typical.   The small colony looks normal microscopically and is coagulase positive even without added CO_2.   If cultures are made as recommended on blood agar incubated in air and incubated anaerobically plus CO_2 these strains will be easily identified, even though the aerobic culture is unrecognizable at first sight, because, since there is CO_2 in the anerobic jar, they grow like a normal strain under these conditions.   This variant is not harmless, it has been isolated in pure culture from clinically typical staphylococcal lesions (U.C.H. laboratory records, 1951–2).

Further tests which may occasionally be necessary are for fermentation of mannite, liquefaction of gelatin and toxin production (see page 361).

Enterotoxin-producing strains of *Staph. aureus* may be isolated from vomit, faeces and food in outbreaks of food poisoning.   Bacteriophage typing is employed in epidemiological investigations of staphylococcal infection.   Almost all enterotoxic strains belong to phage groups III and IV, but not all strains in these groups are enterotoxic (Williams *et al.*, 1953).   Reference laboratories now have antisera to two toxins, A and B, and strains can be tested for toxin production *in vitro*. Moreover, a tentative diagnosis can sometimes be confirmed by demonstrating toxin in the food when the staphylococci have been killed in cooking.

Coagulase negative staphylococci are labelled *Staph. albus*. They can be differentiated from micrococci by being homogeneous in size, strongly catalase positive and able to ferment glucose anaerobically. *Staph. albus* is almost always sensitive to novobiocin, which is seldom used therapeutically, whereas micrococci are normally resistant, (Mitchell and Baird-Parker, 1967).

Although *Staph. albus* is a normal skin commensal it can be pathogenic, particularly in endocarditis following cardiac surgery (Cunliffe *et al.*, 1943), and in the urinary tract; it should not therefore be dismissed as harmess when isolated in pure culture from an infected site.

Staph. aureus causing endocarditis which has received prolonged treatment with bacteriostatic drugs may temporarily lose its power to produce coagulase and the cocci may resemble micrococci. When sub-cultured or stored in antibiotic-free medium it reverts to the typical form and becomes coagulase positive (Clinicopathological Conference, 1966).

Small colony variants can be distinguished from streptococci by a catalase test, see page 129.

Micrococci

This name is given to a group of microbes often encountered as commensals, until recently differentiation from *Staph. albus* has not often been attempted. All Gram-positive aerobic cocci which are not regular in size, like the staphylococci, and which do not form chains or ferment glucose anaerobically and which are resistant to novobiocin are included in the group. When they appear to be causing infection, e.g. in the urinary tract or in endocarditis following heart surgery, full identification should be undertaken, for classification see Baird-Parker (1965). There is evidence that micrococcus sub-group 3 is a urinary pathogen (Roberts, 1967, and Kerr, 1973).

Strains resembling staphylococci are coagulase tested. Those forming comparatively small colonies may be mistaken for streptococci, they can be differentiated by the catalase test (page 129).

STREPTOCOCCI

Gram-stained smear of infected discharge

The stained smear may show Gram-positive cocci in short or long chains or in pairs.   Slightly elongated cocci have their long axis in line, not in parallel like staphylococci and Neisseria.   They may be intra- or extra-cellular.   Capsules are sometimes seen.

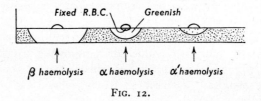

FIG. 12.

Culture :   18 to 24 hours

(*a*) *Blood agar aerobic.*   Colonies 0·1 to 1·0 mm. diameter and variable in appearance.   They may be non-haemolytic, beta-haemolytic, alpha-haemolytic or show alpha-prime haemolysis (see Fig. 7). They may be smooth, domed, entire and translucent, mucoid, or tough, opaque and triangular with a rolled edge.   Their surface may be matt or glossy.

(*b*) *Blood agar anaerobic plus* 10 *per cent* CO_2.   All streptococci grow well anaerobically.   The growth is similar to that on the aerobic culture but the colonies tend to be a little larger and haemolysis more marked.   Strains of *Strept. pyogenes* showing alpha-prime haemolysis aerobically and beta-haemolysis anaerobically are not at all uncommon, occasionally the aerobic culture shows alpha-haemolysis or is non-haemolytic.

(*c*) *MacConkey's bile salt agar.*[1] All strains of *Strept. faecalis* grow on this medium as 0·1 to 0·5 mm. lactose-fermenting colonies. *Strept. pyogenes* is completely inhibited by it.   Most group B streptococci and some group C and G strains, although they are inhibited by 40 per cent bile salt are capable of growth on this medium.   One can therefore exclude *Strept. pyogenes* when growth is present and *Strept. faecalis* when growth is absent.   It may be particularly valuable in examining vaginal swabs from cases of puerperal sepsis.   When a heavy growth of a beta-haemolytic streptococcus is found on blood agar and streptococcal growth is present

[1] See Note, p. 110.

on MacConkey's medium it is almost certainly not *Strept. pyogenes* and although the infection must be treated seriously and the streptococcus identified, a comparatively optimistic preliminary report may be sent. When colonies are large and the Gram-stain shows no chains, a catalase test will differentiate between streptococci and staphylococci (page 129).

1. Beta-haemolytic streptococci

Haemolysis of blood agar by colonies of streptococci is remarkably variable. It is important to realize that it does not necessarily imply pathogenicity. One may be misled in either direction. A strain showing clear zones of haemolysis may prove to be *Strept. faecalis* which in some situations is comparatively harmless, and produces little, if any, soluble haemolysin ; or it may be a member of a group of soluble haemolysin-producing streptococci harmless to man. Alternatively colonies showing very poor haemolysis of blood agar may be found to yield a soluble haemolysin when grown in serum broth and to belong to Lancefield's group A. Some of the factors influencing haemolysis of blood agar are ; the atmosphere in which growth took place, the incubation time, the age of the horse blood, the thickness of the medium and the variety of peptone used in its preparation. The term " haemolytic streptococcus " means to the clinician a microbe capable of causing serious infections and epidemics. A preliminary report is sent on the evidence of blood agar beta-haemolysis and Gram-stain that such an organism may be present, but this is not enough for final identification and after further tests the microbe may prove to be harmless. It may be argued that if a pure or heavy growth of a beta-haemolytic streptococcus is isolated from the infected discharge of a gravely ill patient, that, in itself is sufficient evidence of its importance as pathogen. Such evidence is not however valid for a *laboratory* diagnosis (see Chapter 1). Moreover the prognosis for streptococcal infection is not the same for all groups of streptococci, therefore further tests are essential.

Although the soluble haemolysin test gives a much closer indication of the potential pathogenicity of the microbe than does haemolysis of blood agar, it does not differentiate between soluble-haemolysin-producers pathogenic to man and animal pathogens which are sometimes found as contaminating organisms in human lesions. Lancefield's precipitin test divides streptococci into groups which have quite different clinical significance, therefore all strains of streptococci showing beta-haemolysis on blood agar and also strains of streptococci which fail to show beta-haemolysis but which appear to be playing a major part in infection are grouped.

Lack of haemolysis does not imply low pathogenicity. Epidemic wound infection by non-haemolytic *Strept. pyogenes* has been recorded (Colebrook *et al.*, 1942).

The groups known to be pathogenic to man are as follows:

GROUP A. *Strept. pyogenes.* Almost always beta-haemolytic when incubated anaerobically. It causes erysipelas, epidemic wound infection, puerperal fever, scarlet fever, tonsilitis, otitis media, meningitis, endocarditis etc. It is associated with nephritis and acute rheumatism.

GROUP B. *Strept. agalactiae.* Many strains non-haemolytic. Sometimes a human pathogen especially in neonates (Eickhoff *et al.*, 1964); causes bovine mastitis. Common in the healthy human vagina in small numbers, also found in faeces.

GROUP C. *Strept. equi.* Some strains non-haemolytic or alpha-haemolytic. Sometimes a human pathogen. Found in equine infections. Found commonly in small numbers in the human mouth and throat and in faeces.

GROUP D. *Strept. faecalis.* All strains are members of this group, irrespective of the type of haemolysis or absence of haemolysis on blood agar. They may be beta, alpha or, most commonly, non-haemolytic. *Strept. bovis* is a member of this group, it differs from *Strept. faecalis* in being more sensitive to penicillin.

GROUP F. Occasionally pathogenic in man; the colonies are minute and beta-haemolytic.

GROUP G. Capable of causing severe human infection but mild infection is the rule. Considered in medical bacteriology to be next in importance to group A.

GROUP H and K. *Strept. viridans* isolated from the blood in sub-acute bacterial endocarditis sometimes belong to group H. Group K streptococci have also been isolated from human blood. Beta-haemolytic strains are also encountered.

PROCEDURE FOR IDENTIFICATION OF BETA HAEMOLYTIC STREPTOCOCCI

1st Day. After recording the colony description and result of Gram-stain make the following subcultures from single colonies; into 5 ml. of 0·2 per cent glucose broth in a tube which can be centrifuged, for Lancefield grouping ; on to aesculin-bile medium, for identification of *Strept. faecalis* ;[1] on to blood agar for bacitracin-penicillin sensitivity test, for rapid presumptive identification of group A strains ; on to blood agar for preservation of the strain.

Maxted's bacitracin sensitivity test for group A streptococci

Bacitracin sensitivity is a useful rapid test for presumptive diagnosis of group A strains which are almost all sensitive to an appropriate concentration whereas haemolytic streptococci of other groups are resistant. The test is useful in routine work because it enables a rapid presumptive diagnosis to be made even when the subculture is not pure and grouping must therefore be delayed. It has some pitfalls and the addition of a penicillin sensitivity test is worth while because all group A strains are at present sensitive to this drug.

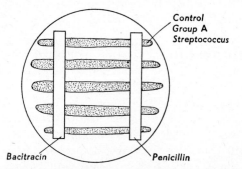

FIG. 13. Inoculation method for bacitracin/penicillin test

METHOD. Soak strips of blotting paper in a 5 units per ml. solution of bacitracin, blot and dry rapidly in a desiccator. Store in the refrigerator at about $-2°$ C. Prepare penicillin-impregnated strips similarly using a 100 units per ml. solution and store at $4°$ C. Streak test strains and a control on blood agar, placing the strips on the surface after seeding (Fig. 13). Incubate in air at $37°$ C. The control group A streptococcus is best maintained by daily subculture on blood agar. Bacitracin in weak solution is unstable but dried strips will keep for some months. Some bought discs and strips tend to give false positive results. This can be overcome either by measuring the zone and comparing it with that given by a control Group A strain, which should always be included to check activity,

[1] See page 372.

or by preparing the discs or strips in the laboratory and discarding any remaining after 2 months.

When typical haemolytic strains are tested the results are usually clear cut, although allowance must be made for heaviness of inoculation which reduces the zone size.   Identification by this method is only presumptive because approximately 1·7 per cent of group A strains are bacitracin resistant and 2·5 per cent of strains of other groups are sensitive (Maxted, 1953).   When haemolytic streptococci are selected by anaerobiosis, strains of *Strept. viridans* are sometimes tested when they are more than usually haemolytic on the anaerobic culture. These may be bacitracin sensitive and it is impossible without grouping to distinguish them from the fairly common group A variant which is alpha haemolytic in air and beta haemolytic anaerobically.   The test is of greatest value in epidemics where the behaviour of the epidemic strain is known and other haemolytic colonies can be rapidly eliminated by their different behaviour in this test.

2nd Day.   Examine the aesculin-bile slope.   Strains showing growth and blackening, due to aesculin fermentation can be reported forthwith as *Strept. faecalis* (page 126).

Strains which are bacitracin or penicillin resistant are most unlikely to belong to group A; they can be identified by Lancefield grouping.   Strains which are beta-haemolytic when incubated in air and sensitive to both bacitracin and penicillin are presumed to belong to group A ; a preliminary report can be sent.   Identification should be checked by grouping.

Lancefield grouping (Fuller's formamide method)

PROCEDURE

Centrifuge all the glucose broth cultures showing good growth at about 3000 r.p.m. for 5 to 10 minutes.   Discard the supernatant fluid.   Check the deposit for purity by Gram-stain and by seeding on blood agar.   One blood agar plate divided into sectors is usually sufficient for all streptococci grouped on one day.

Fuller's method of antigen preparation

1. To the deposit of streptococcal growth add 0·1 ml. formamide (boiling point 155° C.).   Boil in an oil bath [1] at 155 to 170° C. until the growth dissolves.   This takes about 20 minutes.   Cool.

---

[1] Liquid paraffin in a large enamel mug.

2. To the clear fluid add 0·25 ml. acid alcohol (95 parts of anhydrous alcohol, 5 parts 2N HCl). This precipitates the protein fraction. Add 0·5 ml. acetone to precipitate the polysaccharide.

Centrifuge and remove the supernatant fluid very carefully using a pipette.

The precipitate is a mixture of protein and polysaccharide. The

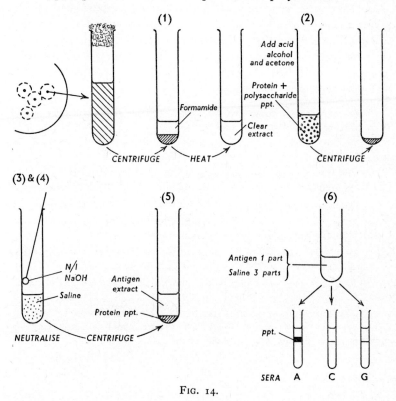

FIG. 14.

polysaccharide is soluble in normal saline at pH 7·4 to 7·6, the protein is insoluble.

3. Add 0·5 to 1 ml. of saline according to the amount of growth originally dissolved.

4. Neutralize. Add a loopful of phenol red indicator. Add small loopfuls (1 to 2 mm. diameter) of normal soda until the solution just turns pink. (It is impossible to add small enough quantities with a pipette.)

5. Centrifuge to obtain a clear supernatant—a few minutes.

6. Set up narrow capillary tubes each containing a small amount of specific group serum. Dilute the antigen ¼ in saline. Layer the diluted antigen carefully on the surface of the serum so that the line of demarcation between the two fluids can be easily seen. The fluids must be clear and free from bubbles.

7. Read. A ring of precipitate should appear within five minutes if good sera are used. Leave negative tests for 30 minutes and if still negative discard them and re-test, using undiluted antigen. If cross reactions occur, i.e. precipitation in more than one tube, dilute the antigen 1 : 8 with saline and test again. The cross reactions will then disappear.

Note 1. Deposit from glucose broth cultures will turn dark brown after heating, this does not matter.

2. Do not allow the temperature to rise above 170° C. because the antigens are not stable at higher temperatures.

3. Once the preparation has begun do not leave it until the extract has been neutralized ; the antigens are not stable in acid solution.

4. Neutralize very carefully or false negative results may be obtained.

5. It is useless to attempt precipitation tests with cloudy fluids. Centrifuge the serum to clear it if necessary.

6. Both false negative and false positive results are seen with extracts of mixed cultures.

7. Readings are best made in the viewing box used for agglutination tests.

8. The minimum number of sera for routine tests is three A, C and G. Cross precipitation occurs between these groups which will not be recognized as such if only one of them is used (see Table 5).

TABLE 5

Serum	A	C	G
Antigen dilutions { 1 : 1	−	−	+
1 : 4	+	−	+
1 : 8	+	−	−

("A" reaction inhibited by excess antigen.)
(Non-specific "G" reaction avoided by dilution.)

Result : Group A streptococcus.

9. Undiluted extract may contain a very large amount of specific antigen and a moderate amount of non-specific antigen which will lead to results as seen in Table 5.

When no precipitation occurs with these sera it may not be necessary to investigate further. For example, if the specimen was a throat swab from a streptococcal carrier, the presence of streptococci other than group A may be of no interest. If however complete identification is required, test the extract against B and D sera and E, F, H and K if necessary. If no reaction is obtained and haemolysis of blood agar was clearly marked, examine a serum broth culture for soluble haemolysin.

Soluble haemolysin test

Take equal volumes of overnight 15 to 20 per cent serum broth culture and 5 per cent thrice washed horse or rabbit blood cells in saline, mix them in a sterile tube and incubate at 37° C. for an hour. Include a control tube with extra saline to replace the culture. If soluble haemolysin is present the cells will lyse and the control tube will be unaltered. *Strept. viridans* cultures will alter the colour of the blood but the fluid will remain opaque.

This is a test for streptolysin S which lyses horse and rabbit cells. It is not formed in broth which contains no serum and it is rapidly destroyed at 37° C. so that it is important to test a very young culture, about 8 to 12 hours old. Growth can be retarded for convenience by incubating as late in the day as possible with the culture tube standing in a beaker containing cold water. Most strains however give a satisfactory result after overnight incubation without this precaution. When the test is negative the strain is reported as *Strept. viridans*.

Almost all beta-haemolytic streptococci encountered in hospital laboratories in Britain can be grouped by Fuller's method using sera A, B, C, D and G, and Fuller's antigen diluted 1 : 4. It is seldom necessary to use sera of other groups or the soluble haemolysin test.

2. Alpha-haemolytic streptococci

There are two important groups of streptococci which produce green discoloration of blood agar, the pneumococci and the large heterogenous group labelled *Strept. viridans*. In addition to these, some strains of enterococci are alpha-haemolytic. Descriptions of typical morphology of members of each of these three groups can be found in any text-book of bacteriology, but unfortunately many atypical strains are encountered. Colonies which at first sight seem to be *Strept. pneumoniae* may prove to be *Strept. viridans* and the reverse is even more common.

Alpha-haemolytic streptococci resemble other streptococci in their ability to grow well anaerobically. Their appearance on aerobic and anaerobic blood agar cultures is very similar, colony size being perhaps a little larger on the anaerobic culture. Microscopically their appearance is very variable. *Strept. pneumoniae*, although typically a capsulated lanceolate diplococcus, may appear non-capsulated in short chains. *Strept. viridans* and *Strept. faecalis* sometimes show very elongated bacillary forms and when these are predominant they may be mistaken for a diphtheroid. Alpha-haemolytic diphtheroids are however very rare and the true nature of the microbe can be revealed by subculture into broth which will show chains of streptococci after overnight incubation. The main constant differences between the three kinds of alpha-haemolytic streptococci can be easily tested and may be tabulated (Table 6).

PROCEDURE FOR IDENTIFICATION OF ALPHA-HAEMOLYTIC STREPTOCOCCI

Seed single colonies heavily on blood agar for bile solubility or for " Optochin " sensitivity tests, and to aesculin-bile medium. Minute colonies showing a clear zone of alpha-haemolysis on the

TABLE 6

	Bile solubility or " Optochin " sensitivity	Growth on MacConkey's bile salt agar[1]	Aesculin bile	Heat resistance
Strept. pneumoniae . .	+	—	—	—
Strep. viridans . . .	—	∓	—	—
Strep. faecalis . . .	—	+	+	+

original plate-culture often fail to grow on MacConkey's medium and then need not be tested for aesculin fermentation.[1]

Bile solubility test

This test depends on a lytic enzyme which is active in the presence of bile salt at pH 7–8. Wash the growth off the surface of

[1] Bile salt varies and each new batch of MacConkey's medium should be tested for its ability to support the growth of *Strept. faecalis*. See Note, page 110.

the blood agar into 1 to 2 ml. saline using a Pasteur pipette bent
to a right angle, being very careful not to scrape up any of the
medium. Distribute this suspension into two sterile plugged
Durham tubes. Add a drop of phenol red to check the pH and
then add a drop of sterile 10 per cent sodium desoxycholate to one
tube, the other is a control. Compare the two fluids and add a
drop of saline to the control tube if necessary to equalize the opacity.
Incubate in a 37° C. water bath and read after 10 minutes and one
hour. If the test is positive, the suspension treated with desoxycholate will be clear and the control will be unaltered. Growth in
5 per cent serum broth is also suitable but it is necessary to centrifuge the culture and resuspend the growth in saline, or to test and
adjust its pH before adding the bile salt because the presence of acid
may inhibit the reaction. All batches of desoxycholate are not
equally effective.

" Optochin " sensitivity

" Optochin " (ethyl hydrocuprein hydrochloride) inhibits the
growth of pneumococci much more than that of other streptococci.
Optochin sensitivity has proved as reliable as the bile solubility test
and much more convenient for routine work (Bowers and Jeffries,
1954). Strips of blotting paper about 8 mm. wide are soaked in
1 in 4000 " Optochin " in distilled water, blotted and then dried in
the incubator. Colonies resembling pneumococci are picked from
the original plate cultures and streaked across a blood agar plate. A
known pneumococcus is included as a control and the strip is laid
on the culture at right angles to the streaks immediately before
incubation. About twelve colonies can be tested on one plate.
Pneumococci show a zone of inhibition about 7 mm. wide, other
streptococci grow up to the strip except for occasional strains which
show very slight inhibition. The test may also be employed for
rapid identification on original cultures. The control is best
maintained by daily subculture on blood agar.

Aesculin-bile test for enterococci

Lancefield grouping is the most reliable single test for enterococci;
it is however time consuming. A much simpler test for hydrolysis
of aesculin in a medium containing 40 per cent bile is almost as
reliable (Swan, 1954) and has proved valuable for rapid identification
(for preparation see page 372). Those which hydrolyse aesculin

blacken the medium after overnight incubation.   Colonies typical
of *Strept. faecalis* which are catalase negative, fail to grow on aesculin
bile but grow on MacConkey's medium (which contains less bile),
can be further investigated by Lancefield grouping or heat resistance.
Aesculin-bile positive strains are presumed to belong to group D
and are reported as *Strept. faecalis.*   *Streptococcus bovis* resembles
Strept. faecalis in being aesculin-bile positive and belonging to
Lancefield's Group D.   It is, however, more sensitive to penicillin
and this is important especially when it causes subacute bacterial
endocarditis.   Detailed examination of streptococci isolated in this
condition is always desirable, for further tests see Cowan and Steele
(1974).

Heat resistance

Mark a blood agar plate into four sectors labelled 0, 10, 20, 30.
Subculture the broth culture on to the area of medium marked 0
then place it in a 60° C. water bath.   After 10 minutes subculture on
to the area marked 10.   Repeat the process after 20 and 30 minutes
exposure to this temperature.   *Strept. faecalis* survives a temperature
of 60° C. for 30 minutes ; other streptococci do not.   The 10 and
20 minutes subcultures may be omitted but as the 30 minutes
exposure to heat is an arbitrarily chosen time, occasional strains may
not withstand it.   Other streptococci do not usually survive 10
minutes at 60° C.   It may be desirable to test by Lancefield grouping,
strains which survive 20 but not 30 minutes.   The differentiation
of *Strep. faecalis* from other streptococci is of considerable practical
importance, because of its different response to treatment with
antibiotics; it is invariably resistant to sulphonamide.

3. Non-haemolytic aerobic streptococci

A non-haemolytic streptococcus may prove to be *Strept. faecalis*,
a non-haemolytic variant of a haemolytic streptococcus, or an
" intermediate streptococcus ".   After overnight culture on blood
agar the colonies vary from minute to 1 mm. in diameter.   *Strept.*
faecalis colonies are comparatively large and show very little
evidence of chain formation when Gram-stained from blood agar.
They may be mistaken for *Staph. albus* or a micrococcus.   To avoid
this error either examine a stained film from negative coagulase-test
broth cultures when chains will be seen in the liquid culture, or do

a catalase test (see below). For identification of *Strept. faecalis* see above.

Non-haemolytic variants of haemolytic streptococci are recognized by Lancefield grouping. All non-haemolytic streptococci isolated in pure culture or found in mixed culture from an infected discharge in the absence of any other likely pathogen, are tested by this method. The "intermediate" streptococci are all those which are not heat resistant and which fail to group by Lancefield's method. As far as is known they are harmless. They can be isolated from the upper respiratory tract, faeces and vagina of healthy people.

Catalase test

Differentiation between staphylococci and streptococci by their morphology is not always possible. In such cases a test for catalase production is of value, since staphylococci and micrococci produce catalase whereas streptococci do not (Isaacs and Scouller, 1948).

Seed a single colony on a nutrient agar slope. After overnight incubation pour several drops of a 10-volume solution of hydrogen peroxide down the slope. When catalase is present bubbles of gas will be seen. (Test for *Mycobacteria*. See page 377.) Colonies which can be easily picked from the plate without scraping off medium with them can be tested rapidly by emulsifying them in a small loopful of water on a glass slide. The addition of a large loopful of hydrogen peroxide will cause very vigorous bubbling when the strain is catalase positive. Some bubbling is caused by catalase in blood from the medium and a doubtful result must be disregarded and the strain retested by the overnight method.

4. Anaerobic streptococci

The anaerobic streptococci and other anaerobic cocci grow more slowly and form smaller colonies than the aerobic cocci. Some are strict anaerobes, others are capable of growth in a microaerophilic atmosphere. Microscopically they are Gram-positive and may show long or short chains. The cocci may be large, elongated or very small. On blood agar incubated anaerobically plus about 10 per cent CO_2 growth is sometimes visible after overnight incubation, but more often 48 to 72 hours' incubation is required before it appears as minute, shiny, smooth, non-haemolytic colonies. Some strains show slight greenish discoloration of the medium after prolonged incubation but true haemolysis of horse blood is rare. They can be isolated from the mouth, vagina and faeces of healthy people. Some strains produce gas and some have a foul odour. Some of the strict anaerobes are very sensitive to oxygen and die after a few hours' exposure on the bench. For this reason cervical and uterine swabs are taken with specially prepared swabs and plunged in transport medium which prevents exposure to oxygen (see page 362).

This group of cocci is not well differentiated but some are undoubted human pathogens and cause severe puerperal sepsis with invasion of the bloodstream. Gas-forming strains sometimes give rise to pseudo gas-gangrene when they invade wounds. They are also commonly found in association with other anaerobes in chronic infection of the mouth and they can be recovered from pus in chronic empyema, from lung abscesses, abdominal abscesses and from pilonidal sinuses and umbilical sepsis. They are also found in tissue at the edge of chronic ulcers of the skin and in infected sebaceous cysts.

If these delicate microbes are not to be missed in routine work the following procedure must be followed : When Gram-positive cocci are seen in the stained film of the original material and do not appear on the plate cultures within 48 hours' incubation, reincubate for two more days. The cocci will probably appear on the anaerobic plate. If cultures of purulent material, in which no acid-fast bacilli have been seen, are sterile after 48 hours' incubation, continue incubation for a week if necessary. Subculture the original cooked meat broth culture after 48 hours on blood agar and incubate the plate anaerobically for at least 48 hours. The gas-forming strains show obvious signs of growth in cooked meat medium but non-gas producers may show none for several days. The cooked meat broth should therefore be subcultured on the fourth day whether there is evidence of growth by this time or not.

Strains which form chains and fail to grow in air plus 5 per cent CO_2 are reported as anaerobic streptococci. Strains which fail to form chains, even in liquid media, are called anaerobic Gram-positive cocci. Further tests of identification are not made routinely because they throw no further light on pathogenicity. This is however a promising field for investigation (Thomas and Hare, 1954, Weinberg, 1974).

NEISSERIA

Members of this genus are all aerobic Gram-negative diplococci. Numerous different species have been described, but apart from the two important pathogens *N. meningitidis* and *N. gonorrhoeae* they are not satisfactorily classified.

Gram-stained films of infected discharges show Gram-negative oval diplococci with the long axis of the cocci parallel. Many of them are intracellular.

Culture : 18 to 24 hours

(*a*) *Blood agar aerobic plus 5 to 10 per cent CO_2.* 0·2 to 1 mm. colonies smooth, shiny, grey or colourless with entire edge, opaque, usually no haemolysis. Various other types of colony are encountered. They may be rough and dry and may show yellow pigmentation.

(*b*) *Blood agar anaerobic plus 10 per cent CO_2.* Colonies very small. Some strains are strict aerobes.

(*c*) *Microscopic examination.* Large Gram-negative cocci, many of them round and single, unlike the bean shaped diplococci seen in the original material. The cocci are fragile and some disintegrating ones are almost always seen, they are particularly numerous in films of gonococcal cultures.

Some strains of *N. meningitidis* and *N. gonorrhoeae* are difficult to maintain in culture and are sensitive to cold. Therefore keep the cultures in the incubator except during actual examination, and subculture on alternate days, while tests are being made, to keep the strain alive. For primary isolation the blood agar must be fresh and moist. An open test-tube containing water and a blotting paper " wick " will help to keep the atmosphere moist in the CO_2 jar. (" Chocolate " blood agar is often recommended for *Neisseria* and *Haemophilus* and it supports their growth well. It is not, however, much superior to a highly nutrient defibrinated horse blood agar, which will support the growth of small inocula equally well although colonies are not quite so large. The advantage in diagnostic work of relying on blood agar is that growth of fastidious organisms will be obtained even when it was not anticipated. Logically one must either include " chocolate " blood agar for all primary cultures or have a blood agar medium sufficiently good to make it unnecessary.)

In routine hospital work it is very important to identify for certain *N. meningitidis* and *N. gonorrhoeae*, but there is no need to attempt detailed investigation of the other members of the group. Those which grow on plain nutrient agar and ferment sugars are labelled *N. pharyngis*, regardless of colonial morphology, and those which grow on nutrient agar but which fail to ferment sugars are called *N. catarrhalis*. Seed pure growth from a blood agar plate or slope on to nutrient agar and on to hydrocele fluid sugar slopes and read the results after 1 to 4 days' incubation in air plus 10 per

TABLE 7

Nutrient	*N. meningitidis*	*N. gonorrhoeae*	*N. pharyngis*	*N. catarrhalis*
Agar slope . . .	–	–*	±	±
Glucose	A	A	A	–
Sucrose	–	–	A	–
Maltose	A	–	A	–
Pigment	∓	–	+ or –	+ or –

A = acid formation   + in line 1 = growth
* See text below

cent CO_2. Colonies showing much pigment formation or a very dry and irregular surface are not meningococci or gonococci. Most members of the genus other than these two will grow on plain nutrient agar without difficulty so that it is a useful preliminary test, but occasional strains fail to do so. Gonococci occasionally grow comparatively well on nutrient agar (Hansman, 1970), but even such strains cannot be maintained on this medium without enrichment. Sugar fermentation must be tested on solid media because *Neisseria* do not grow easily in liquid culture. The medium can be made with rabbit or human but not with horse serum because it contains maltase which may break down the sugar and give a false positive reaction. Hydrocele fluid gives better growth than serum and it is well worth making an effort to obtain it ; the slopes must not be dry (for preparation, see page 375). Some strains of *N. meningitidis* and *N. gonorrhoeae* fail to grow without added CO_2 and almost all grow better in its presence. The carbonic acid formed by CO_2 absorption turns the indicator slightly pink. When growth has occurred, leave the slopes overnight in the incubator without CO_2 and the unfermented slopes will lose their colour.

The identification of *N. meningitidis* is usually straightforward.

Sugar fermentation can be confirmed by agglutination tests if necessary. Agglutination is, however, of limited value only in differentiating meningococci from gonococci because they possess common antigens.

Some strains of *N. gonorrhoeae* ferment glucose with difficulty but they can usually be shown to do so after several subcultures if incubated for 4 days or more.

Rapid identification by fluorescent antibody technique

Staining by fluorescent specific antiserum is a reliable method of identification even when cultures are not pure, and it is possible to distinguish *N. gonorrhoeae* and *N. meningititis* from the other Neisseria by this method within 24 to 48 hours of receiving the specimen. Fermentation tests may be needed to differentiate the gonococcus from the meningococcus because either will react with the antiserum to some extent. In practice, however, there is no great urgency for final identification once the presence of one or other of these species has been confirmed.

The reliability of the test, as in other antigen-antibody reactions, depends on the specificity of the serum. A highly specific fluorescent antiserum combined with a contrasting fluorescent background stain can be successfully employed for rapid diagnosis on smears of the purulent discharge (Fry and Wilkinson, 1964) but in less skilled hands using commercially prepared antiserum it is safer to employ the delayed method, staining Gram-negative cocci which resemble Neisseria on primary culture plates with the fluorescent antibody. The fluorescin-conjugated antigonococcal serum obtainable commercially (Difco Laboratories), is used at its working titre. Two-fold dilutions of each batch are tested on smears of gonococcal cultures. The working titre is one dilution stronger than the highest dilution showing maximal fluorescence. After titration the undiluted serum is kept frozen at $-20°$ C. in convenient small aliquots for appropriate dilution on the day of the test.

METHOD. Make a smear in a drop of tap-water on a clean glass slide as for Gram-stain and allow it to dry in air without heating. Mark round the area with a grease pencil and place a drop of fluorescent antiserum over it. Leave at room temperature for 5 minutes, wash well in two changes of buffered saline, pH 7·2. Mount in buffered glycerine, pH 9, and examine under the optical system described on page 359. A positive control smear from a culture and a negative control *N. catarrhalis* smear must be included on each occasion.

Do not report a strain as *N. gonorrhoeae* or *N. meningitidis* until a satisfactory *laboratory* diagnosis has been made. The knowledge that the organism was recovered from an infected urethra or cerebrospinal fluid must not be used as evidence in identification (see Chapter 1). *N. meningitidis* has been known to cause urethral infection and gonococci may be isolated from cerebrospinal fluid. Moreover some harmless Neisseria closely resemble the pathogens in the early stages of identification.

Gram-negative anaerobic cocci

Members of this group, the *Veillonella*, are sometimes met with in chronic infective discharges. They have not so far been proved pathogenic. The cocci are small and round, unlike *Neisseria*.

CORYNEBACTERIA

In clinical bacteriology the main problem is to distinguish as quickly as possible *C. diphtheriae* from other members of the genus. The main points of difference after overnight culture may be tabulated thus :

C. diphtheriae	Other Species
Bacilli very pleomorphic with many long slender forms.	Bacilli fairly regular often short.
Many metachromatic granules.	Few granules.
Typical grey-black colonies on tellurite medium.	Pale grey or lampblack colonies on tellurite medium.
Some strains haemolytic.	Haemolysis very rare.

None of these findings is significant when taken singly, but consideration of all of them usually makes it possible to send a preliminary report.

The preliminary stages of identification have already been described in Chapter 4. The suspected colonies from tellurite blood agar were seeded into Hiss' serum water and on to a Loeffler slope for sugar inoculation and a virulence test, when required. Liquid serum sugar media are difficult to maintain in good condition, therefore the Hiss serum water culture is seeded heavily, about 5 drops, into each of the appropriate peptone water sugars. The serum carried over in the inoculum is sufficient to ensure good growth. Starch hydrolyses easily in solution and should not be stored in this

TABLE 8

	Glucose	Sucrose*	Maltose	Starch	Haemolysis
C. *diphtheriae gravis* . .	A	—	A	A	±
C. *diphtheriae mitis* . .	A	—	A	—	+
C. *diphtheriae intermedius* .	A	—	A	—	—
C. *xerosis* group . . .	A	A	A	—	—
C. *hofmanni* group . .	—	—	—	—	—

* See text below.

state for longer than a few days. The results are almost always clear after overnight incubation. Occasionally with slow-growing strains it is necessary to reincubate and read after 48 hours.

Virulence test

When a gravis or intermedius strain has been isolated from a patient showing symptoms of diphtheria, virulence is not usually tested because almost all these strains are virulent. Non-virulent *C. diphtheriae mitis* however is common and its pathogenicity needs to be confirmed even when it has been recovered from a sick patient. All strains from carriers are tested because there is always a chance that even gravis or intermedius strains will prove avirulent.

Atypical sucrose-fermenting strains of *C. diphtheriae* are sometimes encountered (Mauss and Keown, 1946). When a " diphtheroid " appears to be playing a major role in any infection it is advisable to test its virulence regardless of the sugar reactions.

METHOD. Add about 2 ml. peptone water to an overnight Loeffler slope culture and scrape the growth from the slope into it carefully without removing pieces of the medium. Transfer this heavy peptone water suspension to a sterile tube. (Do not make the suspension in saline which will kill some of the bacilli within half an hour. Do not use a liquid culture because the growth will not be sufficiently heavy.)

Select two guinea-pigs of approximately equal size. Inoculate one of them intraperitoneally with 100 units of antitoxin and then both of them subcutaneously with about 0·2 ml. of culture. The antitoxin is rapidly absorbed from the peritoneal cavity and provided this large dose is given, it will protect the pig even if the culture is given immediately afterwards.   The unprotected pig will die within 48 hours if the strain is virulent.   Autopsy reveals haemorrhage in the suprarenals and oedema at the site of inoculation.   The protected pig will remain healthy and can be used after a suitable interval for tests other than diphtheria virulence.

Toxin production test

The virulence test described above is very extravagant if many strains have to be investigated.   An alternative is to test them for toxin production on serum agar (Elek, 1949).

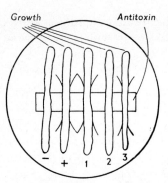

FIG. 15.—Unknown strains 1 and 3 are positive.   Unknown strain 2 is negative.

METHOD.   Liquefy 10 ml. of medium (preparation, see page 373) in a boiling water bath or steamer and when sufficiently cool add 2 ml. of normal horse serum, pour it into a sterile Petri dish.   Dip a strip of sterile blotting paper into a solution containing 1000 units per ml. of purified anti-toxin, drain it of excess fluid and place it across the middle of the liquid medium in the Petri dish.   It will settle below the surface as the agar solidifies.   Dry the plate well in the incubator and then streak the strains to be tested heavily across it at right angles to the antitoxin strip.   Include a control positive and negative strain on each plate.   Incubate the cultures and examine them after 24 and 48 hours.   If toxin is produced it will diffuse into the agar and lines of precipitation will be seen where it meets, in optimal concentration, the antitoxin diffusing from the paper strip (Fig. 15).   That the line of precipitation is caused by toxin is confirmed by comparing its position with lines formed by the positive control.   When they are in identical relative positions it may be assumed that they are caused by the same substances and that the test strain is a toxin-producer.

Since the method has been introduced there have been very few freshly isolated strains of *C. diphtheriae* in Britain with which to test it.   Experience has shown that false negative and false positive results are seen (Bickham and Jones, 1972).   Even lines of identity with the control may be due to non-specific substances. When strains are few guinea-pig inoculation is preferred. Public Health measures should not be instituted on the evidence of a diffusion test alone.

COLIFORMS

This name includes all aerobic non-sporing Gram-negative bacilli, capable of profuse growth in peptone water without the addition of blood or serum and capable of forming colonies about 2 mm. in diameter after overnight incubation on MacConkey's bile salt agar. The medically important coliforms include the *Escherichia, Klebsiella, Salmonella, Shigella, Proteus, Pseudomonas* and other groups

TABLE 9
Lactose fermenters

	Primary tests						Additional tests							Species
Motility	Indole	Lactose	Sucrose	Inositol	Urea	Adonitol	Dulcite	Citrate¹	V.P	Gluconate	Malonate	Gelatin		Species
+	+	(AG)	(AG)	−	−	−	(AG)	−	−	−	−	NL	*Esch. coli*
+	−	AG	AG	−	(+)	(AG)	(AG)	+	+	+	−	(L)	*Ent. cloacae*
+	−	(AG)	AG	AG	(+)	−	−	+	(+)	+	−	L	*Ent. liquefaciens*
+	−	AG	(AG)	−	(+)	−	(AG)	+	−	−	(+)	NL	*Citrobacter freundii*
+	−	AG	AG	AG	−	AG	(AG)	+	+	+	+	(L)	*Ent. aerogenes*
+	−	(AG)	−	−	−	−	−	+	−	+	+	(L)	*Arizona group*
−	−	AG	AG	AG	+	AG	(AG)	+	+	+	+	NL	*Klebs. aerogenes*
−	+	AG	AG	AG	+	AG	(AG)	+	+	+	+	NL	*Klebs. oxytocum*
+	−	(AG)	AG	(AG)	−	(AG)	−	+	+	+	(+)	L	*Serratia marcescens*
+	−	(AG)	(AG)	−	(+)	−	(AG)	+	−	+	−	NL	*Ballerup-Bethesda group*

AG = acid and gas. NL = not liquefied. L = liquefied.
Entries in parenthesis = variable result.

(see Tables 9, 10 and 12). They are found in normal and abnormal intestines and from there may invade other parts of the body to cause severe or even fatal infection.

Identification of atypical members of the common medically important genera sometimes necessitates numerous biochemical tests outside the scope of this book. *Identification of Medical Bacteria* by Cowan and Steele is recommended as a further guide.

In diagnostic work, tests are needed to decide as quickly as possible whether non-lactose fermenters isolated from faeces are

¹ Citrate utilization in Koser's or Simmons' medium.

Salmonella or *Shigella* and it is also necessary to identify coliforms from urine, wounds and other specimens sufficiently to be able to decide whether, when a patient suffers repeated infection, the episodes are caused by the same or by a different species.   Moreover, coliforms are important in hospital infection and an epidemic may be either missed or wrongly suspected if strains are inadequately identified.

The tests described here have been selected to fulfil these needs. In laboratories dealing with large numbers of faecal strains, the list of preliminary tests for non-lactose-fermenters may be abbreviated and Kligler's or other composite medium may be employed in the first instance, (see page 96).

Coliforms are inoculated into the media for primary tests listed in Tables 9 and 10 according to their ability to ferment lactose on the primary MacConkey plate.   When this is not known they are assumed to be non-lactose-fermenters and inosite is added to the preliminary set of tests.   When after overnight incubation identity is still in doubt, the tests on the right of the line in the appropriate Table (9, 10, or 12) are carried out and all tubes are reincubated. When identity is still obscure always check Gram-stain and purity before proceeding further.

Ability to ferment lactose is no longer considered fundamental in classification.   Some non-lactose fermenters are now included in the *Escherichia* genus.   Nevertheless lactose fermentation is a useful guide to further tests.   For example, all lactose fermenters also ferment glucose and mannite, therefore there is little point in testing them.   Adonitol and inosite are better for further differentiation but both need not be employed primarily and as inosite is much less costly it is preferred.   The majority of lactose fermenters prove to be typical *Escherichia* or *Klebsiella*: the short list of primary tests is the minimum for their identification.

Lactose-fermenting coliforms

1. *Isolated from faeces.*   In faeces from individual adults investigated because of enteritis these can be disregarded, but in infants up to the age of 3 years some strains of *Esch. coli* cause acute diarrhoea.   These are usually non-motile, ferment lactose, glucose and mannite with acid and gas overnight, and sucrose and salicin later.   Strains are agglutinated by specific antisera ; there is as yet no satisfactory selective medium.   Traveller's diarrhoea may be caused by unfamiliar types of *Esch. coli* (Rowe *et al.*, 1970) but

investigation of *Esch. coli* isolated from individual patients is un-rewarding. When an outbreak has occurred and no other cause can be found *Esch. coli* strains should be sent to a reference laboratory for typing.

METHOD (Taylor and Charter, 1952)

Using a polyvalent serum[1] pick at least 5 colonies including all morphological types and test each for slide agglutination (see page 141). Test also the confluent growth if these are negative. When the confluent growth but no single colony is positive, replate from the agglutinated particles. Test agglutinating colonies in acriflavine (page 143) because pathogenic coli do not agglutinate in this or in saline.

When slide agglutination is positive with the polyvalent serum and negative in acriflavine, test further with specific O sera to dis-cover to which serotype the strain belongs.

Slide agglutination must always be checked by tube agglutination. Inoculate broth with a single colony and incubate overnight. Heat the culture in a boiling water bath for half an hour, adjust the opacity with saline to about 250×10^6 organisms per ml. and test as des-cribed for Salmonella (Table 12, page 147).

2. *Isolated from specimens normally sterile.* In order to assess the part that these microbes play both in individual infection and in epidemiology they must be identified ; this is particularly important in urinary infection.

A well isolated colony is seeded into peptone water and after a few hours' incubation the media listed for primary tests in Table 9 are inoculated from it. After overnight incubation the majority of strains can be identified but some may need further tests as indicated in the Table. The " sugars " are reincubated and examined daily until identification is established (see Non-lactose-fermenting coli-forms, 2).

Non-lactose-fermenting coliforms

1. *Isolated from faeces.* Those which ferment sugars or give other positive reactions unlike *Salmonella* or *Shigella* can be dis-carded without further tests. Those which fail to ferment glucose overnight or ferment it with a large amount of gas and do not ferment mannite can also be discarded. For interpretation of growth on Kligler's medium see page 96.

[1] Sera are obtainable from Wellcome Reagents Ltd., London.

Typical strains of the Salmonella and Shigella genera can be relied on to ferment glucose, and mannite (when the species ferments it) after overnight incubation. Gas-producing species are listed in Table 10, non-gas-producing species in Table 12. All strains resembling *Salmonella* and *Shigella* must be identified serologically to confirm without doubt a clinical diagnosis of infection, to recognize carriers of these potentially dangerous organisms and for epidemiological purposes.

2. *Isolated from specimens normally sterile.* *Salmonella* are often isolated from these sites and *Shigella* may be. They must be fully identified for the reasons given above. Other non-lactose fermenters from these sites must also be identified. Those commonly encountered are described below. To assess progress and advise treatment for individual patients, identification must be taken to the point where the species can be recognized again in recurrent infection. In most cases the genus and species can be named but sometimes, even when the second line tests have been done and the strain checked for purity, identification may still be obscure. Strains thought from other evidence to be the prime cause of infection should be further pursued. When in doubt, for example an apparently pure growth from a urine with few pus cells, it is usually more rewarding to check evidence of pathogenicity (i.e. isolation from further specimens, symptoms related to the site of isolation, etc.) before proceeding further with identification tests outside the scope of this book.

Salmonella Group [1]

Members of this important group of pathogens are usually isolated from faeces, urine or blood. They may however be found in other sites such as pus from osteitis or from a chronically discharging wound in a patient who is a salmonella carrier. Cultures of non-lactose-fermenting coliforms which give the reactions of this group in the routine primary tests shown in Table 10, or of *Salm. typhi* (see Table 12) need further investigation.

PROCEDURE. The strain is tested by slide agglutination with the polyvalent O serum which reacts with strains in groups A to G, provided agglutination is not blocked by Vi antigen, which can be checked by testing all negative strains against *Salm. typhi* Vi serum. The Vi antigen is possessed by various *Enterobacteriacae*, a positive

[1] See Note, page 148.

TABLE 10

Non-lactose-fermenting gas producers

Indole	Lactose	Glucose	Mannite	Sucrose	Dulcite	Urea	Agar	Salicin	Citrate	V.P.	Gluconate	Malonate	Phenylalanine	Gelatin	H₂S	Species
−	−	AG	AG	−	AG	−	NP	−	+	−	−	−	−	NL	+	*Salmonella* (see *Salm. typhi*, Table 11)
−	(AG)	AG	AG	−	−	−	NP	−	+	−	−	+	−	(L)	+	Arizona group
−	−	AG	AG	(AG)	−	−	NP	(AG)	+	+	−	+	−	NL	+	*Ent. hafnia*
+	−	AG	(AG)	(AG)	−	−	NP	AG	+	−	−	−	+	NL	−	Providence group
−	−	AG	−	(AG)	−	+	NP	−	(+)	(+)	(+)	−	+	L	+	*Proteus mirabilis*
+	−	AG	−	AG	−	+	NP	AG	(+)	−	−	−	+	(L)	+	*Proteus vulgaris*
+	−	AG	−	−	−	+	NP	−	−	−	−	−	+	NL	−	*Proteus morgani*
−	(AG)	AG	AG	−	(+)	−	NP	(AG)	+	(+)	+	−	−	L	−	*Ent. liquefaciens*
+	(AG)	AG	AG	(AG)	(AG)	−	NP	(AG)	−	−	−	−	−	NL	−	*Esch. coli*
−	(AG)	(AG)	AG	AG	−	−	(P)	(AG)	+	+	−	+	(+)	L	−	*Serratia marcescens*
−	(AG)	AG	AG	(AG)	(AG)	(+)	NP	(AG)	+	−	+	−	−	−	+	Ballerup-Bethesda group

AG = acid and gas. NP = no pigment. P = pigment.
NL = not liquefied. L = liquefied.
Entries in parenthesis = variable result.

result in the absence of other diagnostic criteria is therefore of little significance. Strains which prove negative to both polyvalent O and Vi serum are unlikely to be members of the salmonella group.

METHOD. Emulsify a small amount of growth from the agar slope in a large loopful of saline on a clean glass slide without spreading the drop. The emulsion must be absolutely smooth and of medium opacity. Add one small loopful of specific serum and mix without spreading. Positive agglutination should be visible to the naked eye by the time the fluids are well mixed ; late results are disregarded. Discard the slide immediately into disinfectant.

In practice strains are tested against several specific sera for slide agglutination, some of which prove negative so that there is no need for an additional saline control of the suspension. Such a control must, however, be included when all tests prove positive. A small additional loopful of saline is added instead of serum. Acriflavine can also be employed to reveal an unstable suspension when tests prove difficult to interpret (see page 143).

Positive results indicate a *Salmonella* and the next step is to test with sera prepared against O antigens of the individual Salmonella groups. In Britain the most useful sera are those reacting with factors 4 and 5, group B ; with 6 and 7, group C ; with 8, group C_2 ; with 9, group D ; and with 3, 19, 15, and 19, group E.

When biochemical reactions are typical of *Salm. typhi* (Table 12) the polyvalent sera are omitted and the strain is tested directly against *Salmonella typhi* O serum, factor 9, and salmonella H specific serum which reacts with flagella antigen d. It must be remembered that a non-motile Vi strain may react with neither of these, but such strains are rare. When positive results are obtained test it next against polyvalent H non-specific and H specific and non-specific sera. When a good result is obtained with the specific serum and a feeble or negative one with the non-specific serum, the organism is in the specific phase and will react with the appropriate H specific serum. Turn to the Kauffmann-White table of salmonella antigens and test it against H specific serum of the organisms in the appropriate group.

Example : the following results might be obtained by slide-agglutinating a strain isolated from faeces.

Polyvalent *Salmonella* species " O "	+	Polyvalent " H " (specific and	
Group B factor 4	+	non-specific)	+
,, C ,, 7	−	Polyvalent " H " non-specific	−
,, C_2 ,, 8	−	*Salm. paratyphi B* H antigen b	−
,, D ,, 9	−	*Salm. typhi murium* H antigen i	+

The tests first made against O sera have placed the organism in group B. The polyvalent H serum reactions show that it is in the specific phase, therefore it must be tested with H sera prepared for the specific antigens in group B, the commonest in Britain being *Salm. paratyphi B* and *Salm. typhi murium*. It is a waste of time and serum to test it in the first place with sera prepared for H antigens not found in members of this group.

False positive slide agglutination

Slide agglutinations will not always be satisfactory for final identification unless all non-specific antibody has been absorbed from the sera to prevent false reactions. Tube agglutinations with each serum which gives a positive slide test should be set up. If the strain is agglutinated by the specific serum diluted to the titre marked on the bottle, and the saline control of the suspension is satisfactory, identification has been made. When, in the example given, the polyvalent sera only agglutinate the bacillus to titre, it is probably a member of one of the rarer groups with an antigen in common with those of group B.

Diagnostic sera sometimes give a positive slide test with rough non-virulent variants and with other coliforms which are not stable in suspension, moreover this instability may not be demonstrable in a saline suspension on the slide. When this occurs with the polyvalent O serum, a false positive preliminary report may be sent. To avoid this a control suspension is made in 1 : 500 acriflavine solution in saline. The growth is emulsified in this solution on a slide in the usual way without adding serum. When the result is negative it lends weight to the positive serological test. When it is positive a guarded preliminary report must be sent because the organism is likely to prove unstable in suspension and the slide test result to be falsely positive. Agglutination in acriflavine is however sometimes encountered with Vi strains of *Salm. typhi* and other virulent *Salmonella* in the non-specific phase. The test can be used to differentiate " rough " and " smooth " colonies when there is no morphological difference between them. The acriflavine should be stored in the dark and renewed monthly.

Tube agglutination tests

APPARATUS REQUIRED

> Standardized killed bacterial suspension.
> Specific antisera.[1]
> 50 dropping pipettes (or Pasteur pipettes with the *external* diameter of their cut ends approximately equal.)
> Dreyer's agglutination tubes, or round-bottomed 1 × 7·5 cm.
> Water bath controlled at 50° C.

SUSPENSIONS. Living suspensions give good reactions in agglutination tests but they are dangerous to handle. There is no rapid way of killing a suspension which leaves it agglutinable by both H and O sera. It is therefore customary to make two killed suspensions, one for H and one for O agglutination tests.

For H agglutination add a few drops of formalin to an overnight broth culture of the organism. Adjust the suspension to the required density about 250×10^6 per ml., with saline.

For O agglutination wash the growth off the agar slope with 1 ml. of saline and transfer the fluid to a clean tube or screw-capped

---

[1] Obtainable from Wellcome Reagents Ltd., Beckenham, Kent, England.

bottle.   Heat in a boiling water bath for 5 to 10 minutes, then add sufficient saline to make a standard suspension as above.[1]

METHOD.   Set up five Dreyers tubes for each serum to be tested and add the appropriate fluids to them as shown in Table 11.   One control tube is required for each suspension, to check auto-agglutinability of the bacilli.

Add saline first to all tubes.   Using a fresh pipette add the serum to tube 1 and discard the pipette.   Take the saline pipette again, mix the contents of tube 1 well and transfer 10 drops to tube 2, continue in this way making twofold dilutions, and discard ten drops from the fifth tube.   Take a fresh pipette and add ten drops of suspension to each tube.   Incubate in a 50° C. water bath for four hours and read.   When the results are negative, stand on the bench overnight and read next morning.

H agglutinations can be read after two hours in the water bath but it is worthwhile to reincubate for the full time if the two

TABLE 11

Tubes	1	2	3	4	5	6	
Saline drops	18	10	10	10	10	10	10 drops
Serum drops	2	10	10	10	10	0	discarded
Suspension drops	10	10	10	10	10	10	
Final dilution	1/20	1/40	1/80	1/160	1/320	CONTROL	

hour reading is negative.   It is important to read agglutination tests in a viewing box in a bright light with a dark background.   Under these conditions standard agglutination is seen naked eye as clumping of the bacilli, which fall to the bottom of the tube, with clearing of the supernatant fluid.   Traces of agglutination and very fine complete O agglutination may be seen only with the aid of a hand lens.   The test is a quantitative one and a false negative result may be seen unless the suspension is of standard opacity and the dilutions are made carefully with pipettes of comparable size held vertically while dropping.

The titre of the serum is marked on the bottle, it is normally at least 250.   In most cases freshly isolated strains react well and tubes 1 to 5 will show standard agglutination.   If the titre is less than 160 the result is unsatisfactory.   (See page 146.)

[1] Compare with Brown's opacity tube No. 1.   The opacity is such that print is just readable through the suspension.

Difficulties encountered in the identification of Salmonellae

1. *Positive O and negative H agglutination in a non-motile culture.*

PROCEDURE. Inoculate from the peptone-water culture a semi-solid (about 0·5 per cent) nutrient agar plate near the periphery and incubate. Flagellated bacteria will spread across it and subcultures made from the spreading edge will be motile. It may be necessary to repeat the subcultures several times before a satisfactory result is obtained. Alternatively, a Craigie's tube (see below) without antiserum can be employed.

2. *Culture in the non-specific phase.*

This is shown by positive agglutination with the polyvalent H non-specific serum and a negative or feeble result with H specific serum. It is then necessary to set up a Craigie's tube.

PROCEDURE. Pour some sloppy (0·5 to 1 per cent) nutrient agar into a sterile 15 × 2·5 cm tube. Before it sets, mix with it three or four drops of H non-specific serum and place in the middle a piece of sterile glass tube (see Fig. 16). When the agar has set, seed the culture into the middle of the glass tube and incubate. As the majority of bacteria in the culture are in the non-specific phase they will be agglutinated by the serum, but those in the specific phase will be unaffected and will multiply and find their way to the surface of the agar outside the central tube. Subcultures made from this surface will give positive results with specific sera. It may be necessary to repeat the Craigie tube culture several times before the non-specific culture will give rise to bacilli in the specific phase.

FIG. 16.—Craigie's tube.

3. *Positive slide agglutination with more than one group serum.*

There are antigens common to several groups, for example factor 12 is common to groups A, B and D, therefore a strain of *Salm. typhi murium* possessing this antigen may give positive slide agglutination with three group sera. The results of tube agglutination tests will indicate to which group the organism belongs since only an organism possessing a large proportion of the factor specific to the group is likely to be agglutinated to titre.

Sometimes, however, positive results are obtained with O sera from groups with no diagnostic common antigen. In this case the reaction may be due to a common antigen not recorded in the Kauffmann-White scheme or to a non-specific antigen reacting with non-specific antibody originally present in one of the sera (as a result of naturally occurring coliform infection in the rabbit from which the serum was taken). Tube agglutination almost always shows a low titre with such reactions but occasionally high titres are met with. It is therefore very important to perform tube tests to check all positive slide tests and not to be satisfied unless both H and O sera react to titre with the unknown strain. Other enterobacteria giving positive non-specific reactions can easily be mistaken for members of the Salmonella group.

4. *Tube agglutination negative or positive in low titre only.*

PROCEDURE. Retest from a fresh agar slope subculture, using a carefully standardized suspension.

While agglutination tests are proceeding reincubate the " sugars " and set up the " Additional tests " listed in Table 10. If the repeated tube agglutination tests are still unsatisfactory the organism is most unlikely to be a member of the group but it cannot be excluded on this limited serological evidence alone. Much more detailed examination, using sera not available in routine laboratories, is necessary to exclude rare *Salmonella*. A preliminary report may be sent that a bacillus, probably not a *Salmonella*, is under investigation, and the biochemical test cultures are kept in the incubator and examined daily. They may remain unchanged for some days but after 3 to 4 weeks' incubation, and often sooner, one of them will reveal the genus to which the strain belongs. In addition to the routine tests, it is worth while to inoculate equally two peptone-water tubes, incubate one and leave the other in the dark on the bench. If growth at room temperature (as judged by opacity) is heavier, the strain is not a *Salmonella*. A room temperature agar slope culture will sometimes reveal pigment formation when the incubated cultures remain unpigmented, which again excludes a *Salmonella*.

5. *Tube agglutination satisfactory with O but not with H sera when the culture is motile and in the specific phase.*

The strain is probably one of the rare members of the group and is sent to a reference laboratory for further tests.

6. *Biochemical tests positive, slide agglutination tests negative.*

Freshly isolated strains of *Salm. typhi* often have Vi antigens which may mask the O agglutination. Occasional strains therefore which give the biochemical reactions of *Salm. typhi*, are non-motile and are not agglutinated by specific O and H sera, are tested by slide agglutination with Vi serum. Vi antigen is not specific and its presence cannot be taken as evidence that a coliform bacillus is *Salm. typhi* unless biochemical tests are typical. Confirmation of identification by a reference laboratory will then be required.

Shigella Group

Table 12 lists the typical biochemical reactions which are given by the dysentery bacilli and other non-gas-producing coliforms after overnight incubation in the routine biochemical test media. It must be emphasized that at this stage no report can be made. Any of these cultures may prove on further investigation to be a harmless coliform.

PROCEDURE. Test the growth on the agar slope culture for slide agglutination with shigella antisera. The minimal sera required for routine use are polyvalent flexner (composite specific and group) polyvalent boyd, *Sh. sonnei*, *Sh. schmitzi* and *Sh. shigae*. (*Sh.*

TABLE 12

Non-lactose-fermenting non-gas producers

	Primary tests									Additional tests								Species
Motility	Indole	Lactose	Glucose	Mannite	Sucrose	Dulcite	Urea	Agar	Salicin	Citrate	V.P.	Gluconate	Malonate	Phenylalanine	Gelatin	H₂S	
+	−	−	A	A	−	(A)	−	NP	−	−	−	−	−	−	NL	+	*Salm. typhi*
−	−	−	A	−	−	−	−	NP	−	−	−	−	−	−	NL	−	*Sh. shigae*
−	+	−	A	−	−	−	−	NP	−	−	−	−	−	−	NL	−	*Sh. schmitzi*
−	(+)	−	A	(A)	−	−	−	NP	−	−	−	−	−	−	NL	−	*Sh. flexner boydi*
−	−	(A)	A	A	(A)	−	−	NP	−	−	−	−	−	−	NL	−	*Sh. sonnei*
−	(+)	(A)	A	A	(A)	(A)	−	NP	A	−	−	−	−	−	NL	−	Alkalescens dispar group
+	+	−	A	A	(A)	−	+	NP	+	−	−	−	−	+	−	−	*Proteus rettgeri*
+	+	−	A	(A)	(A)	−	−	NP	A	+	−	−	−	−	+	NL	−	Providence group
+	−	−	A	−	−	−	+	(P)	−	+	−	+	−	−	−	L	−	*Pseudomonas*
+	−	−	−	−	−	−	−	NP	−	−	−	−	−	−	(−)	−	*Alkaligenes*
−	−	−	(A)	−	−	−	(+)	NP	−	+	−	−	−	−	(L)	−	Acinetobacter group

A = acid. NP = no pigment. P = pigment.
NL = not liquefied. L = liquefied.
Entries in parenthesis = variable result.

[1] See Note, page 148.

newcastle is agglutinated by the polyvalent flexner serum. *Sh. alkalescens* and *Sh. dispar* are of very doubtful pathogenicity and are now excluded from this group.)

When slide agglutination is strongly positive with one of the appropriate antisera, send a preliminary report and check the result with a tube agglutination test as described for the salmonella group. The suspension is made by scraping growth from the agar slope into saline, adjusting the opacity to about 250×10^6 organisms per ml.[1] and adding a drop of formalin. If the strain is agglutinated to titre the final report can be sent. If slide agglutination is feeble or negative the strain is probably not a dysentery bacillus. Reincubate the original tests and set up the "Additional tests" shown in Table 12. After five days' incubation it is almost always possible to exclude a culture from the dysentery group on the evidence of this augmented series of biochemical tests.

Difficulties encountered in the identification of Shigellae

1. *Positive slide agglutination but low titre agglutination in the tube test* ($\frac{1}{80}$ *or less.*)

Subculture to broth and incubate overnight. Then heat in a boiling water bath for 30 minutes; adjust the opacity of the suspension as before and re-test.

2. *Biochemical tests persistently positive in the absence of agglutination.*

PROCEDURE. Seed the peptone water culture on blood agar and MacConkey agar to test for purity. One or both of these plates will probably reveal a contaminant. Even if none appears, pick a single, well isolated colony from the blood agar plate and repeat the biochemical tests. A contaminant may fail to survive competition with a coliform bacillus and die, having caused confusion by fermenting some of the sugars during its short lifetime.

Repeat the Gram-stain from growth on the blood agar plate. Members of the Bacillus genus, the aerobic sporebearers, can be mistaken for dysentery bacilli if they lose the power of retaining Gram's stain. They can usually be excluded from the dysentery group by their motility or spore formation, but a poorly staining, non-motile, non-sporing strain may cause confusion for some days.

Note. The identification of members of the salmonella and

[1] See footnote, page 144.

shigella groups described above is valid for typical strains of the common pathogens. Rare *Salmonella* have been reported which are indole positive, ferment lactose (late) and even liquefy gelatin. Dysentery bacilli occasionally form gas. When therefore a coliform appears to be the important pathogen in a case of enteric-like disease or in an outbreak of diarrhoea, its agglutinins should be investigated even if its biochemical reactions are unlike those of the majority of members of these groups (Felsenfeld, 1945).

Proteus Group

Numerous species have been described. The biochemical reactions of typical strains are given in Tables 10 and 12. The two most commonly encountered varieties are *Proteus mirabilis* and *Proteus morgani*.

Proteus mirabilis usually swarms on blood agar but some strains show entire colonies. *Proteus morgani* will spread on semi-solid nutrient agar at room temperature but does not swarm on routine culture media.

Inhibition of swarming. Numerous methods have been described to prevent Proteus and other highly motile bacteria from swarming and obscuring other colonies. They depend on two factors, drying the surface of the medium and inhibition of motility by chemical agents; neither is satisfactory. Even very high concentrations of agar in blood agar with over-drying of the plates sometimes fails to inhibit swarming and although added chemicals are effective they inhibit fastidious bacteria which may be important pathogens in specimens contaminated with swarmers.

It is now known that medium containing charcoal inhibits swarming without spoiling its nutrient properties for fastidious organisms (Alwen and Smith, 1967). Charcoal horse-red-cell nutrient agar is the medium of choice for primary culture from sites likely to harbour *Proteus* or *Clostridia*. Whole blood cannot be used because it reduces the adsorbing power of charcoal; some nutrient agar bases are also unsuitable (for preparation, see page 365). Although very effective this medium is not easy to prepare. In practice the problem is often solved by culturing on CLED medium on which most common pathogens will grow, see page 369. Many fastidious anaerobes can be selected by incorporating an aminoglycoside antibiotic in blood agar as an additional culture medium. The selective medium for gonococci (page 366) inhibits Proteus

sensitive to trimethoprim and can be used to isolate meningococci, although some strains of these pathogenic Neisseria fail to
grow on it. Charcoal blood agar should therefore be employed to
separate anaerobic cocci (aminoglycoside sensitive), Haemophilus,
some strains of Neisseria and rarely other fastidious pathogens from
swarming Proteus. It is indicated when bacteria resembling these
are seen in the Gram-stained smear of pus and cultures are spoiled
by a swarmer originating from a few colonies. CLED can be used
to separate swarming Clostridia from each other. If MacConkey
agar is used routinely in addition to blood agar for primary culture
it may be possible to recover some strains of streptococci and staphylococci from it when the blood agar cultures are spoilt by the unexpected presence of Proteus. In this case however the cultures
should be repeated, using charcoal red-cell agar, to reveal any bacteria
which are incapable of growth on MacConkey's medium.

Klebs. friedländeri. Typical strains show large mucoid colonies
on blood agar and MacConkey agar. Some strains ferment lactose
overnight, the majority do not. The stickiness is due to the formation of capsules and sometimes of free slime. Non-capsulated
non-mucoid strains are sometimes encountered. It is non-motile.
Some strains are strict aerobes. The differentiation of this organism
from other Klebsiella and slime-producing coliforms is not at all
easy. Biochemical tests are very variable, those given in the table
are true of typical strains only. The two species most often isolated
from urine are *Klebs. aerogenes* and *Klebs. oxytocum* (Table 9) which
because it is indole positive may easily be mistaken for *Esch. coli.*
Differentiation is worthwhile because, in contrast with Escherichia,
Klebsiella are normally resistant to ampicillin. Neither of these
species are likely to be primary respiratory pathogens although they
are often isolated from the sputum of ampicillin-treated patients.
Klebs. friedländeri can be recognized because like other respiratory
Klebsiella it is MR positive, VP negative but differs from them in
being malonate, citrate and urease positive. The most reliable test
is mouse pathogenicity. Diluted overnight peptone water culture
is injected intraperitoneally (0·1 ml. of a 10^{-5} dilution). If the
strain is virulent *Klebs. friedländeri*, the mouse will die within three
days and the organism will be recoverable from the heart blood.

Primary *Klebsiella* pneumonia is extremely rare but secondary
invasion by antibiotic-resistant strains in patients particularly at risk
is common. Any of the biotypes may be implicated.

Ps. aeruginosa is found frequently in pus, urine and faeces ; it is relatively resistant to most antibiotics and is often isolated from treated wounds. All strains encountered so far are sensitive to polymyxin. When found in treated wounds or urine after catheterization hospital infection should be suspected because it is the most common non-sporing organism to be found contaminating antiseptic solutions and primary infection in healthy people is probably very rare. In epidemological surveys a selective medium employing cetrimide has proved valuable (page 369). In the routine test media it acidifies glucose by oxidation, it liquefies gelatin, usually within 3 days' incubation, and it forms a green pigment which diffuses into the media and makes it easily recognizable. Some strains do not produce pigment easily and may be persuaded to do so by growth in the dark at room temperature or by cultivation on special medium. Fluorescin is also produced which can be seen when the culture is exposed to ultraviolet light but this, unlike pyocyanin, is produced by other pseudomonads. For an account of culture methods, see Brown and Lowbury (1965).

When cross-infection by *Escherichia, Klebsiella* or *Pseudomonas* is suspected evidence for or against it can be greatly strengthened by typing; methods employing bacteriocins, serology, and bacteriophages are used. A method for *Proteus* employing the Dienes phenomenon and a resistogram has also been described (Kashbur *et al.*, 1974).

Non-fermenting Gram-negative bacilli

The identity of many of these strains is still in doubt, but they are tentatively classified according to the morphology especially of their flagellae. Most of them are oxidase positive. The following groups are included ; *Acinetobacter, Achromobacteria, Alkaligenes, Lophomonas, Pseudomonas, Flavobacterium*, and *Moraxella*. Some of them are very short fat bacilli easily confused with Neisseria. *Mima polymorpha*, classified in the *Acinetobacter* group, is notorious in this respect. For further information, the reader is referred to Wilson and Miles (1975, page 836) and to Cowan and Steele (1974).

PARVOBACTERIA

This name is given to five genera of small non-sporing mainly non-motile Gram-negative aerobic bacilli, which grow comparatively poorly on blood agar and poorly or not at all on MacConkey's bile salt agar.   The classification of some of the species is undetermined, different authorities include them in different genera.   It is convenient therefore to have a name comparable with " coliform " to include them all.

An account of preliminary tests which indicate to which genus an unknown strain belongs will be followed by a description of the commonly encountered species.   The reader must then turn to a text-book of systematic bacteriology for further details.

The parvobacteria are all human or animal obligate pathogens.

TABLE 13

Genus	Colonial morphology overnight. Blood agar* aerobic plus 5–10% CO_2	Haemolysis	Motility	Anaerobic growth overnight	Growth in peptone water	Growth on MacConkey agar[1]	Fermentation of carbohydrates (acid only)
Haemophilus	Minute—0·8 mm. transparent	(+)	—	+	—	—	+
Bordetella	No growth	—	—	—	—	—	—
Brucella	Minute—1 mm. Individual colony size variable	—	—	—	+	(+)	—
Pasteurella	0·1—1 mm.	—	—	+	+	—	+
Yersinia	0·1—1 mm.	—	(+)	+	+	+	+
Francisella tularensis	Grows on Dorset's egg medium but not on other ordinary routine media.						

* 10% defibrinated horse-blood agar, see page 363.
Parentheses signify variability.

They are isolated from blood, serous fluids and infected discharges. Differentiation from coliforms is usually clear after examination of the overnight plate cultures.   Many species, particularly haemophilus, grow with difficulty on culture media.   To maintain viability of the strain while tests are being made, the primary cultures are reincubated and daily subcultures are made on blood agar incubated in the atmosphere in which growth first occurred until it has been demonstrated that such frequent subculture is unnecessary.

Occasionally small colony variants of coliform bacilli are encountered, which at first sight may be mistaken for Parvobacteria. Their true nature is revealed by their ability to form gas in sugar fermentation and on repeated subculture they yield a few large typical colonies on MacConkey agar.

[1] See Note page 110.

Haemophilus

The important human pathogens are *H. influenzae* and *H. parainfluenzae*, which normally inhabit the nasopharynx, and *H. ducreyi*, the cause of soft sore. They are all dependent on one or both of the growth factors X and V. X factor is haemin and has to do with iron metabolism of the organism under aerobic conditions. V factor is a codehydrogenase present in many tissues and produced normally by most other bacteria.

Haemophilus has been studied extensively and species can be subjected to a variety of tests and divided into subspecies. *H. influenzae* which causes meningitis is usually encapsulated and belongs to Pitman's type b. There is no evidence that encapsulated strains are better able than others to initiate infection (Turk and May, 1967). Some strains are haemolytic; most of these need V but not X factor and would therefore be classified as *H. parainfluenzae*. The Koch-Weeks bacillus, *H. aegypti*, closely resembles non-capsulated *H. influenzae*. Since *H. influenzae* and *H. parainfluenzae* can also be associated with acute conjunctivitis there seems little point in differentiating them except for epidemiological purposes. In individual patients the pathogenic role of an *Haemophilus* is judged by its presence in large numbers associated with pus cells in the Gram-stained smear of an infected discharge. No *Haemophilus*, whatever its laboratory reactions, can be ignored in these circumstances. They are most commonly found in sputum, in acute bronchitis and bronchopneumonia and in infections of the ear and nasal sinuses. Occasionally they are encountered further afield in abdominal pus and even in urinary infection. They are fairly common in the vagina.

When strains are isolated from unusual sites, *H. parainfluenzae* is more often found than *H. influenzae*. " *H. vaginalis* " needs neither X nor V factor and is probably misnamed.

Until further tests are proved of value in prognosis, prophylaxis or treatment, there is much to be said for retaining a simple classification based on the need for growth factors and leaving further subdivisions to those specially interested.

Chocolate blood agar is often recommended for primary isolation but defibrinated horse-blood agar preparation (page 363) as recommended here for primary culture, is almost as good and has the advantage that *Haemophilus* will be isolated even when its presence was not anticipated (see page 131). When expense is no object both media can be employed routinely. Beta-haemolytic strains isolated from throat swabs may easily be mistaken for haemolytic streptococci. Gram-stain, however, reveals either typical Gram-negative coccobacilli or extremely pleomorphic bacilli with serpent and club forms. The colonies are often transparent and the more usual non-haemolytic form may be missed unless the culture is examined carefully by reflected light.

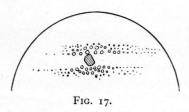

FIG. 17.

The need for X and V growth factors can be conveniently demonstrated in two ways, the simplest is by a satellitism test. Blood agar [1] and nutrient agar plates are inoculated with the test organism in parallel streaks about 1 cm. apart. A spot of a Staphylococcus culture is then seeded between the streaks. (The standard *Staphylococcus* employed as control for antibiotic sensitivity tests is suitable.) The plates are incubated overnight. On blood agar there is plenty of X factor. If the strain needs V factor the bacteria near the Staphylococcus will be favourably placed ; they will therefore form larger colonies than those at the periphery (see Fig. 17).

In plain nutrient agar there is no haemin, so that strains needing X factor will not grow ; those needing V but not X will again show large satellite colonies round the staphylococcal growth. In mixed cultures satellitism may be seen on the original blood agar culture. When the colonial and microscopic appearance of the satellite is typical there is no need to test it further, it can be labelled an haemophilus.

Blood agar contains small amounts of V factor, enough for the isolation of *H. influenzae* in pure culture. If the medium is unusually rich in it, or if the strain needs comparatively little, the satellitism test may fail. Need for the growth factor is then tested by inoculating four peptone-water tubes, one containing V factor, one with X factor, one with both, and one with neither. The tubes are inoculated from growth on blood agar, care being taken not to scrape up any of the medium. They are incubated overnight. (For pre-

---

[1] Routine defibrinated horse-blood agar (page 363) is so highly nutritive for haemophilus that satellitism may not be seen. Oxalated horse-blood agar is more suitable for this test. Strains needing traces only of X factor may grow on some kinds of nutrient agar.

paration of the growth factors see page 378.) As the opacity after overnight culture is not always easily seen a control set of uninoculated tubes is incubated for comparison. This also checks the sterility of the added growth factors.

TABLE 14

	X	V	X and V
H. *influenzae*	—	—	—	+
H. *parainfluenzae* . . .	—	—	+	+
H. *ducreyi*	—	+	—	+

+ = growth.

H. ducreyi. Some strains grow on horse blood agar, but if a clinical diagnosis of soft sore has been made, scrapings from the ulcer floor are seeded on 3 per cent nutrient agar containing 20 to 30 per cent defibrinated rabbit blood in addition to the routine cultures.

Bordetella

Bord. pertussis is the only medically important species.

Whooping cough is sometimes caused by *Acinetobacter parapertussis* (not so far in Britain) and perhaps by *Alkaligenes bronchisepticus*. The association of these organisms with the same clinical condition is no longer thought to be sufficient reason for including them in the genus *Bordetella*. Unlike *Bordetella pertussis*, which is highly fastidious, both will grow on nutrient agar. They differ from each other in motility, growth on MacConkey agar, oxidase and nitrate reduction, *Alkaligenes bronchisepticus* being positive and *Acinetobacter parapertussis* negative in these tests. Adenovirus infection is also associated with a whooping cough and the idea that pathogens which cause similar clinical conditions must be similar is clearly untrue in this case. Although *Bord. pertussis* is not dependent on X and V growth factors it needs a highly nutrient medium without peptone for primary isolation (see page 367). On Bordet-Gengou or Lacey's medium the colonies are o 5 to 1 mm. in diameter after 4 days' incubation and are pearly in appearance. Identification is by agglutination with specific antiserum. Specific antigens are unusually labile ; they can be altered by culture on different media. Therefore a negative agglutination test is not significant unless the strain used in preparation of the specific antiserum was cultured on medium similar in composition to that on which the strain to be tested was grown (Lacey, 1951).

Difficulty may be encountered with autoagglutinable strains. On subculture such strains may become smooth, or it may be possible to suspend the organism in saline, allow agglutinated particles to settle and then obtain a satisfactory tube agglutination test (see page 143) with the smooth supernatant.

Brucella

In Britain *Br. abortus* is much the commonest cause of undulant fever. The organism can be isolated from the blood of patients during one of the bouts of pyrexia. On first isolation growth in an atmosphere containing 5 to 10 per cent CO_2 is essential. Identification of the organism as a member of the genus is easily made by lack of sugar fermentation, inability to grow under strict anaerobic conditions and agglutination with specific antisera. Differentiation into the different species, however, is not a simple matter because antigenically they are very similar. The main points of difference are shown in Table 15.

TABLE 15

	Animal Host	Need for 5–10 per cent CO_2	Growth in the Presence of Standard Dyes		
			Thionin	Basic Fuchsin	Pyronin
Br. melitensis .	Goats and sheep (mainly)	—	+	+	—
Br. abortus . .	Cattle	+ +	—	+	+
Br. suis . . .	Swine	—	+	—	—

In addition, tests for H_2S production and agglutination with absorbed sera are of value. It is possible to divide these species into a number of biotypes which aids epidemiological studies (Wilson and Miles, 1975). Treatment of the patient is the same for all brucella infections so that identification of the species is not a matter of urgency.

Brucella may be isolated from contaminated material by culture on blood or liver agar plates screened with penicillin, or a guinea-pig may be inoculated intramuscularly with the test material; after

6 weeks, autopsy will reveal signs of infection and the organism can be recovered in pure culture from the spleen. The guinea-pig serum will usually show brucella agglutinins from the second week after inoculation.

Pasteurella and Yersinia

Pasteurella septica is fairly commonly encountered in specimens from patients in Britain. Colonies on blood agar resemble Neisseria and may be mistaken for them when bipolar staining is marked. It is carried by animals, including dogs and cats, and is found in bite wounds. It may also be isolated from sputum in mixed culture, but probably comes from the upper respiratory tract in patients in close contact with animals and plays no part in chest infection.

TABLE 16

	Motility 18 hr. at 22° C.	Growth on MacConkey agar[1]	Indole	Sucrose	Maltose	Salicin
Past. septica . . .	−	−	+	A	(A)	−
Y. enterocolitica . .	+	+	−	A	A	−
Y. pseudotuberculosis .	+	+	−	(−)	A	A
Y. pestis	−	+	−	−	A	A

All species ferment glucose and mannite.

Yersinia pseudotuberculosis is well known as a guinea-pig pathogen giving rise to autopsy findings resembling those which result from injection of *Myco. tuberculosis*, as its name implies. Natural infection of these and other rodents kept as pets, may lead to accidental ingestion and human abdominal lymphadenitis (Mair *et al.*, 1960). The bacillus can be cultivated from the cut surface of glands removed at operation, the patient having been suspected of acute appendicitis. The glands are reddish in colour, swollen and soft and can be recognized by the surgeon as likely to be infected with *Yersinia*. Cultures overnight on blood agar yield a few colonies 0·5 to 1 mm. in diameter, non-haemolytic. Biochemical reactions are given in Table 16. Identification can be confirmed by agglutination with specific antisera. There are several types, but almost all human infections are due to type 1. The patient's serum may also agglutinate a known suspension. Some types of *Y. pseudotuberculosis* have antigens in common with *Salmonella*

[1] See Note, page 110.

of groups B and D.   *Yersinia enterocolitica* which closely resembles *Y. pseudotuberculosis* has caused outbreaks of diarrhoea.   Infection with *Yersinia* may simulate Crohn's disease and acute rheumatic fever and is probably more common than is generally supposed. (Chessum *et al.*, 1971, Toivanen *et al.*, 1973).

Plague no longer occurs in Britain, but *Y. pestis* may be imported from endemic areas.   The bacillus is most likely to be isolated from a bubo, i.e. pus from incised inguinal glands from a patient recently landed by air from India or China.

Ps. mallei

Glanders is extremely rare in Britain.   When it is suspected, cultures are made on 4 per cent glycerol blood agar in addition to routine media.   *Ps. mallei* should be considered as a cause of any granulomatous lesion, or generalized infection, when the more common pathogens have been excluded.   Human infections usually arise from contact with horses suffering from glanders.   The organism grows on blood agar but somewhat slowly on first isolation. Plate cultures should not be discarded until they have been incubated for five days ; there is no growth on MacConkey's medium.   Identification is by cultural characteristics and intraperitoneal inoculation of a male guinea-pig which is followed by enlargement of the testicles after two or three days (the Straus reaction), by ulceration of the scrotum and finally death, usually within ten days.   It is included under parvobacteria because of its morphological and cultural characteristics although it is now classified as a pseudomonad.

SPORE-BEARING AND ANAEROBIC BACILLI

The spore bearers are divided into two genera, *Bacillus* which is aerobic and *Clostridium* which is anaerobic.   They are found in air, water, soil, house dust, faeces and wounds.   The anaerobes are more important in medicine and will be considered first.

Clostridia

The degree of anaerobiosis necessary for growth varies with different species.   Some, such as *Cl. tetani*, are very strict anaerobes ; others, such as *Cl. welchi*, are capable of growth in the presence of very small amounts of oxygen.   All the medically important species except *Cl. welchi* are highly motile and tend to swarm on blood agar.   Swarming clostridial growth is sometimes very difficult to see, particularly if the organism has spread all over the plate so that there is no sterile area of medium for comparison. They are usually found in mixed infection, either in pus from deep and lacerated wounds or from the uterus. Many discharges infected with anaerobes have a foul odour. The odour is caused by the breakdown of dead tissue by comparatively harmless proteolytic species. The four most dangerous species, *Cl. welchi*, *Cl. oedematiens*, *Cl. septicum* and *Cl. tetani* produce no foul smell and may be found in odourless discharges.

Swarming and spore formation can be used to aid isolation from mixed cultures. About 0·5 ml. of culture taken with a pipette from the bottom of a cooked meat broth, or a drop of original material, is seeded at the bottom of a blood agar slope and incubated in an anaerobic jar. After 48 hours smears are examined from the slope above the inoculum. It may then be possible to obtain a pure culture of a spreader from the top of the slope. On the whole, however, swarming is a disadvantage because more than one spreading *Clostridium* may be present and then they are very difficult to separate ; moreover, the spreader may obscure the presence of other anaerobes such as anaerobic streptococci and *Bacteroides*. It may be prevented by culture on charcoal agar as already described for *Proteus* (page 149). Cysteine-lactose-electrolyte-deficient medium (CLED) incubated anaerobically can also be employed to separate swarmers from each other but will not support the growth of the fastidious non-sporing anaerobes.

Spore formation aids isolation because other bacteria can be killed by heat, leaving the sporebearers viable. Moreover, since all spores are not equally heat sensitive, it is sometimes possible to separate one sporebearer from another less resistant by suitably heating either liquid cultures or the original material. If the presence of *Clostridia* is suspected either from the appearance of the stained film or from the nature of the lesion, the routine method is amplified. The original material is seeded on to charcoal red-cell agar, to avoid the risk of damage to the cultures by swarming, and also on to Nagler's medium (see below). When all cultures and smears have been made, the remainder of the material is heated in a tube in a boiling water-bath, cultured on blood agar (in air and anaerobically) and into cooked meat broth after 5, 10 and 20 minutes' heating.

Cl. welchi rarely spores and is more likely to be recovered from the original anaerobic plate culture or from the cooked meat broth. It can be identified after overnight incubation by the Nagler reaction. The original material, or a suspected colony, is seeded on a clear medium composed of broth, Fildes' extract, human serum and agar (see page 373). When *Cl. welchi* grows on it the α-toxin, which is a lecithinase, diffuses into the medium and changes the fat in the serum forming a cloudy halo round each colony. This reaction can be specifically inhibited by antitoxin.

METHOD (Hayward, 1941)

Dry a Nagler plate in the incubator for an hour and spread about five drops of antitoxin [1] over one half of its surface. Seed the material for culture so that the pool of inoculum covers part of both

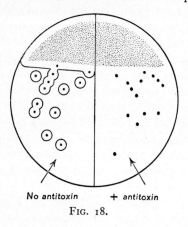

No antitoxin   + antitoxin

FIG. 18.

areas, and spread it out on both halves of the plate, then incubate it anaerobically overnight.

If toxin producing *Cl. welchi* is present colonies about 1 to 2 mm. in diameter of Gram-positive non-sporing bacilli will be found on both halves of the plate. Those on the untreated half will be surrounded by cloudy haloes due to the presence of α-toxin. No haloes will be seen on the antitoxin-treated half because the toxin is neutralized. Provided there are no spores and the organism is an anaerobe the reaction is specific and the presence of *Cl. welchi* can be reported without further tests. *Cl. bifermentans* gives a positive reaction but can be recognized by its spores. Some aerobic bacilli show cloudy zones and closely resemble *Cl. welchi* when grown anaerobically. When growth of the organism is profuse on the original anaerobic plate culture it can be assumed to be an anaerobe if no similar bacilli are found on the aerobic culture. When growth is scanty, lack of it on the aerobic culture may be due to sampling error and it is then necessary to subculture to prove that the organism is incapable of aerobic growth, before a final report can be sent.

[1] Polyvalent therapeutic antitoxin can be used, provided there is not too much preservative in it.

All these *Clostridia* produce toxins which are lethal to mice and guinea-pigs. With the exception of *Cl. welchi*, which is identified by the Nagler reaction, they are finally identified and proved virulent by inoculating two animals with cooked meat broth cultures, one being protected with specific antitoxin. It is usual with the gas gangrene organisms to inoculate guinea-pigs intramuscularly : 2 per cent calcium chloride is added to the culture to produce a nidus of necrosis so that infection can be established. *Cl. tetani* and *Cl. botulinum* are usually identified by intraperitoneal mouse inoculation.

TABLE 17

Clostridia which are known human pathogens

	Spores	Lactose	Glucose	Sucrose	Maltose	Pathogenicity
Cl. welchi . .	OS	AG	AG	AG	AG	⎫
Cl. oedematiens .	OS	—	AG	—	AG	⎬ Gas-gangrene
Cl. septicum . .	OS	AG	AG	—	AG	⎬
Cl. histolyticum .	OS	—	(A)	—	(A)	⎭
Cl. tetani . . .	RT	—	—	—	—	Tetanus
Cl. botulinum .	OS	(AG)	AG	—	AG	Food poisoning

OS = oval subterminal.　　　RT = round terminal.

Enteroxtoxic Cl. welchi

Food poisoning is caused when meat contaminated with *Cl. welchi* spores is cooked insufficiently to kill the spores and is then allowed to stand at room temperature for some hours. The spores germinate to form a " cooked-meat culture " which, when eaten, causes abdominal cramps and diarrhoea (Knox and MacDonald, 1943 ; Hobbs *et al.*, 1953). The strains are poorly haemolytic on horse-blood agar, but are Nagler positive. For primary isolation, see page 5. Heat-resistant strains which will withstand several hours simmering are more often implicated in outbreaks than the common faecal strains which are less resistant.

Note.—Although cooked meat medium is very useful in anaerobic work, its role should be no more than that of broth for aerobic cultures. Solid medium is just as important for the primary culture of anaerobes as it is for aerobes. The practice of seeding material into liquid medium and then searching for anaerobes only when growth or gas formation has occurred is deplorable. As in mixed aerobic broth cultures, the species which grow easily, multiply at the expense of the others and important pathogens may fail to appear. Incubation in jars is essential for anaerobic plate cultures (see page 351). In an emergency the specimen can be taken on a swab in transport medium (page 362) and left on the bench until anaerobic plate cultures can be made. Delicate anaerobes survive well in this medium and as it is non-nutrient there is no danger that they will be rapidly outgrown. Commensals capable of growth in poorly nutrient fluids may spoil the specimen if culture is delayed for more than a few hours. Liquid media can be made anaerobic by reducing substances, the most convenient is sterile strips of sheet iron, even very strict anaerobes such as *Cl. tetani* will flourish in peptone water sugar media containing them. A disadvantage is that iron discolours the medium on prolonged incubation, which obscures colour change of the indicator. When prolonged incubation is likely a jar can be used, or Brewer's medium, with sugar and indicator added, is satisfactory.

Bacillus

Members of this genus are often encountered in wounds, particularly after road accidents, in discharges from chronic ulcers, and from other sites likely to be contaminated with soil and dust. They also occur as airborne contaminants on solid media. Microscopically they are large, usually square-ended Gram-positive bacilli, sometimes showing spores. Only one species, *B. anthracis*, is an habitual human pathogen, the others are usually considered to be harmless to man. They should not, however, be light-heartedly dismissed on this account as they have been known to cause human infection. Easily tested points of difference between *B. anthracis* and other members of the group are as follows. *B. anthracis* is non-motile, capsulated, grows in long chains, forms no turbidity in broth, forms inverted fir tree growth in gelatin, with slow liquefaction, and is pathogenic to laboratory animals.[1] Other sporebearing aerobes may show one or two but not the majority of these characteristics. Occasionally they lose their ability to retain Gram's stain and may be mistaken for coliforms. They form acid but no gas in a variety of sugar media and all the common species are indole negative. Although *B. cereus* is no more infective than other species it has been implicated in outbreaks of food poisoning. An enterotoxin is produced when it is allowed to grow in cooked rice which, having been exposed to air, is kept warm, sometimes for many hours to serve to customers in restaurants (Mortimer and McCann, 1974).

---

[1] Virulence tests should not normally be undertaken because of the danger of laboratory infection and contamination of the animal house with anthrax spores.

Bacteroides and Fusiformis

This group includes a number of non-motile, non-sporing, anaerobic or microaerophilic Gram-negative bacilli. They are very pleomorphic and although they may appear fusiform in smears of infected discharges, for example from Vincent's angina, in stained films from cultures they may be long and slender, or short cocco-bacilli, or in serpent and club forms, or large and similar to the coliforms. Colonies on blood agar are seldom visible after overnight incubation and may not appear for five days or more ; they are usually small, translucent and non-haemolytic. They are found in pus from suppurative appendicitis, from lung abscess and cerebral abscess, also from the cervix in puerperal fever and may be recovered from the blood in generalized infection arising from any of these conditions. They are also found in mixed cultures from chronic suppurating wounds. Their natural habitat is the mouth, intestinal tract and vagina. They are successful parasites in that they live happily with their host for the most part and only seldom cause infection and very seldom death. It is as a rule unnecessary to attempt to differentiate the species since identification will not affect treatment or prognosis and the bacilli are too fragile to be of epidemiological importance ; since they grow so slowly this is in any case a lengthy process. Nevertheless, when time permits identification should be attempted in order to learn more about pathogenicity and frequency of different species isolated from different sites. *Bacteroides fragilis* is very commonly encountered in pus from abdominal abscesses and in high vaginal swabs from patients with uterine sepsis. It is not a very strict anaerobe and tends to grow more rapidly than other non-sporing anaerobes. It normally has reduced sensitivity to ampicillin ; for methods of identification, see Sutter and Finegold (1971). Primary isolation is aided by adding about 10 per cent CO_2 to the anaerobic atmosphere (Watt, 1973). Some strains are very oxygen sensitive and die quickly unless replaced, immediately after examination, in an anaerobic jar. Many of them survive well in Robertson's cooked meat broth.

They are often found growing in symbiosis with anaerobic cocci. There is evidence that the two organisms together do more damage than either separately (Hayward, 1947; Gibbons, 1974). Discharges containing them are almost always malodorous.

MISCELLANEOUS

1. *Acinetobacter* and *Alkaligenes*

A number of Gram-negative bacteria which occasionally cause infection and are found as secondary invaders are classified in these two genera. Acinetobacter are non-motile and non-fastidious, they oxidase carbohydrates. Among them are included *Acineto-bacter parapertussis* previously classified with *Bordetella pertussis* and the organism known as *Mima polymorpha Acinetobacter anitra-tus* which, although a bacillus, can appear coccoid and may be confused with *Neisseria*.

Alkaligenes (Table 12) are also non-fastidious, but show some motility and do not attack carbohydrates. *Alk. bronchisepticus* previously classified with *Brucella* or *Bordetella* may be isolated from sputum where it is usually considered to be a secondary invader.

2. *Actinobacillus moniliformis*

This organism is a branching, Gram-positive, non-motile, pleo-morphic bacillus. It can be isolated from the blood in one type of rat-bite fever (the other type is caused by *Spirillum minus*) and from the blood of patients suffering from Haverhill fever. In liquid culture it shows long branching filaments, some of which break up to form bacilli or cocci in chains which stain irregularly, Gram-positive and Gram-negative forms appearing in the same field. Most strains grow with difficulty on blood agar but when sub-cultured from the original blood-broth will appear after 3 to 4 days' incubation as minute translucent colonies. Growth on the solid medium is usually found to consist of comparatively short and almost diphtheroid forms. Sometimes large rounded bodies are found, and the bacilli may show central or terminal swellings ; a pleuropneumonia (*Mycoplasma*) type of organism may be associated with it. In 30 per cent serum broth growth appears after 3 to 4 days as a granular fluffy deposit. *Actinobacillus muris* is an aerobe but most strains prefer an anaerobic or microaerophilic atmosphere plus 5 to 10 per cent CO_2. It is recognized by its extreme pleomorphism and slow growth. It is most difficult to maintain in culture. Pathogenicity of cultures is variable. Intraperitoneal mouse inoculation usually proves fatal within 2 days, but there are no very characteristic post-mortem appearances. The organism can be recovered from the upper respiratory tract of healthy rats.

3. *Erysipelothrix* and *Listeria*

Erysipelothrix rhusiopathiae and *Listeria monocytogenes* are small Gram-positive bacilli which will grow on ordinary laboratory media, without enrichment, to form minute clear colonies. The smooth form resembles small diphtheroids microscopically except that there is a tendency to chain formation. The rough form yields slightly larger opaque colonies about 0·2 mm. in diameter after overnight growth and microscopically shows long filaments and branching. Their growth is improved by small amounts of glucose and slightly improved by blood. They ferment sugars without gas formation. They are widespread animal pathogens. Man is occasionally infected through the skin when handling infected material. The organisms survive in dead animal and vegetable matter for long periods.

Ery. rhusiopathiae causes erysipeloid in man. It can be recovered from tissue, or tissue fluid, taken from the edge of the lesion. When the condition is suspected the material is seeded into 0·2 per cent glucose broth in addition to the routine media. Septicaemia due to either species is very common in animals and is occasionally encountered in man. Inoculation of rabbits causes monocytosis. Intraperitoneal inoculation of mice causes death within 2 to 4 days with conjunctivitis and focal necrosis of the liver, the organisms can be recovered from the bloodstream. Pathogenicity tests are made as soon as possible after isolation because virulence falls after subculture. Agglutination tests are of value in diagnosis.

Listeria monocytogenes is very similar in colonial and microscopic morphology. Easily tested differences between the two species are that *Listeria monocytogenes* shows beta-haemolysis on horse-blood agar, is motile, grows at 4° C., hydrolyses aesculin and is catalase positive, whereas *Ery. rhusiopathiae* is alpha-haemolytic on horse-blood agar and gives negative results in the other tests. Two sites are especially liable to infection in man. The uterus may be either acutely or chronically infected and this probably causes sterility in some patients. Acute or subacute meningitis may also be due to *Listeria* and since a monocytic cell reaction is seen, tuberculous meningitis may be wrongly diagnosed. A correct diagnosis is important since the organism is penicillin sensitive and treatment is rapidly effective in most cases. Colonies may be scanty on the primary plates, but the organism can be recovered from the cooked-meat culture if the method recommended (page 49) is employed.

When isolation from contaminated material is needed, for example in attempting to find an infective cause for abortion, ability to grow at 4° C., or screening the medium with nalidixic acid may be employed for selection (Kampelmacher, 1967).

4. *Francisella*

Fr. tularensis (Table 13) needs to be considered when a febrile patient has arrived ten days or less from America, Russia or other country where tularensis is endemic.

5. Lactobacilli

These organisms live in the intestinal canal and vagina and appear to be harmless. Most of them prefer a medium more acid than that used in routine work so that they are seldom isolated. Döderlein's bacillus is a member of this group. It is often seen in cervical and vaginal smears as a large Gram-positive bacillus and it may appear on blood agar cultures as minute non-haemolytic colonies. Other varieties of lactobacilli are found in cervical and wound swabs, they are non-motile, slender, Gram-positive bacilli which tend to form chains. Some species are strict anaerobes.

6. *Moraxella*

The Morax-Axenfeld bacillus causes one kind of human conjunctivitis. Stained films of the discharge show a Gram-negative bacillus resembling *Klebs. friedländeri* but non-capsulated. Growth on blood agar is poor and for this reason cultures of discharge from acute conjunctivitis should include a Loeffler slope and Dorset's egg medium which both favour the growth of these organisms. They are identified by their morphology, by the small pits of liquefaction which they show on Loeffler's medium and by their inability to grow on nutrient agar.

7. *Mycoplasma*

These differ from bacteria in having no rigid cell wall, and from viruses in being able to grow in cell-free medium. They resemble the L-forms of bacteria described by Kleineberger (1936). They have been known to cause fatal pleuropneumonia in cattle for many years. There are various species, markedly host specific.

Mycoplasma pneumoniae (Eaton agent), the cause of primary

typical pneumonia, is very difficult to isolate and culture as a method of routine diagnosis is not suitable with the methods at present available. A sample of blood must therefore be taken early in the infection if a reliable serological diagnosis is to be made without undue delay, see Chapter 8. In man the genital species *Mycoplasma hominis* is most easily grown. It can be isolated on defibrinated horse-blood agar provided the sample is not allowed to dry (Stuart's transport medium preserves them well). The medium must be fresh and remain moist, and be incubated from 3 to 5 days. They have been found in pure culture in the blood in puerperal sepsis, in pus from Bartholin's abscess, in pleural fluid (Stokes, 1955) and in the vagina, urethra and throat of healthy people. They are no longer believed to be an important cause of Reiter's syndrome and non-specific urethritis. They grow with difficulty on artificial media appearing as very minute, non-haemolytic, transparent colonies on blood agar after three to four days' incubation ; when examined under a plate microscope they have a " fried egg " appearance. The blood agar must be fresh and moist and prepared from defibrinated horse or human blood ; oxalated blood is not suitable. Nagler's medium (20 per cent human-serum Fildes' extract agar) used for the identification of *Cl. welchi* supports their growth better than blood agar or plain 20 per cent serum agar. Culture in liquid medium is difficult and growth is insufficient to cause opacity until the strain has been subcultured many times. *Mycoplasma hominis* is resistant to penicillin and erythromycin and can be isolated from mixed cultures by screening the medium with these antibiotics; thallous acetate is also an efficient selector at a concentration of 1·25 mg. per 100 ml. medium. *In vitro* it is sensitive to lincomycin, streptomycin, chloramphenicol and the tetracyclines. Aerobic and anaerobic growth is approximately equal but the addition of 5 to 10 per cent CO_2 appears to aid primary isolation. It is essential to incubate cultures in a jar, or to seal the plates, to prevent undue drying during incubation.

Mycoplasma hominis would probably be recognized much more often if it was easier to examine microscopically. When colonies are scraped off the surface of the medium with a loop in the usual way their structure is destroyed and staining reveals nothing recognizable. In order to recognize them it is necessary to adopt a technique such as that described by Ledingham (1933). A glass coverslip is dropped on to the surface of the medium over the colonies and gently pressed down.   After a second or two it is lifted with forceps, great care being taken not to slide it on the surface of the medium, the colonies adhere to the glass leaving minute pits in the agar.   The coverslip is then lightly flamed to fix the preparation and laid on a glass slide, colonies uppermost.   It is fixed to the slide, for convenient handling, by sealing the edges with molten paraffin wax and the colonies are then stained by Gram's method or by Leishmann's stain.   They appear as Gram-negative rather sponge-like structures with minute bacilliform and coccoid bodies at the periphery, and are quite unlike ordinary bacterial colonies prepared for examination in the same way.   It is common in the vagina and urethra of healthy people and its pathogenic role must be judged according to the circumstances of isolation.   When it grows profusely on primary plate culture from a purulent specimen from which no other pathogen can be seen in the direct Gram-stained film, or in the culture, it is reasonable to assume a mycoplasmal infection, especially if there is a rise of antibody in the patient's serum.   When it can only be isolated in highly selective liquid culture, in small numbers on solid selective medium or in tissue culture, its pathogenic role is extremely doubtful.   In mycoplasmal urinary infection a profuse growth is obtained from a loopful of uncentrifuged urine (U.C.H. laboratory records);   a scanty growth from the deposit is consistent with a normal urine.   Identification is made by growth-inhibition tests using paper discs impregnated with specific antisera.

T-mycoplasmas (T=tiny) produce colonies too small to see with a hand lens and no opacity in liquid medium, their growth is recognized by a change in pH shown by indicator incorporated in the culture medium (Taylor Robinson *et al.*, 1969).   Until there is firm evidence of their importance in medicine there is no need to attempt culture in diagnostic laboratories.

Mycoplasma are easily recognized from their morphology which is maintained on medium without selective agents.   L-forms which resemble them tend to revert to the parent strain.

8. Pseudomonads

Ps. aeruginosa (page 151) is most commonly isolated and is an important hospital pathogen (page 324).   Other pseudomonads particularly *Ps. cepacia* are sometimes encountered (Phillips *et al.*, 1971).   All are motile and oxidative.   Probably only two members of the genus *Ps. mallei* and *Ps. pseudomallei* are primary pathogens in healthy people. *Ps. pseudomallei* causes melioidosis and may be isolated from the blood of febrile patients arriving by air from India or Asia. Although it grows on MacConkey agar it differs from other pseudomonads by being inhibited by desoxycholate citrate and centrimide. *Ps. mallei* is fastidious and at first sight unlike other pseudomonads it is described under parvobacteria, page 158.

REFERENCES

ALWEN, J. and SMITH, D. G. (1967). *J. appl. Bact.*, **30**, 389.
BAIRD-PARKER, A. C. (1965). *J. gen. Microbiol.*, **38**, 363.
BARBER, M. and KUPER, S. W. A. (1951). *J. Path. Bact.*, **63**, 65.
BICKHAM, S. and JONES, W. (1972). *Amer. J. clin. Path.*, **57**, 244.
BLAIR, E. B., EMERSON, J. S. and TULL, A. H. (1967). *Amer. J. clin. Path.*, **47**, 30.
BOWERS, E. F. and JEFFRIES, L. R. (1954). *J. clin. Path.*, **8**, 58.
BROWN, V. I. and LOWBURY, E. J. L. (1965). *J. clin. Path.*, **18**, 752.
CHESSUM, B., FRENGLEY, J. D., FLECK, D. G. and MAIR, N. S. (1971). *Brit. med. J.*, **iii**, 466.
Clinicopathological Conference (1966). *Brit. med. J.*, **i**, 93.
COLEBROOK, L., *et al.* (1942). *Lancet*, **ii**, 30.
COWAN, S. T. and STEEL, K. J. (1974). *Identification of Medical Bacteria.* University Press, Cambridge.
CUNLIFFE, A. C., GILLAM, G. G. and WILLIAMS, R. (1943). *Lancet*, **ii**, 355.
EICKHOFF, T. C., KLEIN, J. O., DALY, A. K., INGALL, D. and FINLAND, M. (1964). *N. Engl. J. Med.*, **27**, 1221.
ELEK, S. (1949). *J. clin. Path.*, **2**, 250.
FELSENFELD, O. (1945). *Amer. J. clin. Path.*, **15**, 584.
FRY, C. S. and WILKINSON, A. E. (1964). *Brit. J. vener. Dis.*, **39**, 190.
GIBBONS, R. J. (1974). *Anaerobic Bacteria.* Charles C Thomas, Illinois.
HANSMAN, D. (1970). *J. med. Microbiol.*, **3**, 359.
HAYWARD, N. J. (1941). *Brit. med. J.*, **i**, 811.
—— (1947). *Recent Advances in Clinical Pathology*, ed. S. C. DYKE, Churchill, London.
HOBBS, B. C., SMITH, M. E., OAKLEY, C. L., WARRACK, G. H. and CRUICKSHANK, J. C. (1953). *J. Hyg. Camb.*, **51**, 75.
ISAACS, A. and SCOULLER, J. M. (1948). *J. Path. Bact.*, **60**, 135.
KAMPELMACHER, E. H. (1967). *Lancet*, **i**, 165.
KASHBUR, I. M., GEORGE, R. H. and AYLIFFE, G. A. J. (1974). *J. clin. Path.*, **27**, 572.
KERR, H. (1973). *J. clin. Path.*, **26**, 918.
KLEINEBERGER, E. (1936). *J. Path. Bact.*, **42**, 587.
KNOX, R. and MacDONALD, E. K. (1943). *Med. Offr*, **69**, 21.
LACEY, B. W. (1951). *J. gen. Microbiol.*, **5**, vi.

LEDINGHAM, J. C. G. (1933). *J. Path. Bact.*, **37**, 393.
MAIR, N. S., MAIR, H. J., STIRK, E. M. and CORSON, J. G. (1960). *J. clin. Path.*, **13**, 432.
MAUSS, E. A., and KEOWN, M. J. (1946). *Science*, **104**, 252.
MAXTED, W. R. (1953). *J. clin. Path.*, **6**, 224.
MITCHELL, R. G. and BAIRD-PARKER, A. C. (1967). *J. appl. Bact.*, **30**, 251.
MORTIMER, P. R. and McCANN, G. (1974). *Lancet*, **i**, 1043.
PHILLIPS, I., EYKYN, S., CURTIS, M. A. and SNELL, J. J. S. (1971). *Lancet*, **i**, 375.
ROBERTS, A. P. (1967). *J. clin. Path.*, **20**, 631.
ROWE, B., TAYLOR, J. and BETTELHEIM, K. A. (1970). *Lancet*, **i**, 1.
STOKES, E. J. (1955). *Lancet*, **i**, 276.
SUTTER, V. L. and FINEGOLD, S. M. (1971). *Appl. Microbiol.*, **21**, 13.
SWAN, A. (1954). *J. clin. Path.*, **7**, 160.
TAYLOR, J. and CHARTER, R. E. (1952). *J. Path. Bact.*, **6** 729.
TAYLOR ROBINSON, D., ADDEY, J. P. and GOODWIN, C. S. (1969). *Nature*, **222**, 274.
THOMAS, C. G. A. and HARE, R. (1954). *J. clin. Path.*, **7**, 300.
TOIVANEN, P., TOIVANEN, A., OLKKONEN, L. and AANTAA, S. (1973). *Lancet*, **i**, 801.
TURK, D. C. and MAY, J. R. (1967). *Haemophilus Influenzae*. English Univ. Press, London.
WATT, B. (1972). *J. med. Microbiol.*, **6**, 307.
WEINBERG, A. N. (1974). *Anaerobic Bacteria*. Charles C Thomas, Illinois.
WILLIAMS, R. E. O., RIPPON, J. E. and DOWSETT, L. M. (1953). *Lancet*, **i**, 510.
WILSON, G. S. and MILES, A. A. (1975). *Topley and Wilson's Principles of Bacteriology, Virology and Immunity*, 6th edn. Edward Arnold, London.

6

Investigation of chronic infections

Laboratory diagnosis of these infections is considered separately because, with the possible exception of actinomycosis, moniliasis and torulosis the routine methods so far described will not reveal their nature. Examination for tubercle bacilli and fungi is often initiated in the first place by the clinician. But if routine blood agar cultures of an inflammatory fluid are sterile or yield no likely pathogen, or if microscopy reveals acid-fast bacilli or structures resembling fungi, further cultures should be made using the methods now to be described.

TUBERCULOSIS
Prevention of laboratory infection

When ordinary laboratory precautions are practiced and work is carried out on the open bench there is some risk of infection to the laboratory staff. Sputtering of living tubercle bacilli from flamed loops has long been recognized as a hazard. Prevention by using hooded burners or removing large particles before flaming deals with this but infection can be acquired in other ways. The following rules for handling specimens likely to be tuberculous are therefore followed :

1. The laboratory worker must be tuberculin positive.
2. All specimens, cultures and infected animal cadavers must be handled in a safety cabinet. (Collins *et al.*, 1974.)
3. Specimens must be shaken without wetting the caps of bottles, e.g. on a vortex mechanical mixer (page 61).
4. A tube or bottle to be centrifuged must first be wrapped in a self-seal plastic bag to contain spilt material if the tube should break.
5. Acid-fast bacilli are resistant to disinfectants, therefore glassware must be heat treated. Boiling for 5 minutes is effective but autoclaving may be convenient. Pipettes must be discarded with their tips under fluid (hypochlorite cleaning fluid) to prevent an aerosol when the teat is removed.

6. The cabinet must contain a discard jar for pipettes and a burner. Other equipment should be kept outside until required.

7. When possible a separate room should be provided for work on tuberculous material.

8. Autopsy on patients with known active tuberculosis should not be undertaken. Staff working in the autopsy room must be tuberculin positive.

Four types of investigation are available to aid diagnosis. They are, microscopic examination, culture, animal inoculation and a skin test for hypersensitivity to tuberculin.

MICROSCOPIC EXAMINATION

This is a test to aid *clinical* diagnosis, the bacteriologist does not assume that the bacilli he sees are *Myco. tuberculosis* (see Chapter 1). *Laboratory* diagnosis can only be made by culture or animal inoculation and takes several weeks. Because of this delay microscopy plays an unusually important part. The value of a positive finding depends on the probability that there are no acid-fast bacilli, other than tubercle bacilli, in the specimen ; this varies with different types of specimen. The chance of finding harmless acid-fast bacilli in sputum is slight and microscopy is therefore of great value ; gastric juice, urine and faeces fairly often contain them and a diagnosis made on the strength of positive microscopy alone is therefore not satisfactory and needs confirmation by further tests. A source of error which must be avoided is contamination of solutions or glass-ware, either with saprophytic acid-fast bacilli, which are sometimes found in dust and tap water, or with the dead bodies of tubercle bacilli from previous positive specimens and cultures. Acid-fast bacilli, unlike most other bacteria, retain their structure and staining properties after death. They may adhere to glass, survive the cleaning process, and be washed off into subsequent specimens. The following rules are adopted to avoid the grave risk of false positive reports from this cause.

1. As far as possible all specimens are sent in disposable cartons.

2. New glass slides are used for smears.

3. Stains are made up in distilled water and dropped from bottles on to the slides. (The use of jars of stain for immersion of the slides is indefensible.)

4. Blotting paper is cut in small squares, each piece is discarded after the whole of its surface has been used.

5. Centrifuge tubes, test tubes, and bottles used for tuberculous material are immersed in 50 per cent nitric acid [1] for two hours after sterilization and ordinary cleaning. They are then washed free of acid, dried, sterilized and marked to be used for T.B. only.

6. Marked bottles and caps are used to hold media for the culture of tubercle bacilli. They are kept entirely separate, and should be of a different shape, from those in which specimens are collected.

Methods

There are two methods in general use at the present time, the classical Ziehl-Neelsen method followed by examination under ordinary illumination, and the fluorescent method in which the bacilli are stained with auramine phenol which fluoresces in short-wave light. The principle of both methods is to demonstrate the unique power of acid-fast bacilli to retain strong stains after decolorization with acid and alcohol. In laboratories where ten or more examinations are made daily there is no doubt that the fluorescent method is to be preferred. The staining process is quicker and simpler than Ziehl-Neelsen's method, the bacilli appear bigger and the whole area of the smear can be scanned in less than five minutes with the 16 mm. objective once experience has been gained. The requirements for fluorescent microscopy of acid-fast bacilli and for immunofluorescence (see Chapter 8) are different. The same microscope and light source can be employed, but in searching for acid-fast bacilli a wide field is essential in order to cover a large area quickly under low magnification and therefore a bright-field condenser must be employed. In immunofluorescence the microbes are less brilliant and therefore a darkground condenser gives better contrast. An iodine quartz 100 watt lamp is a suitable light source for both purposes and is much more convenient to use and lasts longer than a mercury vapour lamp. For details of the equipment required, see p. 359. A known positive specimen is kept in the refrigerator and a control smear is made and examined with each batch to check the stain and to ensure that the microscope and lamp are arranged so that fluorescence is optimal Negative results should not be reported without this precaution.

[1] The acid lasts indefinitely, more is added to replace gradual loss.

Staining Methods

Ziehl-Neelsen stain

1. Flood the slide with strong carbol fuchsin and heat gently till steam rises. Add more stain when necessary and keep steaming for 5 to 10 minutes but do not allow it to boil.
2. Wash in tap water (from a freely flowing tap which is in frequent use).
3. Decolorize alternately with 25 per cent sulphuric acid and 95 per cent alcohol until no more colour comes away.
4. Wash in tap water.
5. Counter stain with Loeffler's methylene blue for 3 minutes.

Auramine-phenol stain for fluorescent method

1. Flood the slide with auramine phenol, 4 minutes.
2. Wash well in tap water.
3. Flood with acid-alcohol-decolorizer, 4 minutes.
4. Wash.
5. Counter stain with potassium permanganate, 30 seconds.

Concentration of specimens

Concentration for microscopy only

Liquid, non-purulent specimens are centrifuged at 3000 r.p.m. for 30 minutes. The deposit is examined as concentrated as possible because acid-fast bacilli are likely to be few, see also under individual specimens. Sputum, thick pus and gastric juice : Add 2 parts of 4 per cent sodium hydroxide to one part of the specimen. Incubate for one hour or more, until fluid. Centrifuge.

Faeces : Emulsify a lump of faeces in 1 ml. of broth, add 3 ml. of ether and mix well. Centrifuge for 20 to 30 minutes at 3000 r.p.m. Remove the ether layer with a pipette and make films from the layer immediately below this, which contains very large numbers of the faecal organisms.

Concentration for culture and microscopy

None of the many concentration methods is ideal because all the chemicals employed to kill contaminating organisms also damage living tubercle bacilli to some extent. Strong chemicals such as sodium hydroxide and sulphuric acid kill contaminants in a short time but the tubercle bacilli will also suffer unless exposure to them is accurately timed. Sputum is particularly difficult to treat because time must be allowed for liquefaction, and for the chemical to penetrate to all the contaminating organisms. As the consistency of specimens is extremely variable no rule can be made for the time each is to be treated. It is a matter of judgement, and the tendency is to over-treat the specimen so that contaminated cultures are avoided. Under experimental conditions the original method described by Petroff using 4 per cent NaOH has proved as good or better than any (P.H.L.S. Working party, 1952) but in routine work careful timing is very difficult to achieve and a slower acting substance less lethal to tubercle bacilli is to be preferred for sputum, gastric juice and large samples of pus. Treatment of different types of specimens is considered in detail later.

CULTURE

Culture of the tubercle bacillus is an essential part of the investigation of tuberculosis. A laboratory diagnosis, independent of clinical findings, cannot be made until the bacillus has been cultured or has produced typical lesions after animal inoculation. Since culture usually gives a more rapid result and is more economical it is the method of choice.

A negative culture is of more value than negative microscopy alone because, as in other infections, cultures are frequently positive when no organisms can be found in the stained film. Moreover when the bacillus has been cultured it can be tested for sensitivity to therapeutic substances.

Culture of uncontaminated material presents no difficulty. Contaminated specimens such as sputum, laryngeal swabs, gastric juice, faeces, pooled specimens of urine and pus from secondarily infected wounds are treated to kill the non-acid-fast bacilli and to concentrate the specimen before culture. Media for isolation of tubercle bacilli have been improved to the point where animal inoculation is seldom if ever necessary (Marks, 1972). Even slightly contaminated specimens can now be cultured successfully on selective medium (Mitchison *et al.*, 1973), see page 370. Lowenstein-Jensen's medium is still employed but Stonebrink's pyruvate egg medium has replaced Dorset's egg medium for bovine strains and will support the growth of human strains which fail to grow on L-J medium (Marks, 1963). Hughes (1966) found that cultures of sputum from 5 of 99 tuberculous patients yielded growth only on pyruvate egg and confirmed that when, for economy, only one medium can be used this is the medium of choice.

Specimens needing treatment to get rid of contaminating organisms and others such as cerebrospinal fluid which the bacteriologist has reason to think might be accidentally contaminated should be cultured on selective medium. Non-purulent fluids which are unlikely to be contaminated should in addition be cultured in Kirschner's liquid medium which can also be introduced into the original container to allow any bacilli remaining there to grow when the bottle is incubated. Culture of heavily infected tuberculous material on good batches of media takes longer, but is no more difficult, than culture of the common pyogenic organisms on blood agar, but with specimens containing relatively few viable bacilli two difficulties arise. First the bacilli tend to remain localized in clumps, even in concentrated specimens, which makes the sampling error great, and second the egg media are more variable in composition than blood agar. Even when the medium has been carefully prepared from a large batch of fresh eggs it can be shown that occasional slopes are incapable of supporting the growth of small inocula. This is probably due to the presence of inhibitory substances, occasionally found in eggs, which have not been evenly distributed throughout the medium. For this reason large batches at a time are made from fresh eggs so that the inhibitory substances, if present, will be well diluted; the mixture is passed through a homogenizer.

Blood agar prepared from human blood discarded from the hospital blood bank when out of date has proved simple to prepare and as effective as Lowenstein-Jensen's medium for primary culture of human strains. Contaminants are suppressed with penicillin (Tarshis *et al.*, 1953). For method of preparation, see page 370.

The media are seeded heavily and the caps are screwed on tightly to prevent evaporation; moist medium is essential for the growth of tubercle bacilli. The cultures are inspected weekly and discarded after 8 weeks' incubation if they are sterile. When acid-fast bacilli have been seen in the original smear and cultures yield no growth, they are incubated for a further month. Occasionally growth appears very late, particularly from specimens from treated patients. Growth of human strains on egg media can be expected within 2 to 3 weeks but may be delayed for much longer, even from material not exposed to chemotherapy. Negative 3-month cultures indicate either that the bacilli seen are dead, which is likely in a treated patient, or that they are opportunist mycobacteria incapable of growth at 35–37° C. Culture from untreated patients should therefore be transferred to a 30° C. incubator for a further 3 months with weekly inspection. Equipment contaminated with dead but visible bacilli (see page 172) inferior culture media, or too rigorous pre-culture treatment are likely causes of failure to isolate when acid-fast bacilli have been seen.

Procedure for different types of specimen

The methods for specimens other than sputum and faeces are adapted from Marks (1972).

1. *Sputum.* Examine each specimen for acid-fast bacilli before treatment. When they are seen in one sample culture this sample ; it is a pity to dilute it with further samples which may be negative. Deposits from up to three samples with negative microscopy can be pooled for culture. The concentrated pool should then be examined for acid-fast bacilli.

METHOD. Add an equal volume of 10 per cent crystalline sodium triphosphate to sputum in its container. Shake on a vortex mixer and incubate overnight (18 hours). Measure 2 ml. of the now liquefied sputum into a centrifuge bottle or tube, preferably disposable, containing 20 ml. acidified sterile distilled water (35 ml. N HCl in 1 litre distilled water). Mix again on the vortex mixer then place the bottle or tube in a self-seal plastic bag and centrifuge at 3000 r.p.m. for 20 minutes. Decant the supernatant and using a pipette distribute the deposit over the surface of one pyruvate egg and one Lowenstein-Jensen slope. Incubate in the sloped position for the first week. When three samples are to be pooled mix the deposits from each of the three bottles in one of them before inoculating the medium and prepare a slide of the material for microscopic examination.

2. *Cerebrospinal and non-purulent serous fluids.* Break up clot when necessary by shaking with sterile glass beads. Centrifuge at 3000 r.p.m. for 30 minutes ; retain the original bottle. Remove the supernatant fluid with a sterile pipette. Make two smears, one as concentrated as possible, for examination for acid-fast bacilli and another for Gram-stain.

When no bacteria are seen in the Gram-stained smear and there are few or no polymorphonuclear cells it can be assumed that other bacteria are absent.   Culture the remains of the deposit without treatment on the following media ; Lowenstein-Jensen's, selective Middlebrook's medium, pyruvate egg, Kirschner's liquid medium. Add Kirschner's medium, using sterile precautions, to the specimen bottle and incubate it with the other media because acid-fast bacilli cling to the glass and may be lost in the specimen bottle.

When bacteria are seen in the Gram-stained smear it is not often necessary to culture for tubercle bacilli but when this is required add 1 drop 5 per cent sulphuric acid to the deposit, hold for 20 minutes at room temperature and inoculate media as above using a pipette and sterile precautions.   Wash out the centrifuge tube with Kirschner's medium which should be inoculated last.

3. *Purulent serous fluids and pus.*   Gram-stain and culture overnight for other microbes refrigerating the specimen meanwhile ; examine a second smear for acid-fast bacilli.

When the Gram-stained smear and culture show no contaminants inoculate the refrigerated sample directly to the media described above for non-purulent fluids.   When bacteria are seen, the overnight cultures yield growth and culture for Mycobacteria is still required, add an equal volume of 5 per cent sulphuric acid to up to 2 ml. pus in a centrifuge tube.   When a swab only is provided submerge it in 2 ml. of the acid, rubbing as much of the specimen as possible off the swab against the side of the centrifuge tube. Mix on a vortex mixer and stand at room temperature for 20–40 minutes depending on the degree of contamination.   Add 16 ml. sterile distilled water, squeeze as much liquid as possible out of the swab against the side of the tube before removing it and centrifuge at 3000 r.p.m. for 20 minutes.   Remove the supernatant fluid and culture the deposit on the media recommended above but omit Kirschner's medium because the risk of contamination of a liquid culture is great.

4. *Urine.*   (a) Early-morning specimens.   Three consecutive early morning specimens are collected in wide-mouthed screw-capped jars which have either been specially treated and marked for this purpose (page 172) or are disposable.   Each is delivered to the laboratory on the day of collection. (If delay is unavoidable they are refrigerated.)   Mix well and centrifuge 20–30 ml. at 3000 r.p.m. for 30 minutes ; discard the supernatant fluid.   Make two smears from the deposit, one for methylene blue stain and the other, a concentrated smear, for auramine or Ziehl–Neelsen stain.

Examine the methylene blue-stained smear. Add to the deposit 2 ml. 5 per cent sulphuric acid and hold at room temperature for 10, 20, 30 or 40 minutes according to the number of bacteria seen in the methylene blue-stained smear. Add 16 ml. sterile distilled water, mix and then centrifuge for 30 minutes. Distribute the deposit between 1 pyruvate egg slope, 1 Lowenstein–Jensen slope and one selective medium slope.

(*b*) Ureteric catheter specimens: treat as recommended for cerebrospinal fluid.

5. *Tissue.* Grind in a Griffith's tube, using sterile sand when necessary, or process in a stomacher. Then treat the emulsion as described for pus.

6. *Gastric juice.* The examination is made when patients cough but can produce no sputum, the small quantity coughed up is involuntarily swallowed and can be recovered from the stomach. The resting gastric juice is sucked out through a stomach tube early in the morning. If little fluid is available the stomach may be washed out with a small amount of water or sodium bicarbonate solution. On reaching the laboratory the gastric juice is treated as sputum. All specimens are cultured because of the unreliability of diagnosis by microscopy alone. When delay in examination is unavoidable, the specimen is neutralized and refrigerated because tubercle bacilli do not survive well in the presence of acid and active enzymes.

FIG. 19.

7. *Laryngeal swabs.* These are taken as an alternative to the examination of gastric juice from patients who cough but can produce no sputum. It is a particularly useful method for outpatients. Swabs are prepared on strong wire applicators as in the diagram and moistened in sterile water just before use. (Fig. 19).

METHOD. Wrap a strip of lint round the patient's protruded tongue, hold it gently and ask him to breathe through his mouth. Pass the tip of the swab through the mouth, over the back of the tongue, until the long straight part of the applicator lies along its surface. The sampling end should now be in the larynx and the patient will cough. It is best to take two swabs on each occasion because the material for culture is very scanty. The specimen is treated immediately by the method described for pus swabs omitting the Gram-stain and blood agar cultures. Microscopy for acid-fast bacilli is also omitted because all the material on the swab is needed for culture.

Note. The sampler protects his face with a transparent visor because he is in the direct line of fire of the patient's cough. One can be cheaply made by clipping washed, old X-ray film to a head band ; after sampling the film is discarded and replaced with a fresh piece.

8. *Faeces.* Handle the specimen in the cabinet *with the burner turned off.* Emulsify a pea-sized portion of faeces in 5 ml. distilled water in a glass bottle using a short swab. Add about 2 ml. ether. Place the bottle in a self-sealing transparent plastic bag and with the cap screwed on firmly shake well manually. Centrifuge at 3000 r.p.m. for 5 minutes. Remove the cap while the bottle is still inside the plastic bag. Open the bag but do not remove the bottle or lid from it. Using a pipette transfer the layer immediately below the ether, which will contain the bacteria, to a centrifuge tube and treat it as a heavily contaminated urine deposit. Place the glass bottle and lid, still within the plastic bag, to be autoclaved.

Examination of cultures

The human type of tubercle bacillus grows on egg medium as pale cream-coloured, rough, crumb-like colonies which are difficult to emulsify. Bovine strains grow slowly and often appear on pyruvate egg medium only, their growth is usually smoother and more confluent. Slopes should always be examined with a hand lens to avoid missing growth which is sometimes difficult to see through the bottle glass.

It is important to check that colonies of acid-fast bacilli are not saprophytes. The saprophytic bacilli have the following characteristics. They grow rapidly, usually within a few days ; they are often pigmented and they are capable of growth at room temperature and on plain nutrient agar. The growth is therefore subcultured to another egg medium slope and left in the dark at room temperature and also to a nutrient agar slope which is incubated. If these cultures are sterile after a week's incubation and the microscopic morphology is typical of *Myco. tuberculosis* a preliminary report may be sent. On first isolation antibiotic sensitivity tests are needed and are best carried out by a reference laboratory who will also confirm that the organism is indeed *Mycobacterium tuberculosis*. Cultures must

TABLE 18

| | Niacin | Rapid growth <7 days | Pigmentation | | Growth stimulated by glycerol | Growth on nutrient agar | Growth at | | INAH sensitive | Catalase | Pathogenicity | | | |
			in light	in dark			37° C	25° C			Human	Guinea-pig†	Mouse†	Rabbit†
Myco. tuberculosis														
human	+‡	−	−	−	+	−	+	−	‡+++	(+)	++++	++	++	−
bovine	−	−	−	−	−	−	+	−	++	(+)		+	++	++
Myco. ulcercans	−	−	−	−	+	−	−	+	−	++	−	−	++	++
Myco. balnei	−	+	+	−	−	−	−	+	−	++	+‡	+‡	+	+
Opportunist mycobacteria	−	(+)	(+)	(+)	(+)	(+)	++	(+)	−	(+)	+‡	−	+	
Saprophytes	−	+	(+)	(+)	(+)	+	+	+		++	−	−	−	−

† Diluted culture injected.
Parentheses indicate variable result.
‡ Except in treated patients and occasionally those in contact with them.

not leave the laboratory until growth is present on two slopes, one to remain in case the one sent is lost.

Opportunist mycobacteria

Since chemotherapy stimulated activity in culturing acid-fast bacilli in order to test their sensitivity, a number of strains have been isolated from patients who are not suffering from typical tuberculosis, indeed their symptoms may be minor or absent. These organisms are often resistant to antituberculous drugs and it is important to differentiate them from resistant strains of *Myco. tuberculosis*. The infectivity of these organisms is much lower than that of *Myco. tuberculosis* and patients harbouring them should not have to undergo the rigours of antituberculosis treatment. Their frequency varies in different parts of the world, being more common in India than in Britain, for example (Mitchison *et al.*, 1960).

There have been various attempts to classify them, none altogether successful. Runyon (1959) divided them into four groups and his is still the most widely accepted classification. Collins (1962) followed Runyon with modifications, and Marks and Richards (1962) divided them into seven groups. In practice human strains of *Myco. tuberculosis*, including those resistant to chemotherapy, can be separated from the rest by the niacin test (see page 376) which should be done on all strains, particularly on those which are resistant, and on acid-fast bacilli isolated from patients without typical signs of infection.

Of the niacin-negative strains there are three important groups: those which are not capable of infecting man, i.e. the saprophytes ; those which cause skin ulceration and grow at lower temperatures and those which cause comparatively mild, deep-seated infections. It is this last group which has proved difficult to classify ; among them are well-recognized species, *Myco. fortuitum*, *Myco. kansasi* and *Myco. avium*. While not attempting to classify them, Table 18 lists the minimal tests necessary to assess the significance of mycobacteria isolated from patients (for the catalase test method, see page 377). Further classification is desirable but not essential as it will not affect the management of the patient.

Animal Inoculation

Guinea-pigs are highly susceptible to both human and bovine types of tubercle bacilli. The material is inoculated without preparation, other than centrifuging, whenever possible. Heavily contaminated specimens are prepared as described for culture. The guinea-pig is a better selector of tubercle bacilli than the culture medium because contamination of the injected material with organisms non-pathogenic to guinea-pigs does not matter, whereas if harmless organisms survive treatment or are accidentally introduced, cultures will be ruined. On the other hand even though the animals are highly susceptible to tuberculosis they are not absolutely helpless against the attack of a few bacilli ; less than ten viable bacilli cannot be relied on to cause infection in every animal (Schwabacher and Wilson, 1937). Another disadvantage is the risk of death from intercurrent disease before the bacilli have produced recognizable tuberculosis. The guinea-pig is mainly of use as an additional selector when an important specimen is heavily contaminated and is of sufficient volume for animal inoculation in addition to the media recommended. The culture methods described above can be expected to do better than guinea-pig inoculation for uncontaminated or lightly contaminated specimens.

Method

Cerebrospinal and serous fluids, ureteric urine deposits and uncontaminated pus are inoculated without preparation. Unfortunately guinea-pigs are easily killed by penicillin ; it cannot therefore be given as a prophylactic against contaminants. Other specimens are treated as described for culture. They are neutralized carefully because injection of acid is painful. As much of the fluid as possible, is inoculated into the right thigh muscles of an adult guinea-pig. A 350 g. animal receives 0·5 ml. without distress. The subcutaneous route leads to the formation of ulcers and intraperitoneal inoculation causes post-mortem findings which may be difficult to interpret. If two guinea-pigs are used the chance of a positive result is greater, provided there is adequate material ; it is unlikely that both will die of intercurrent disease. When one guinea-pig only has been inoculated it is killed after six weeks. If two have been used they are examined weekly, from the third week on, for enlargement of femoral lymph glands and when this sign appears, the animal which shows it best is killed. This may enable a quicker report to be sent. The second animal remains for later examination if the glandular enlargement proves to be non-tuberculous.

Animal autopsy

A thorough post-mortem examination is always made to establish the cause of a death in the animal house.   Death occurring before six weeks in guinea-pigs inoculated intramuscularly with tuberculous material is not often due to tuberculosis.   If death from intercurrent infection occurs more than three weeks after inoculation it may be possible to diagnose tuberculosis at autopsy.   A negative finding indicates that the inoculated material was not very heavily infected. When the animal dies sooner than this, diagnosis of tuberculosis is not often possible and a negative finding is valueless.

MATERIALS REQUIRED FOR ANIMAL AUTOPSY

Tray (enamel or metal).
Cork or soft wood board.
Pins.
Instrument sterilizer.
Bunsen burner.
Searing iron (soldering iron will do).
Jar containing 5 per cent Sudol.[1]
Discard bin with lid.
Two sets of instruments, one clean and one sterile.
   (*a*) Clean scissors, scalpel and one pair of forceps.

(*b*) *Sterile* scissors and two pairs of forceps.   Bone cutters if the skull is to be opened.
Glass slides.
Sterile pipettes.
Platinum loop.
Two sterile Petri dishes.
Sterile saline.
Pyruvate egg and Lowenstein-Jensen's slopes.
2 blood agar plates.
2 tubes of cooked meat broth.
1 bottle containing 10 per cent formaldehyde in saline.

If a table with cleats to hold strings be available the animal can be tied out on the tray with strings attached to the limbs, the board and pins are then unnecessary.

PROCEDURE (*a*) *Spontaneous death of animal*

Remove the animal from the cage with forceps and place it in a small covered bin, or wrap it in at least four sheets' thickness of newspaper.   When death has occurred after inoculation with tuberculous material the autopsy should be carried out on a tray in the inoculation cabinet.

Arrange the apparatus listed above on the autopsy table and heat the searing iron in the Bunsen flame.   Unwrap the body on the tray and discard the paper into the bin (which is placed under the table). Using forceps, dip the body in the Sudol and then pin or tie it out firmly in the supine position.   (The Sudol serves to prevent the spread

[1] or equivalent disinfectant.

of infected particles of fur and to kill some of the superficial bacteria.)
Taking the first set of instruments, pick up the skin in the mid-line
over the abdomen and make an incision with the scissors from pubis
to chin and then radial incisions along all four limbs. Dissect the
flaps of skin from the body wall and proximal part of the limbs very
thoroughly so that they lie flat and have no tendency to curl back into
position ; take great care not to penetrate the abdomen or thorax.
Discard this set of instruments and sear with the hot iron the whole
of the exposed area. Take the sterile second set of instruments and
using the scissors make a mid-line incision through the abdominal
wall from pubis to xiphisternum, then cut along each costal margin
as far posteriorly as possible. Make two similar incisions above and
parallel to the inguinal ligaments so that the abdominal wall is
reflected in two flaps. Be very careful not to perforate the gut. As
the abdominal cavity is opened, look for excess peritoneal fluid, if it is
present suck some up with a sterile pipette and culture on blood agar
and in cooked meat broth and make smears on glass slides for
microscopic examination. Then with the points of the scissors
towards the left axilla make a cut through the left side of the chest
wall and diaphragm as posteriorly as possible. Insert the points of
the scissors through the hole thus made into the thorax and separate
the diaphragm from the ribs along the costal margin. Then cut
through the ribs continuing the first incision as far as the clavicle.
Cut across the manubrium sterni and continue down the right side
of the thorax posteriorly so that the whole of the front of the chest
wall can be lifted off to give a good view of the thoracic viscera. If
pleural exudate is found treat it by the method described above for
peritoneal fluid. Next sample the heart blood by plunging the
end of a sterile Pasteur pipette into the ventricle, holding the heart
firmly at its base with forceps ; culture it on blood agar and
in cooked meat broth and make films. Now examine the viscera
more closely, starting as far as possible from the site of inoculation
so as to avoid contamination of distant sites with bacteria from the
neighbourhood of the local lesion. Look for enlargement of the
bronchial lymph glands and for lesions in the lungs, then examine
the liver, spleen, kidneys and portal lymph gland, then the sublumbar
and inguinal glands and finally the femoral glands and the site of
inoculation.

In widely disseminated tuberculosis all the lymph glands are
enlarged, there are small necrotic patches and sometimes tubercles

on the lungs, numerous greyish yellow necrotic patches on the liver and on the spleen. The femoral and inguinal glands are much bigger on the side of inoculation, and at its site, in the right thigh muscles, a deep caseous abscess may be found; often the local lesion is healed. Smears and cultures are made as the examination proceeds from macroscopic lesions. For evidence of tuberculosis make smears, and cultures when required, from the cut surface of the spleen and sublumbar glands when they show evidence of disease. Heavy inocula of saprophytic acid-fast bacilli sometimes cause a local lesion and enlargement of the femoral lymph gland, but after intramuscular injection they do not penetrate as far as the sublumbar glands or spleen. If acid-fast bacilli are found in these smears the diagnosis of virulent *Myco. tuberculosis* is made.

In mild cases or when the animal has died prematurely the only evidence may be enlargement of the femoral gland near the inoculation site, with or without a local lesion. Even if acid-fast bacilli are found this is not sufficient for a diagnosis. Remove the gland, grind it in a sterile grinding tube or stomacher, culture on egg medium and inoculate the remainder into another animal.

Disease caused by salmonella infection, the pyogenic cocci, *Brucella, Y. pseudotuberculosis* or fungi may be mistaken for tuberculosis. No acid-fast bacilli are found. The liver and spleen show areas of necrosis similar in appearance to tuberculous necrosis, but the lymph glands are often comparatively small. Cultures of spleen and blood yield the pathogen.

When lesions are atypical and several cultures are required, place the organs in separate sterile Petri dishes and make cultures from them at the bench when the autopsy is complete. Finally fix portions of them in formol-saline for section, which may indicate the nature of the disease when cultures and smears prove to be negative.

The autopsy is now complete. Using forceps place the body, with strings attached, in the discard bin to be incinerated, boil all the instruments and pins in the sterilizer for 5 minutes. Flood the tray and board with 50 per cent Sudol, leave them for 2 hours and then wash and dry them.

Note. This method can be applied to the examination of other small laboratory animals. Because of their size mice are always pinned out on a board.

(b) Guinea-pig killed at 6 weeks

The animal should be pinned out on a tray in the inoculation cabinet used for tubercle culture. The autopsy is performed by the method described above except that blood agar and broth cultures are omitted. When the viscera have been exposed, they are examined from the thorax downwards. If a caseating lesion is present in the thigh it is carefully avoided until the end so that organisms from it are not inadvertently spread to distant sites. Smears and cultures are made from the spleen and sublumbar glands. When several autopsies are made in succession it is a wise precaution to remove the spleen and one or two enlarged glands from the infected animals and leave them in sterile Petri dishes until the smears have been examined, because if no acid-fast bacilli are found it will be necessary to examine more material from them, to make cultures on blood agar, to fix part of them for section and perhaps to inoculate another animal with the ground tissue.

When a large number of apparently healthy guinea-pigs are to be examined for tuberculosis, apparatus may be saved by using one set of unsterile instruments only in the first place. When there is no macroscopic evidence of infection no smears or cultures are made and the same instruments and tray may be used for the next animal. When an infected animal has been examined the instruments and tray are removed for sterilization and a fresh set is used for the next animal. A wire loop is employed as much as possible for sampling tuberculous material. Instruments are flamed to a dull red heat after boiling because of the risk of transferring dead, but visible, tubercle bacilli to smears made from subsequent autopsies.

Notes. 1. Sleeves are rolled well back so that they do not touch infected material and so that hands and forearms can be well washed when the examination is complete.

2. Animals inoculated with dangerous organisms, for example, *B. anthracis*, *Y. pestis* and *Fr. tularensis*, need a special technique. Rubber gloves are worn so that the skin never comes in contact with the animal or the contents of its cage (isolation in special cages is necessary). Clothing is protected by a gown and rubber apron. The gloves and protective clothing are sterilized after use. Animal inoculation of such organisms is avoided as far as possible because of the risk of laboratory infection and of an epidemic in the animal house.

Typing cultures of *Myco. tuberculosis*

Since the treatment of human and bovine tuberculosis in man is the same, it is seldom necessary to find which type of bacillus is causing the disease.   When the source of the infection is to be traced, however, typing may be important.   Although there are cultural differences between the types, they are not sufficiently constant for reliable identification and a pathogenicity test in rabbits is required.

METHOD

Seed the culture heavily on a Dorset's egg slope.   When good confluent growth appears, after 1 to 3 weeks' incubation, wash it off in the water of condensation, add 1 to 2 ml. of saline and transfer the suspension to a tube.   Break up lumps of growth by grinding them if necessary.   Make the volume up to 10 ml. with saline and allow any remaining lumps to settle.   Make two consecutive tenfold dilutions in saline from the supernatant fluid (1 ml. of 1 : 100 dilution is equivalent to $\propto$ 0·001 mg. of moist solid growth).   Inoculate 1 ml. of this dilution intravenously into a rabbit.

Bovine strains almost always kill the animal in from 4 to 6 weeks. If it is not dead at the end of 3 months it is killed and an autopsy is made.   Human strains are capable of causing minimal lesions only which will be hard to find after this time; bovine strains cause miliary tuberculosis with numerous tubercles in the lungs and kidneys.   As the susceptibility of animals varies, for really accurate work inoculation of each strain into several rabbits is desirable.

Tuberculin test

Numerous methods for testing hypersensitivity to tuberculin have been described.   The graded intradermal (Mantoux) test is reliable and has been used extensively in epidemiological surveys in many countries.   Two antigens are available.   Old tuberculin has proved very reliable and is comparatively stable, PPD is a purer preparation and theoretically should give fewer false reactions; there is however more chance of variation between different batches.

METHOD

Adults are injected intradermally in the forearm with 0·1 ml. of 1 : 10,000 dilution which raises a bleb about 1 cm. in diameter; no control is necessary.   The result is read between 48 and 72 hours and is positive if there is a patch of *oedema* at the site of inoculation at least 5 mm. in diameter ; erythema alone is of no significance. Reactions occurring on the first day which fade rapidly are non-specific. When the result is negative the test is repeated in the other forearm using the 1 : 1000 dilution. Alternatively a multiple puncture test can be performed see following page.

Patients with erythema nodosum, who are likely to be highly sensitive, are tested initially with weaker dilutions, one in a hundred thousand or even one in a million. The consequences of giving too large a dose to a hypersensitive patient are serious and include necrosis of the skin at the site of inoculation, general symptoms with fever and sometimes increased size of the tuberculous lesion.

It is very difficult to remove traces of tuberculin from glass syringes, therefore those used in the Mantoux test are kept for this purpose only. Syringes must not be used for different dilutions of tuberculin without very thorough cleaning. If a syringe which has previously held a strong dilution is then used for injecting a weak dilution, an unpleasantly strong reaction may follow in a very sensitive patient. Disposable syringes and needles are ideal for hyper-sensitivity tests.

Interpretation. A positive result indicates that the tissues have been invaded by the tubercle bacillus some time previously and is useful in the diagnosis of tuberculosis in children under 5 years who are normally negative. It gives no indication of the activity, size or site of the lesion. In Britain many adults are sensitive so that a positive result is of no diagnostic importance.

A negative result almost excludes tuberculosis of more than 2 months' duration at any age, unless the patient has received steroids. Patients with acute fresh lesions and those with longer standing but rapidly progressive disease may fail to react, but a negative multiple puncture tuberculin test in patients with active chronic tuberculosis is extremely rare.

The multiple puncture test (Heaf and Rusby, 1959) is ideal for screening large numbers of patients attending weekly clinics who are expected to be tuberculin negative, or newly converted after B.C.G. vaccination. One drop of a special preparation of PPD is spread on the forearm and injected into the skin by releasing a 6-bladed " gun " over the area. The method is painless and the result can be read any time between the third and the seventh day. The " gun " is sterilized between patients by flaming. The blades can be set in either of two positions so that they are effective in adults and can be made to penetrate less deeply for children. In highly sensitive patients an unpleasantly severe reaction results and the site of inoculation may be visible as six small red papules for months or even years. Therefore when a positive result is likely, in adults from endemic areas or in suspected infection, the graded Mantoux test is preferred. Those failing to react to the 1 : 10,000 intradermal dose can safely be retested by the multiple puncture technique.

A low-grade positive tuberculin test can result from infection by opportunist mycobacteria. When this is suspected the test can be further evaluated by comparing in parallel tuberculin tests using standard mammalian PPD and PPDs prepared from avian, Battey and other mycobacteria ; see Edwards *et al.* (1970).

Mycobacterial ulcers

Since the original report by MacCallum *et al.* (1948) of chronic ulcers caused by Mycobacteria, a number of other episodes have been recorded in other parts of the world, including Britain (Thomas, 1967). The organisms usually gain entry following accidental grazing of the skin in a swimming bath. Two pathogens are now well recognized ; *Myco. ulcerans* and *Myco. balnei* (see Table 18). When swabs are received from such lesions or when acid-fast bacilli are seen in smears from any superficial ulcer, cultures should be made at 30° C. on pyruvate egg and Lowenstein-Jenson's medium in addition to the routine cultures. When no 30° C. incubator is available, culture should be left in the dark at room temperature. These organisms will grow at the lower temperature, but poorly, if at all, at 37° C.

LEPROSY

Myco. leprae can be seen in scrapings of the nasal mucosa in the lepromatous form of the disease and in sputum when the lungs are affected. They are acid-fast bacilli which can be stained by the Ziehl-Neelsen or by the fluorescent method as described for tubercle bacilli. They tend to appear in very large clumps and many are found within the cytoplasm of pus cells. They are differentiated from other Mycobacteria by their inability to grow on egg medium or to infect animals.

Tests for reagin in the diagnosis of syphilis are often positive in leprosy, particularly in the nodular type of the disease.

ACTINOMYCOSIS

Although it is possible to make a bacteriological diagnosis of actinomycosis by examination and culture of pus (see Chapter 3) when the disease has not previously been suspected, isolation of the actinomyces occurs more often as the result of a deliberate search to confirm a clinical diagnosis.

The cause of typical human actinomycosis is *Actinomyces israeli*. It is an obligate human or animal parasite and has been isolated from the mouth of patients not suffering from the disease, so that its presence in a culture is not proof of actinomycosis in the patient. It is associated with salivary calculi and is probably the most common cause of their formation. The factors which determine the onset of the disease in a previously healthy human carrier are unknown. When the organism multiplies in tissues they respond by forming a granuloma which enlarges, penetrates the surrounding tissues, whatever their nature, and finally reaches the surface, where it appears as a hard mass with discharging sinuses. The usual sites of infection are the jaw, the caecum or the lungs. Spread by the bloodstream is rare but actinomycotic cerebral abscess has been recorded as a complication of the pulmonary form of the disease.

The discharge is seldom profuse and in order to obtain a good sample the whole of the inner dressing is sent to the laboratory in a sterile screw-capped bottle. A swab taken from the sinuses is usually rich in contaminating organisms and poor in material from the deep part of the granuloma which is most likely to yield a positive culture.

Actinomyces bovis, which causes actinomycosis in cattle and occasionally in man, closely resembles *Actinomyces israeli* and is also an anaerobe. The name actinomyces is now limited to the anaerobic and microaerophilic species; the aerobes, some of which are human and cattle pathogens, are termed *Nocardia* (see Table 19).

MACROSCOPIC AND MICROSCOPIC EXAMINATION

Using sterile forceps and scissors, remove the part of the innermost layer of the dressing which is most heavily contaminated and place it in a 30 ml. screw-capped bottle containing 5 to 10 ml. of peptone water. (Saline may interfere with the viability of delicate microbes. At this stage infection by actinomyces cannot be assumed.) Shake well and examine the fluid for actinomycotic granules. They are colonies of *Actinomyces israeli* which have developed in the tissues. They are about 0·2 to 1 mm. in diameter, yellowish and oily in appearance. Remove one with a pasteur pipette and squash it between two glass slides, separate the slides and Gram-stain. Fresh granules are soft and easily emulsified but old ones may be calcified. If no soft granules can be found it is necessary to treat calcified ones with dilute hydrochloric acid before staining. The fungus is seen as a tangled mass of Gram-positive branching mycelium with Gram-negative clubs at the periphery, the appearance is typical and diagnostic of actinomycosis.

CULTURE

Material from old lesions is often sterile but there is usually no difficulty in recovering the organism from young active lesions. When soft fresh granules are found, they are washed three times in saline to remove contaminating organisms and are then cultured on two blood agar plates, incubated in air plus 5 to 10 per cent CO_2 and anaerobically plus about 10 per cent CO_2. When no granules are found the fluid is centrifuged and the deposit is cultured in the same manner. *Actinomyces israeli* is incapable of growth in air, which distinguishes it from the many harmless aerobic species which sometimes contaminate wounds. Growth is usually visible on the anaerobic culture after 3 days' incubation as 0·5 to 1 mm. yellowish crumb-like colonies which sometimes tend to adhere to the medium and may not be very easily emulsified. Gram-stained films from cultures reveal a very different appearance from the colonies found in tissues. The organism is mainly bacilliform and resembles diphtheroids for which it may easily be mistaken. The arrangement, however, is a little different because many bacilli lie in V or Y forms and careful search usually reveals a few filaments showing true branching ; it is non-acid-fast. For a detailed account of strains and their antibiotic sensitivity, see Garrod (1952 a and b).

Nocardia, some of which are acid-fast, have been described as the cause of atypical actinomycosis. There is no doubt that they are occasionally pathogenic and even cause generalized infection and death. When one of them is isolated from a chronic suppurating wound or from sputum very thorough investigation is necessary before it can be regarded as the causal organism (see Chapter 1) and proof is seldom forthcoming unless an autopsy is performed.

Table 19 lists the human pathogens.

The actinomyces are not finally classified. Like the *Mycobacteria* they lie between the fungi proper and bacteria. An anaerobic or microaerophilic actinomycete is likely to be pathogenic. Aerobic species may be mistaken for diphtheroids or erysipelothrix (see Chapter 5) but can be differentiated from them by cultural methods and by agglutination reactions. Animal pathogenicity tests are not of much practical value because, although virulent Actinomyces cultures will cause disease when injected into cattle, they are not

TABLE 19.

Actinomyces and Nocardia

Name	Disease	Acid-fast	Remarks
Actinomyces israeli	Actinomycosis (man)	—	Anaerobic obligate parasite
Nocardia graminis	Actinomycosis (cattle). Various human infections	—	Found on grasses
Nocardia madurae	Madura disease (man)	—	Aerobic, mainly in tropics
Nocardia asteroides	Pulmonary infection (man)	+	Aerobic, rare as pathogen

highly pathogenic for small laboratory animals. A virulent culture injected intraperitoneally into a rabbit may cause some actinomycotic nodules in the omentum after three months but the animal remains apparently healthy.

FUNGOUS INFECTIONS

In addition to ringworm and other well-recognized diseases, fungi cause infections resembling tuberculosis and actinomycosis and they are sought whenever it has proved impossible to see or culture tubercle bacilli or actinomyces from non-syphilitic granulomatous lesions.

The methods used for identification of true fungi differ from ordinary bacteriological technique.[1] Their structure is best seen in wet preparations, stained or unstained, rather than in dried films, and

[1] A more detailed account of mycological technique is given in the *CDC Laboratory Manual for Medical Mycology*, Communicable Disease Centre, Atlanta, Georgia 30333, U.S.A.

identification depends mainly on the appearance of the mycelium and reproductive spores instead of on biochemical and serological tests. Frequently a presumptive diagnosis can be made after examination of a wet preparation of the original material, but confirmation by culture on medium which encourages spore formation is necessary for identification.

CULTURE METHODS

Most fungi which cause human disease will grow on blood agar incubated at 37° C. but they grow more slowly than the majority of bacteria and it is necessary to incubate the cultures, for 1 to 8 weeks, and to seal them so that they do not dry up. Some grow more quickly at room temperature.

Some of the pathogenic fungi exist in two forms. One phase is found on blood agar incubated at 37° C. and in infected tissues and discharges; reproduction is by simple budding, the colonies resemble large smooth bacterial colonies and show simple yeast-like forms without mycelium. The other phase, which is important in diagnosis, shows hyphae and spores microscopically and the colonies are large and fluffy like moulds; they appear on Sabouraud's glucose agar incubated at 18° to 26° C. Almost all the pathogenic fungi are simple in structure and have no sexual spores. They are not susceptible to penicillin and streptomycin, which can be used to inhibit bacterial growth in cultures of contaminated material. If resistant bacteria are present other antibiotics may inhibit them but the presence of chloramphenicol and tetracyclines in the cultures may not be harmless to fungi although these antibiotics are apparently unsuccessful in the treatment of fungous infections.

It is usual to seed the inoculum heavily in one or perhaps two places on the medium without spreading, because one large typical colony will arise from the heavily inoculated area. Sabouraud's glucose agar, sloped in 30 ml. screw-capped bottles and screened with penicillin and streptomycin, is popular and efficient. Material from granulomatous lesions, pus, sputum, cerebrospinal fluid or bone marrow may be cultured when generalized fungal infection is suspected. It should be seeded on Sabouraud's medium and on blood agar sloped in bottles similarly screened. The slopes are examined twice weekly for at least 4 weeks.

Examination of fungous cultures

Note the appearance of the surface of the colony and also inspect it from below when the medium is translucent. Scrape off two pieces of the colony, one from the centre and one from the edge, using a short stout loop or a small thin metal spatula. Place each piece in a drop of saline on a glass slide and tease it out. The material is sometimes tough and mounted botanical needles are most suitable for dealing with it. Cover the preparation with a coverslip and examine it using the 16 mm. and 4 mm. objective.

Generalized fungous infections

(a) *Cryptococcus neoformans* (Torula histolytica)

Infection in man may occur in the skin, subcutaneous tissues or lungs where it often remains undiagnosed until generalized infection with meningitis supervenes. The disease is rare and is usually mistaken for tuberculous meningitis. The laboratory diagnosis is simple because the yeast cells are seen in the cerebrospinal fluid. They are about the same size as red cells or small lymphocytes but unlike blood cells they are thick walled and vary in size. Culture on blood agar yields a growth of gelatinous white colonies, usually within a few days, but incubation for 2 weeks is occasionally necessary. The diagnosis is made either from the original material, or from culture by making a wet preparation in indian ink which reveals the typical thick gelatinous capsule which surrounds the cells. The capsule may be thicker than the diameter of the cells. Reproduction is by budding and no mycelium is formed.

(b) *Candida albicans* (Monilia albicans, Oidium albicans)

Of the *Candida* this is the only important pathogen but it may be found in the mouth, vagina and faeces of normal people. It causes thrush, commonly infecting the mouths of infants and the mother's nipple and occasionally infects adult mouths. It is a common cause of vaginitis, sometimes infects the skin and occasionally causes pulmonary and generalized infection. Patients treated for long periods with broad spectrum antibiotics which do not affect the yeast are especially prone to infection. It can be easily seen either in wet preparations or in Gram-stained films as large Gram-positive oval budding cells sometimes with short strands of pseudomycelium. It grows well on blood agar and Sabouraud's glucose agar, yielding smooth white colonies 1 to 2 mm. in diameter within 1 or 2 days at 37° C. The colonies sometimes show pseudomycelium which appear as short hair-like spikes around the periphery giving them an irregular edge. *Candida albicans* is identified by germ tube formation in serum, by its ability to form chlamydospores on suitable medium, by fermentation tests and carbon assimilation and by agglutination tests.

The germ tube method is rapid, reliable and sufficient for most diagnostic purposes provided the yeast examined has typical morphology and grows readily at 37° C. (Mackenzie, 1962). Species other than *C. albicans* which give a positive test are so rarely encountered in Britain that they can be ignored except in serious generalized infection when full identification should be made. This will include assimilation of carbon from carbohydrates and is best undertaken by a specialist mycology laboratory.

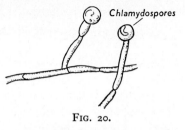

Chlamydospores

Fig. 20.

Formation of germ tubes

Any mammalian serum is satisfactory and human plasma from out-of-date blood transfusion blood can also be used. Inoculum is critical, if it is too heavy germ tube formation will be suppressed. At 10⁷ yeasts per ml. or more in some serum it may be impossible to see germ tubes although the strain is capable of forming large numbers from a lighter inoculum. Since incubation time is short, 3 hours, unsterile tubes can be employed and they need not be covered ; the serum should be sterile.

METHOD. Inoculate 3 small plastic tubes, 6 × 50 mm., each containing about 0·1 ml. mammalian serum or human plasma, from several typical yeast colonies rubbing off the inoculum in each tube consecutively to ensure diminishing inoculum. Incubate at 35°–37° C. for 3 hours. Examine a small loopful of culture in a wet preparation for outgrowth from the yeast cells resembling the handle of a mirror. When no germ tubes are seen in the first tube examine the other two. Cells which appear threadlike are not *C. albicans*. On the rare occasions when no germ tubes are seen and yet the cells are not threadlike fermentation tests may be employed. Media should contain 2–3 per cent sugar instead of the usual 1 per cent in media for Enterobacteriacae. Results for species which may be encountered are seen in Table 20. It is essential to subculture on antibiotic-free medium to ensure purity before fermentation tests are performed. Examination for chlamydospores may also be attempted.

We have found Taschdjian's medium to be superior to corn-meal or Zein agar (Taubert and Smith, 1960). It is inoculated heavily with the original material, or from a colony, and a sterile coverslip is placed over the inoculated area. After overnight incubation at room temperature (18°–25° C.), the culture is ex-

TABLE 20.

	Glucose	Maltose	Sucrose	Lactose
C. albicans	AG	AG	(A)	—
C. tropicalis	AG	AG	AG	—
C. pseudotropicalis	AG	—	AG	AG
C. parapsilopsis	AG	—	—	—
C. krusei	AG	—	—	—
C. guillermondi	AG	—	AG	—
Crypt. neoformans	—	—	—	—
Torulopsis glabrata	—	—	—	—

Bracket indicates variable result.

amined under a plate microscope or low power of an ordinary microscope and chlamydospores (Fig. 20) can be seen. Temperatures higher than this will inhibit chlamydospore formation.

(c) *Coccidioides immitis*

This fungus causes coccidioidomycosis, a disease endemic in California ; cases have been reported from other areas in America and from Europe. It resembles tuberculosis both clinically and, to some extent, epidemiologically. Chronic cases suffer from cough and show chest lesions similar to tuberculosis. Erythema nodosum is common. Glands, joints and bones are sometimes affected with the formation of granuloma and suppuration. The coccidioidin test, similar in principle to the tuberculin test, reveals hypersensitivity to the fungus in a high proportion of apparently healthy people in the endemic area and also in laboratory workers who have handled cultures. The fungus gains access to the lungs by inhalation of the spores which are found in dust. Direct spread of infection from man to man has not been demonstrated. Cultures are dangerous because of the risk of inhaling the spores which are readily detached from them and contaminate the air.   Laboratory infections have often been reported.   Diagnosis is by direct examination, culture and animal inoculation of sputum or pus.

C. immitis appears on Sabouraud's glucose agar within a week. If incubation is continued the colony becomes covered with white mycelium which gradually turns brownish.   Microscopic examination reveals hyphae and the highly infective arthrospores.   Cultures must be handled with the greatest care in the safety cabinet.   In tissues the fungus has a totally different appearance ; no mycelium is seen but there are large thick-walled spherules, about 80 μm. in diameter when fully developed, containing numerous endospores ; their appearance is diagnostic.   When they cannot be demonstrated in the original material it is inoculated intraperitoneally into a guinea-pig which is killed after 6 weeks unless death has supervened. The post-mortem findings resemble tuberculosis but no acid-fast bacilli are found, the spherules can be demonstrated in sections of the liver and spleen and the fungus can be cultured from the lesions. Arthrospores of *C. immitis* inoculated into animals revert to the endospore form in the tissues within a week.   Preparation of sputum and pus for animal inoculation as described for the diagnosis of tuberculosis will kill the endospores of *C. immitis* which are not as resistant as acid-fast bacilli to chemicals.   Contaminating bacteria can be inhibited by treating the specimen with an equal volume of 0·5 per cent copper sulphate solution for 4 hours and then centrifuging before inoculation.

(d) Histoplasma capsulatum

This fungus is the cause of histoplasmosis, a disease of the lungs and reticulo-endothelial system.   It is common in some parts of the United States and has been encountered elsewhere.

The pulmonary form of the disease is usually mild.   Sub-clinical infection, demonstrated by the histoplasmin skin hypersensitivity test and by chest X-ray is common in the endemic areas. Generalized infection is almost always fatal.   The fungus enters through the alimentary tract and becomes disseminated throughout the reticulo-endothelial system, causing enlargement of liver, spleen and lymph glands ; when the disease is well established the fungus can be seen in bone marrow smears and even in peripheral blood films stained by Geimsa's method.

H. capsulatum can be isolated from the bone marrow by culture on blood agar or Sabouraud's agar and sometimes from the peripheral blood. Growth will be seen on the agar slope of the diphasic medium which should be incubated for 6–8 weeks because growth on primary isolation may be slow. On blood agar incubated at 37° C. the colonies appear smooth and white and budding yeast-like forms only are found microscopically. For diagnosis it is necessary to culture on Sabouraud's agar incubated at 18° to 20° C. After 2 to 3 weeks white colonies with a fluffy surface appear which later turn brownish. Microscopically, mycelium is found with hyphae bearing numerous thick-walled chlamydospores covered with short projections. These spores are 8 to 20 μm. in diameter and their appearance is diagnostic. They are highly infective and all cultures must be handled with the greatest care in the safety cabinet. Mice and guinea-pigs are susceptible to the disease.

Ringworm

(*a*) *Microsporon* species

TABLE 21.

Microsporon (ringworm of hair and skin only)

	Animal pathogen	Growth on Sabouraud's medium	Macroconidia
M. audouini . .	—	Slow	Very rare, slender, distorted
M. canis . . .	+	More rapid	Numerous, long, thick-walled with slender tips
M. gypseum . .	+	More rapid	Very numerous small thin-walled with blunt tips

Microsporon audouini causes epidemic scalp ringworm in children. If stumps of broken hair in the bald patch are pulled out with forceps, mounted in 10 per cent potassium hydroxide and examined microscopically under a low power objective the fungus is seen as a mass of small spores enveloping the lower end of the hair. Differentiation between *M. audouini* and the animal pathogens *M. canis* (lanosum) and *M. gypseum* is of considerable importance because the animal pathogens, which only occasionally cause human ringworm, will not give rise to a human epidemic. Identification can only be made by culture and examination of the reproductive spores.

Short lengths of hair (about ½ cm.) are implanted on the surface of Sabouraud's glucose agar containing penicillin and streptomycin to inhibit bacterial growth.   The cultures are incubated in the dark at room temperature or better at 26°–28° C.   The points of difference between the species are listed in Table 21.

(*b*) *Trichophyton* species (ringworm of hair, skin and nails)

Several different species are pathogenic to man, some are animal pathogens.   All grow in culture, show numerous hyphal swellings

(a)   (b)

FIG. 21.

Macroconidia *Microsporon*, 5 to 15   Macroconidia *Trichophyton*, 2 to 6
segments 40 to 50 μm. long.   segments 10 to 50 μm. long.

and chlamydospores.   Macroconidia are usually scanty and may be absent.   Their appearance, see Fig. 21(b), is diagnostic.   *T. inter-digitale* is the commonest cause of foot ringworm.

(*c*) *Epidermophyton* species (ringworm of skin and nails only)

The group consists of one species only, *E. floccosum*.   It is the commonest cause of ringworm of the groin and is not a natural

FIG. 22.

Macroconidia   Epidermophyton
2 to 4 segments.   30 to 40 μm. long.

animal pathogen. Growth on Sabouraud's medium is slow; macroconidia are numerous (see Fig. 22).

All the ringworm fungi are cultured by implanting pieces of infected material on the surface of Sabouraud's medium screened with penicillin and streptomycin to inhibit bacterial growth. They are incubated in the dark at room temperature to encourage the production of spores on which diagnosis depends.

Fungous infections of the subcutaneous tissues

Chronic spreading cellulitis of the exposed part of the body, particularly the fingers, hands and forearms, is occasionally caused by fungi. The most likely fungal cause of such an infection is *Sporotrichum*. The condition is diagnosed clinically by its lack of response to the usual forms of treatment and in the laboratory by microscopic demonstration of the fungus in the lesion and by culture. In pus and tissues *Sporotrichum* appears as a Gram-positive short fusiform body 1·5 to 4 μm. in size. Its appearance is diagnostic. Cultures on blood agar at 37° C. or on Sabouraud's agar at room temperature can be screened with antibiotics to prevent growth of secondary bacterial contaminants. White or cream coloured colonies of the fungus usually appear within a week. Intraperitoneal inoculation of mice or guinea-pigs will cause infection. Mice sometimes show purulent orchitis and the fungus can be recovered from the pus.

References

Collins, C. H. (1962). *Tubercle*, **43**, 292.
Collins, C. H., Hartley, E. G. and Pilsworth, R. (1974). Public Health Laboratory Service Monograph Series No. 6. HMSO, London.
Edwards, P. Q., Furcolow, M. L., Grabau, A. A., Grzybowski, S., Katz, J. and MacLean, R. A. (1970). *Amer. Rev. resp. Dis.*, **102**, 468.
Garrod, L. P. (1952a). *Tubercle*, **33**, 258.
—— (1952b). *Brit. Med. J.*, **i**, 1263.
Heaf, F. and Lloyd Rusby, N. (1959). *Recent Advances in Respiratory Tuberculosis*. Churchill, London, page 62.
Hughes, M. H. (1966). *J. clin. Path.*, **19**, 73.
MacCallum, P., Tolhurst, J. C., Buckle, G. and Sissons, H. A. (1948). *J. Path. Bact.*, **60**, 93.
Mackenzie, D. W. R. (1962). *J. clin. Path.*, **15**, 563.

MARKS, J. (1963). *Mth. Bull. Minst. Hlth Lab. Serv.*, **22**, 150.
—— (1972). *Tubercle*, **53**, 31.
MARKS, J. and RICHARDS, M. (1962). *Mth. Bull. Minist. Hlth. Lab. Serv.*, **21**, 200.
MITCHISON, D. A., ALLEN, B. W. and LAMBERT, R. A. (1973). *J. clin. Path.*, **26**, 250.
MITCHISON, D. A., WALLACE, J. G., BHATIA, A. L., SELKON, J. B., SABBISH, T. V. and LANCASTER, M. C. (1960). *Tubercle*, **41**, 1.
Public Health Lab. Service Working Party (1952). *Mth. Bull. Minist. Hlth Lab. Serv.*, August, Vol, 11.
RUNYON, E. H. (1959). *Med. Clin. N. Amer.*, **43**, 273.
SCHWABACHER, H., and WILSON, G. S. (1937). *Tubercle*, Lond., **18**, 442.
TARSHIS, M. S., KINSELLA, P. C., and PARKER, M. V. (1953). *J. Bact.*, **66**, 448.
TAUBERT, H. D. and SMITH, A. G. (1960). *Lab. clin. Med.*, **55**, 820.
THOMAS, D. T. (1967). *Brit. med. J.*, i, 437.

7

Antibacterial drugs

Since the discovery of antibiotics precise bacterial diagnosis has assumed a new importance. The bacteriologist is now called upon to test the sensitivity of pathogens to the various drugs available and his advice as to the most suitable antibacterial treatment is often sought. In my view the medical bacteriologist should be prepared to take full responsibility for advising treatment. He should familiarize himself with the effect on bacteria and the advantages and possible toxic manifestations of new antibacterial agents. There are now innumerable drugs other than antibiotics which physicians and surgeons have to learn about if they are to treat their patients optimally. Antimicrobial agents are, of necessity, partly the province of the bacteriologist and he should do all he can to help his clinical colleagues in this increasingly complex field. Rational antibiotic treatment, particularly in hospital, is essential not only for the good of individual patients but to ensure prolonged usefulness of the drugs. The reader is referred to Garrod *et al.* (1973) for further information on the principles of antibiotic treatment which is outside the scope of this book.

To meet clinical demands rapid sensitivity tests are essential; the choice of drug, however, does not depend entirely on the result. If an antibiotic, in concentration higher than that which can be obtained in body fluids, fails to impede growth of an infecting microbe in the laboratory, it is extremely unlikely that the drug will be of use in treatment. But unfortunately it by no means follows that because an antibiotic is found to inhibit growth *in vitro* that treatment with it will succeed. Many other factors have to be considered. The subject is much too complex to be discussed here in detail, a few important points, however, will be enumerated to emphasize the limitation of laboratory sensitivity tests as guides to treatment.

1. Type of Infection

(*a*) *Acute generalized infection.* Adequate doses of any of the drugs which prove active against the pathogen in laboratory tests will as a rule cure the infection ; exceptions to this are typhoid fever and tuberculosis. The typhoid bacillus and tubercle bacillus are usually sensitive in the laboratory to various antibiotics for example to tetracycline, but clinical response to treatment with this and some other apparently active drugs is unexpectedly poor.

(b) *Subacute infections with local lesions.* Any of the drugs which prove active in the laboratory will probably improve the patient's condition, at least temporarily. Lasting success is most likely to be gained by giving a drug, such as penicillin or streptomycin, which is likely to penetrate to the local lesion in a concentration sufficient to *kill* the pathogen (see Table 30).

Much has yet to be discovered about the penetration of antibiotics to the bacteria in different kinds of septic lesion, but it has already been shown that minimal concentration of penicillin in tissue fluid from inflammatory lesions is actually higher than the corresponding blood level. As would be expected, penicillin treatment of subacute infection by sensitive bacteria is highly successful. The disadvantage of streptomycin therapy is that the majority of bacteria are capable of producing resistant variants and therefore quickly become resistant unless they are all killed very rapidly ; this can sometimes be prevented by combining it with other drugs. Gentamicin, which is included with streptomycin, kanamycin and tobramycin in the aminoglycoside group of drugs, has not this disadvantage, it is also bactericidal and is preferred for the treatment of severe infections, streptomycin now being reserved for the treatment of tuberculosis.

(c) *Chronic deep-seated infections.* Treatment with antibiotics alone will not as a rule cure these conditions, even if the microbes causing them are highly sensitive. They are useful in combination with other forms of treatment to prevent generalized spread of infection during surgical procedures, or for inserting into infected cavities after these have been opened and drained. For the latter purpose the pH of the fluid in the cavity is of importance in the choice of drug ; for example, the aminoglycosides are relatively inactive in acid solutions and the tetracyclines are unstable in neutral and alkaline solutions.

2. Site of Infection

(*a*) *Meninges.* In meningitis drugs pass to a greater extent than normal from the bloodstream into the cerebrospinal fluid (CSF). In fact in miliary tuberculosis treated by intramuscular injection of streptomycin, the presence of the drug in the cerebrospinal fluid may be an early sign of meningitis. This increased permeability is probably insufficient for treatment and when the diagnosis is established streptomycin is given intrathecally. In pyogenic meningitis the drug of choice for intrathecal use is one which, in addition to its lethal effect on the pathogen, is non-irritant, highly soluble, stable in isotonic solution and which can be prepared as a sterile liquid in sealed ampoules. Sterility of fluids introduced intrathecally is extremely important. It has been shown (Smith and Smith, 1941) that organisms usually considered harmless, and sometimes difficult to demonstrate by routine methods, may cause severe meningitis if they are accidentally introduced into the theca. Such an accident during treatment of meningitis would probably be mistaken for a relapse of the original condition and it may well be more common than is generally supposed.

Chloramphenicol, sulphonamides, trimethoprim, and isoniazid penetrate freely to the subdural space. After the usual oral doses, levels of chloramphenicol in blood and cerebrospinal fluid are approximately equal, even in patients without inflammation of the meninges. These drugs are therefore particularly valuable for treating meningeal infection by organisms resistant to penicillin. Although penetration by penicillin is relatively poor, the drug is harmless except in high dosage given intrathecally and has proved so efficient that it is still the first choice when the microbe is sensitive. A high dose given frequently by injection into an intravenous tube will usually, in acute meningitis, raise the level in the CSF sufficiently to inhibit or kill the bacteria thus obviating the need for intrathecal therapy. By testing the fluid against the organism isolated from it in the manner described for serum in endocarditis, page 245, the sufficiency of penetration can be checked.

(*b*) *Urinary tract.* Most antibiotics are concentrated in the urine, sometimes more than a hundredfold. It is often possible therefore to treat urinary infections successfully when the pathogen is relatively resistant. For example, a low dose, 250–500 mg. ampicillin 6-hourly by mouth is sufficient for the treatment of

Esch. coli cystitis although the minimal inhibitory concentration for this species is 8 *μ*g. per ml. and much higher doses given by injection will be needed for infections elsewhere. Moreover, the pH of urine can be altered when necessary to suit the antibiotic. It should be alkaline for the aminoglycosides and acid when tetracyclines are given.

(*c*) *Superficial ulcers and wounds.* By local application, in addition to systemic treatment, it is sometimes possible to cure these lesions when the pathogens are relatively resistant. Great care is needed to prevent cross infection with highly resistant strains. Local treatment alone is less likely to be effective and may cause sensitization to the drug. Comparatively toxic drugs such as neomycin and bacitracin are usually employed locally so that should the patient develop hypersensitivity he will not be deprived of a valuable drug which he may desperately need in a future systemic infection.

Laboratory Control of Treatment

Laboratory control of treatment is always desirable and often essential, especially in endocarditis, if the best use is to be made of the numerous antibiotics now available. Although some species are always sensitive to certain drugs, in practice, reliable sensitivity tests can usually be reported sooner than the pathogen can be identified. Therefore all specimens received from sites normally sterile are tested against appropriate antibiotics; a separate plate culture for sensitivity tests is included with the primary cultures. Primary sensitivity tests are not made on material from sites with a normal flora except when microscopic examination of the specimen indicates bacterial infection. The chance of a useful result of primary tests of throat swabs, faeces and high-vaginal swabs from gynaecological conditions is so remote that primary testing is omitted.

Complete control with assay of the drug in body fluids is usually only required in endocarditis. In this condition the bacteriologist can make a major contribution and he needs constantly to be prepared to meet demands which may be made on him at short notice. In endocarditis a drug, or more often a combination of drugs, is required which will *kill* the bacteria, not merely inhibit their growth. In addition to sensitivity tests which demonstrate inhibition, tests of bactericidal power are required. Tests with the patient's own serum, after treatment has started, against the pathogen isolated from him, will confirm that the antibiotics given are reaching the blood in sufficient concentration to achieve almost complete killing of a moderately heavy inoculum (see page 246).

The concentration in blood and urine after varying doses of different preparations is well known. In individual patients the levels in these fluids can be assumed to follow, approximately, known excretion curves, provided there is no evidence of renal failure or of lack of absorption. If, however, the pathogen was sensitive in the laboratory to a drug which proves ineffective in treatment, assay of serum, or serous fluid may be required to determine whether the expected levels are in fact attained. Control of dosage by serum assay is necessary when a toxic drug is given to a patient with renal failure; for example, streptomycin in tuberculous pyelonephritis or gentamicin in endocarditis or other serious infection. Even when renal function is normal gentamicin should be assayed in patients who have to receive the drug for longer than one week, which is seldom necessary, because in some people it may accumulate slightly and when this happens over a prolonged period ototoxicity often follows. (Mawer *et al.*, 1974.)

When treatment, which should theoretically be effective, fails the possibility that the organism isolated is not the pathogen, or not the only pathogen, must be borne in mind.

When treatment is urgent every effort should be made to obtain specimens for culture before the first dose is given. If this be omitted the chance of making the diagnosis is greatly reduced and the clinician may be left with no alternative but blind antibiotic treatment which is undesirable both for therapeutic and financial reasons.

It is sometimes considered that routine sensitivity tests are unnecessary because, when the pathogen is known, the drug of choice is clearly indicated. There are, however, good reasons for making such tests whenever antibiotic treatment is contemplated. One reason is that the result of the test is often known sooner than the microbe can be reliably identified. For example *Strept. pyogenes* is almost invariably sensitive to penicillin. An overnight sensitivity test culture of a specimen infected with this organism will show an haemolytic streptococcus sensitive to penicillin, but certain identification will take one more day. Other species always sensitive at present when isolated from untreated patients are listed in Table

30.   In practice the result of antibiotic sensitivity tests is often a guide to identification.   During the last decade *Strept. pyogenes*, *Strept. pneumoniae* and *Haemophilus* resistant to tetracycline have made their appearance and are no longer uncommon.   Possibly resistance of *Strept. pyogenes* to benzylpenicillin may develop, *Strept. pneumoniae* is already occasionally resistant (Hansman *et al.*, 1973), and *Haemophilus* may become resistant to chloramphenicol; we should look out for these changes.

The majority of microbes develop resistance to some drugs very quickly and when they are unable to develop resistance they are replaced by different strains, sometimes of the same species, which are naturally resistant.   Resistant organisms can be expected in all the normal bacterial reservoirs of the body within a day or two of the onset of treatment ; this change of flora sometimes takes place within a few hours.   Staphylococci, coliforms and streptococci other than the two species listed are frequently resistant to penicillin, tetracycline and other drugs.   The frequency with which these resistant strains are encountered is related to the frequency with which the drugs are prescribed in the community sampled.

ANTIBIOTIC SENSITIVITY TESTS

Numerous methods have been described for testing the drug sensitivity of bacteria.   In routine hospital work speed is much more important than a high degree of accuracy.   The method commonly employed is to place blotting-paper discs impregnated with the drug on plates pre-seeded with the original material or with a pure culture.   A separate sensitivity test culture is required because of the danger of altering morphology and making identification difficult if the discs are placed on the diagnostic plate culture and because special sensitivity test medium is required.

Primary tests

There are three good reasons for testing sensitivity on first isolation, instead of after subculture.   First the need for speed is met in many cases ; second the microbes tested are as fair a sample as possible of those at the site of sampling.   It is not uncommon for two strains of the same species to infect a single wound, one sensitive, the other resistant.   If tests are made after subculture the sensitive one only may be found and this may be followed by an unexplained failure of antibiotic treatment.   The third reason is that antibiotic sensitivity may change. This occasionally happens with *Strept. pyogenes* which on subculture is almost invariably sensitive to penicillin but sometimes resists treatment in the patient and in such cases can be shown to be resistant *in vitro* on first isolation by the controlled disc technique (U.C.H. Laboratory records, 1952). This was seen in a patient with streptococcal olecranon bursitis penicillin-treated for a month on board ship with only partial response; on subculture the strain reverted to sensitivity. A more common cause of failure with penicillin is a mixed infection of *Strept. pyogenes* with a penicillinase producer, usually a resistant staphylococcus. On the primary plate culture the streptococcus may appear resistant as it is able to grow near the disc when accompanied by the penicillinase producer. Since this is what is happening in the patient, it is as well to be reminded by the test result that treatment with penicillin is unlikely to succeed even though the streptococcus when tested alone is penicillin sensitive (see Fig. 27).

Testing sensitivity after subculture from the primary plates is sometimes considered to be " more scientific " than testing the total growth from the original specimen. Plates made from pure cultures may be easier to read and have a more elegant appearance, but there is no virtue in an elegant test unless it gives at least as good an answer to the problem which it is designed to solve, and there is no doubt that testing the original material gives a more rapid and reliable result in many cases particularly in urinary tract infection (see page 219). Occasionally a resistant organism may obscure all other growth in the mixed culture, or the inoculum may prove to be either too heavy or too light. The test can then be repeated from the diagnostic plate cultures. Very often, however, it is possible to say after overnight incubation from the controlled test about to be described that all the organisms are sensitive to one or more drugs and this is invaluable information for those faced with the problem of choosing antibiotics for treatment.

Definition of Sensitivity and Resistance

When sensitivity tests are carried out by a diffusion method it is not possible to give precise figures for the sensitivity of a given organism to a given antibiotic, but bacteria can be divided into three groups : sensitive, moderately resistant and resistant.

Sensitive: strains amenable to treatment in ordinary dosage.

Moderately resistant : strains amenable to treatment when large doses are used or when the drug is concentrated at the site of infection, e.g. in the urinary tract or by local application.

Resistant : strains unaffected by high concentration and therefore unlikely to respond to any dosage of the antibiotic.

For clinical purposes the term " sensitive " must be related not only to the minimal inhibitory concentration (M.I.C.) for the organism, but to the concentration of antibiotic obtainable *in vivo*. In most acute infections it can be assumed that if the average blood level exceeds the M.I.C. by a safety factor of 2 or 4 the infection will respond to treatment. When the pathogens reside in dense or avascular tissue, e.g. endocarditis and actinomycosis, a higher safety factor is required.

In practice, grading is achieved by testing in parallel unknown organisms with a sensitive control and comparing their inhibition zones.

1. Impregnated Paper Disc Method

Blotting-paper discs impregnated with antibiotic can be purchased and are likely to give a more constant performance than home-made discs. Even with a properly controlled method the choice of disc content is critical. When it is too high there will be little change of zone over a wide range of difference in minimal inhibitory concentration and unless strains are highly resistant they will appear sensitive. When it is too low the zone will be small and a slightly too heavy inoculum will make a sensitive organism appear resistant. The optimum disc content is usually the smallest quantity which will give a zone radius of about 10–15 mm. when tested against a semiconfluent growth of the appropriate sensitive control organism (see page 215). The disc contents recommended in Table 22 are all commercially available. Two strengths are recommended for some drugs because their high concentration in urine makes treatment of urinary infection possible when the organism in any other site would be regarded as moderately sensitive only or even resistant. It must be understood, however, that infection of the urinary tract involving renal tissue cannot be expected to respond unless the organism is also sensitive when tested with a low content disc or, in the case of a penicillin or cephalosporin, a high dose can be used or excretion can be blocked to ensure a high blood as well as urine level.

Because sensitivity in the report indicates a drug suitable for treatment in particular circumstances it is possible to report *Esch. coli* isolated from blood as moderately sensitive to, say, ampicillin when the same strain isolated from the patient's urine will be reported sensitive. This seems illogical but it is inevitable if the report is to be both rapid and reliable in indicating appropriate treatment. It only happens with penicillins and cephalosporins which are concentrated in the urine and are sufficiently harmless to be given in exceptionally large doses when the need arises ; the term moderate reminds the clinician that a high dose will be needed.

Discs can be stored at $-20°$ C. in unopened vials for a year or more. Some drugs are less stable at $4°$ C. and the manufacturer's expiry date should always be observed. When removed from the refrigerator for use the vials should be kept unopened for sufficient time to avoid condensation. Enough discs for the day's use should be taken out and any not used should be discarded.

Disc content is important and a variety of different strengths is available. Choice is governed by the method of interpretation to be used. Suitable contents for the method described below are given in Table 22. The indiscriminate use of high or very low content discs with unsuitable methods can give grossly misleading results. Discs can be applied with a sharp botanical or hypodermic needle and each must be pressed firmly in position ; even diffusion will take place only when the disc is in close contact with the medium. The needle need not be flamed between each application. The discs must be sufficiently widely spaced (2 cm) to prevent overlapping of the zones.

When no commercial discs are available, discs about 5 mm. diameter are cut with a leather punch from good quality blotting-paper. They are separated and spread out singly in 15 cm. Petri dishes, so that each disc is not less than 2 mm. from its neighbours. They are sterilized in the oven at $160°$ C. for one hour. When they have cooled one drop (0·02 ml.) of a sterile solution of the required antibiotic or sulphonamide is dropped from a standard dropping pipette* on to each disc. They are then dried in the incubator with the lids of the Petri dishes tilted.

The amount of liquid absorbed and hence the strength of solution required can be discovered by weighing dry discs, wetting

* For preparation, see page 346.

them in the manner described and reweighing ; different blotting papers absorb differently. Solutions of antibiotic used for parenteral treatment can be diluted to achieve the desired concentration with the exception of chloramphenicol and clindamycin phosphate where the drug is activated after absorption ; a small and variable amount of free antibiotic is present in the solution before injection.

Paper strips instead of discs can also be prepared. In order to achieve uniform impregnation the strip should be dipped in solution

TABLE 22. Suitable disc contents (μg)

Antibiotic	Organisms from urine	Organisms from other infections
Ampicillin	25	10
Carbenicillin	100	100
Cephaloridine	30	5
Chloramphenicol	30	10
Clindamycin	—	2
Erythromycin	—	15
Fucidin	—	5
Gentamicin	10	10
Kanamycin	30	30
Methicillin	—	10 (strips 25)
Nalidixic acid	30	—
Nitrofurantoin	200	—
Penicillin	—	2[1]
Polymyxin B	300[1]	300[1]
Streptomycin	25	10
Sulphafurazole	100	100
Tetracycline	30	10
Trimethoprim	1·25	1·25

[1] Units.

and then blotted before drying to prevent an excessive amount of solution accumulating at one end.

MEDIUM

Many drugs are affected by various constituents of laboratory media. The most important examples are sulphonamides and trimethoprim, which act by inhibiting folate metabolism. If the medium used for these tests contains end products of folate metabolism (e.g. thymidine), sensitive organisms will be able to bypass the actions of these drugs. On such media inhibition zones are not simply reduced in size, but colonies, often much smaller than normal will grow right up to the disc. Media containing small amounts of such substances can be rendered suitable for use by the addition of lysed horse-blood.

Other ingredients which may affect some antibiotics are salt, which may reduce the activity of some (e.g. aminoglycosides) and enhance that of others (e.g. fusidic acid); carbohydrates, which may enhance the action of nitrofurantoin and ampicillin in some diffusion tests; and some minerals such as calcium, magnesium, and iron (e.g. tetracycline and gentamicin). Many ingredients such as peptone, yeast extract, tryptone, and the agar used for solidifying the medium vary in their mineral content, not only between different manufacturers but even between different batches from the same source (Bovallius and Zacharias, 1971).

Results of sensitivity tests are both more reliable and easier to interpret if a specially defined medium is used. Several are now available in Britain; for example Oxoid Diagnostic Sensitivity Test Agar, to which lysed horse-blood must be added for tests with sulphonamides and trimethoprim, and Wellcotest Agar and Oxoid Sensitest Agar, both of which are free of sulphonamide inhibitors. Five per cent whole blood (in addition to the lysed horse-blood, when this is required) should be added when testing fastidious organisms and plates should be " chocolated " for tests with *Haemophilus*. It should be remembered that zones will be smaller on media containing blood when the drug is highly protein-bound (e.g. fusidic acid and novobiocin).

The medium should be poured into flat-bottomed dishes on a flat horizontal surface, to a depth of 3–4 mm. (20 ml. in an 8·5 cm. Petri dish). An increase in depth of 1 mm. will not significantly affect the results but very thin plates are unsatisfactory.

The antibiotics most affected by pH are tetracyclines which give large zones on acid medium and the aminoglycosides, especially streptomycin, which are most active in alkaline medium. Many other drugs are affected to some extent, for example, the cephalosporins, the macrolides (erythromycin group), and lincomycin are favoured by alkalinity whereas methicillin, cloxacillin and fusidic acid perform better in acid medium. Conditions for sensitivity tests should resemble as nearly as possible conditions in the patient's tissues. The buffering action of blood is beneficial in nullifying the effect of minor changes. The pH of sensitivity test media containing no blood must be very carefully adjusted and conditions such as incubation with added CO_2 which alter pH must be avoided.

INOCULUM

All zones of inhibition are affected by the size of the inoculum and whenever pure cultures are tested adjustment of the inoculum is essential. When this is too heavy zones are always reduced in size and the edges become indistinct, but with sulphonamides the zone will completely disappear, although the organism may be fully sensitive. The inoculum which gives the most consistent results is that which gives dense but not confluent growth (once growth is confluent it is impossible to judge how heavy it is); it is essential that it is evenly distributed over the whole area. Small variations in the weight of growth are unlikely to invalidate the results but the use of uncontrolled and inadequately spread inocula is one of the main sources of error in these tests. In the ACP Sensitivity Test Trial (1965) and in a recent quality control trial with 87 participants (unpublished), there were fewer errors in primary sensitivity tests than in secondary tests, almost certainly because the inoculum is commonly near the optimum in the former.

Choice of drugs to test

Suggestions for suitable drugs to be tested against various bacteria and specimens are given in Table 23. The number of tests can be reduced by including only one representative of closely related drugs amongst which cross-resistance always occurs. Only one tetracycline need be tested; methicillin should represent all penicillinase-resistant penicillins and cephalosporins against staphylococci (see page 222); kanamycin will serve for neomycin, clindamycin for lincomycin, cephaloridine for cephalosporins, and either polymyxin or colistin will serve for both. Only one sulphonamide need be tested and sulphafurazole is satisfactory; some confusion arises because *Ps. aeruginosa* may appear resistant to this but sensitive to sulphatriad used on some discs; this is due to the greater activity of sulphadiazine (included in sulphatriad) against this species, but this is not thought to be clinically significant.

Tests with mandelamine are both unnecessary and misleading; this drug acts because formalin is released in acid urine, all organisms are sensitive to formalin but if they split urea (e.g. *Proteus spp.*)

they render the urine too alkaline for it to be liberated (Waterworth, 1962).

Maintenance and choice of control cultures

Control cultures are essential and should always be included with any method. Difference in species between the control and test strains does not usually matter provided both grow at approximately the same rate, an exception is *Ps. aeruginosa*, see page 224.

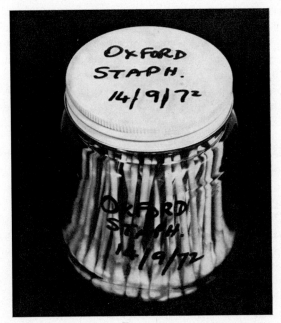

FIG. 23.

Control organisms should be maintained on agar slopes kept in the dark at room temperature. A broth culture is made at the beginning of each week and further subcultures in broth daily from this. The agar slopes are sub-cultured monthly.

When plates are to be inoculated with swabs, a more convenient method of maintaining controls, less liable to contamination, is to impregnate sterile 7·5-cm. swabs in bulk and store these in screw-capped glass or plastic jars (Felmingham and Stokes, 1972). About 20 drops of overnight broth culture in 20 ml. broth, well mixed,

CB—H

will impregnate and be quickly absorbed by about 90 Q-tip swabs[1] (Fig. 23) and will give a suitable inoculum when one swab is used per plate. They can be stored for at least one week at 4° C.

No attempt should be made to purify a contaminated control culture by picking a single colony because of the risk of variation. When the slope is contaminated the control should be revived from a lyophilized preparation.

The control must be an organism known to respond to treatment with normal doses of the drug and the choice depends on the site of the infection and the concentration of drug attainable there. Drugs which are excreted by the kidneys produce high concentrations in the urine and more resistant organisms are likely to respond to treatment. In these circumstances a high content disc and a more resistant control organism can be used. It must be remembered, however, that when coliforms or enterococci are isolated from specimens other than urine, they must be compared with the fully sensitive control. This is particularly important with ampicillin and the cephalosporins because, although urinary infections with these organisms respond well to low doses, coliforms and enterococci are really only moderately sensitive to these drugs and the report must make it clear that higher dosage will be required for infections in other parts of the body, see page 211.

The Oxford staphylococcus (NCTC 6571) can be used as the sensitive control for all drugs except polymyxins. *Esch. coli* (NCTC 10418) should be used for all tests with polymyxins and for all drugs with organisms from the urine.

METHOD

A separate plate for sensitivity tests is set up when the cultures for identification have been made, before the swab is placed in cooked-meat broth or on any antibiotic-containing medium. Blood agar must be employed for primary tests since the organism isolated may be fastidious, and the blood should be added to medium specially prepared for sensitivity tests ; see above. This is essential for sulphonamide and trimethoprim tests and important for other tests also. The disc content recommended is related to the medium employed and may be inappropriate for other media. For example, tetracycline discs will show smaller zones and false resistance may

[1] Obtainable from Chesebrough-Ponds Ltd., Victoria Road, London, N.W. 10.

be reported if the blood agar base recommended here for primary cultures is also employed for sensitivity tests.

The swab, loopful of pus or loopful of *uncentrifuged* urine is seeded heavily and evenly in a band across the middle third of the plate, in the manner described below (see Urine). The control is applied on either side of the test organism either by inoculating one small loopful of overnight broth culture and spreading with a dry sterile swab or by using one pre-impregnated swab per plate (see Fig. 24 and 25). Using sterile forceps or a flamed needle, the 4 discs are now placed on the gap between the inoculated areas making sure they touch the edge of the central band and are about 1 cm. from the edge of the plate.

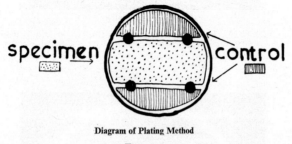

Diagram of Plating Method

Fig. 24.

The control and test material must be inoculated within about 15 minutes. A delay of not more than 2 hours may be allowed before the discs are applied when this is more convenient.

When tests need repeating because, for example, the primary growth of an important pathogen is too scanty, or it has been overgrown by another resistant organism or is itself resistant to all four drugs, the same method can again be employed. The inoculum of the test organism must be adjusted to match the control. See secondary tests below.

Pus swabs and liquid pus

Sensitivity test cultures are normally incubated aerobically but when anaerobes are likely to be present, e.g. septic abortion pelvic or abdominal abscess, it may save time to do, anaerobic sensitivity tests as well. Since by the method described below anaerobic incubation does not invalidate results on facultative aerobes one

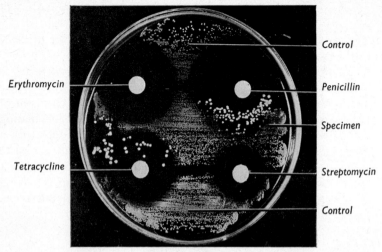

FIG. 25a.

Sensitive control (Oxford staphylococcus) spread over top and bottom thirds of plate. Specimen of pus from an abscess spread over middle third. The specimen yielded two staphylococci; one sensitive to all drugs (i.e. zones equivalent to control), and a smaller number of a penicillinase-producing staphylococcus also resistant to tetracycline.

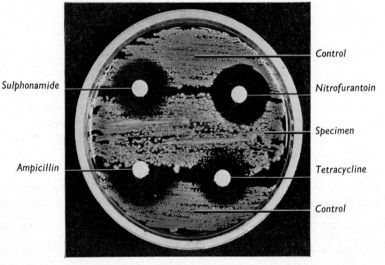

FIG. 25b.

Primary test of infected urine made by student technician. *Esch. coli* (N.C.T.C. 10418) control spread over top and bottom thirds, specimen (uncentrifuged) spread centrally. RESULT: Inoculum equivalent, bacillus sensitive to sulphonamide, nitrofurantoin and tetracycline, resistant to ampicillin.

(anaerobic) test plate only can be employed. Methicillin may be included in primary tests on blood agar when staphylococci are expected, provided the culture is incubated at 30° C. (see page 222).

Urine

Primary sensitivity tests should be done only on specimens likely to be infected as judged by microscopy or other suitable tests (Thomas and Baldwin, 1971). An optimal inoculum of infected

TABLE 23. Suggested choice of drugs for sensitivity tests

	Primary culture			Subculture				
	Pus	Skin and ear	Urine	Staphyl-ococci	Strept-ococci	Haemo-philus	Coli-forms	Pseudo-monas
Penicillin	×	×		×	×			
Methicillin[1]	×			×				
Ampicillin			×			×	×	
Carbenicillin								×
Cephaloridine				×			×	
Erythromycin	∨			×	×	×		
Clindamycin	×			×		×		
Fucidin				×				
Tetracycline	×	×		×	×	×	×	
Chloramphenicol		×				×	×	
Streptomycin							×	
Kanamycin		×		×				
Gentamicin	×		×				×	×
Polymyxin		×						×
Trimethoprim	×		×			×	×	
Sulphonamide	×		×			×	×	
Nalidixic acid[2]			×				×	
Nitrofurantoin[2]			×				×	

[1] See page 222 concerning this drug.
[2] Only applicable to organisms from urine.

urine can almost always be achieved by placing a standard loopful (5 mm. external diameter, held vertically) of uncentrifuged, well mixed urine on a plate of sensitivity test medium, and spreading evenly with a sterile dry swab ; swabs dipped in urine will usually give too heavy an inoculum.

Results are interpreted as for tests with pure cultures, but when zones are significantly smaller than the control, particularly if the inoculum is heavier, tests must be repeated. It is also important to assess the significance of the growth before reporting sensitivity tests. Reporting the sensitivities of commensals may lead to unnecessary treatment in the first place or to a change in established therapy, possibly detrimental to the patient.

If large square dishes are used or if the cultures are spread using a rotator (Pearson and Whitehead, 1974) six discs can be tested on each (Fig. 26) : suitable drugs are given in Table 23.   The extra

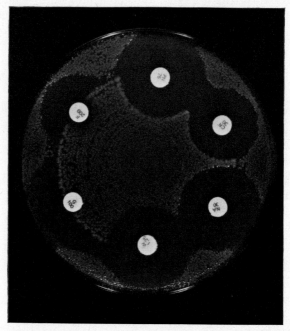

Fig. 26.   Primary sensitivity test of urine using mechanical rotator[1].   A pre-impregnated control swab is inoculated peripherally, one loopful of urine is inoculated centrally, see page 219.

time taken to apply the control cultures to each plate can be greatly reduced if pre-impregnated swabs prepared in bulk are used (page 215), and it is compensated for by the ease with which zones can be compared.

Secondary tests on pure cultures

Inoculum

Even when the primary culture appears to yield a pure growth it is seldom desirable to test a single colony in case there is variation

[1] Obtainable from Denley Ltd., Bolney, Sussex, England.   Rotation speed 150 revolutions per minute.

in sensitivity and part of several colonies should be taken. It is not usually possible to achieve a satisfactory inoculum (semiconfluent growth, see page 214) by transferring solid growth directly to the plate ; a suspension, approximately equivalent in density to an overnight broth culture can be made by emulsifying the colonies in a small volume of broth or peptone water. A small loopful (2–4 mm. diameter) of this or of an overnight broth culture is transferred to the centre of the plate and spread as evenly as possible with a dry sterile swab. When colonies are very small or not easily emulsifiable it may be possible to achieve an optimal inoculum by picking one or two and after seeding them centrally on the plate spreading with a swab, but broth should also be inoculated so that tests can be repeated when necessary from a liquid culture next day.

INTERPRETATION OF RESULTS

Zones of inhibition of the test organism are compared with those of the appropriate control organism, and when there is any doubt as to their relative size, they are measured with calipers or by laying a millimetre rule across the dish. Provided the inoculum is optimal and the control zones 8–15 mm. radius, results are reported as follows :

Sensitive—zone radius equal, wider, or not more than 3 mm. smaller than the control.

Moderately sensitive—zone more than 3 mm. radius but smaller than the control by more than 3 mm.

Resistant—zone of 2 mm. radius or less.

This method of interpretation is not valid for some tests with penicillinase-producing staphylococci nor for tests with polymyxin (colistin). They are interpreted as described below.

The paper strip method

When a number of organisms have to be tested against only one or two drugs, blotting paper strips are more convenient to use than discs. Suitably diluted cultures are streaked across the plate ; eight strains (seven tests and a control) can be accommodated on one plate, and an impregnated strip is placed at right angles across the inocula.

Several drugs are now commercially available in this form.[1]

[1] Obtainable from Mast Laboratories, 38 Queensland Street, Liverpool, L7 3JG, England.

The method is valuable for testing staphylococci with methicillin (see below).

Tests presenting special difficulty

Penicillinase-producing staphylococci

The penicillin inhibition zone of a sensitive organism may be reduced if it is mixed with a penicillinase producer (see Fig. 27). If sufficient penicillinase is formed and the logistics of its production and the growth of the organisms is favourable satellitism of the sensitive organism round colonies of the penicillinase producer will be seen.

If a penicillinase-producer fails to form enough of the enzyme to neutralize the penicillin close to the disc, it will show an inhibition zone. It can, however, be recognized because the colonies at the edge of the zone are large and well developed and there is no gradual fading away of growth towards the disc (see Figs. 25a and 27). Treatment with penicillin is more likely to stimulate penicillinase production than to kill the organism (Geronimus and Cohen, 1957). These strains are therefore reported resistant.

Methicillin-resistant staphylococci

Methicillin-resistant staphylococci will often appear fully sensitive when tested in the ordinary way. Many of these organisms grow more slowly in the presence of methicillin and growth will only appear within the zone when incubation is continued for 48 hours. This difficulty can be overcome either by incubating the culture at 30° C. (Annear, 1968) or by using 5 per cent salt agar (Barber, 1964) when most of the culture will appear resistant overnight.

Although there is cross-resistance between methicillin and all other penicillinase-resistant penicillins both clinically and when tested by a dilution method, many resistant strains appear sensitive when tested against cloxacillin or flucloxacillin discs (Hewitt, *et al.*, 1969 ; Garrod and Waterworth, 1971). Methicillin only should be used in diffusion tests but it should be made clear in the report that results apply equally to the other drugs.

Cephalosporin resistance in staphylococci

Cephaloridine is only moderately resistant to staphylococcal penicillinase and when high penicillinase-producing strains are

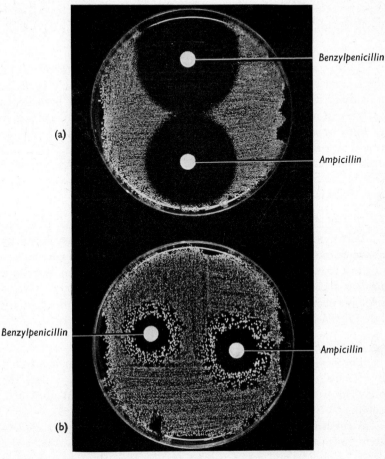

Fig. 27.

Cultures showing the effect of a penicillinase-producing staphylococcus on inhibition zones to penicillin of a sensitive staphylococcus mixed with it.
(a) Sensitive strain alone showing wide zones.
(b) (same strength discs) Sensitive staphylococcus plus penicillinase producer showing : (1) small zone with well defined edge typical of a penicillinase producer ; (2) zone of sensitive strain smaller than in (a).

tested using a heavy inoculum they may appear resistant. There is cross-resistance between methicillin and the cephalosporins and sensitivity to the latter can be assumed from tests with the former. Cephalothin, which has greater resistance to this penicillinase is recommended for the treatment of staphylococcal infections. The disc method is misleading for testing staphylococci to cephalo-sporins and should not be attempted (Blowers *et al.*, 1973).

Co-trimoxazole

This is a mixture of sulphamethoxazole and trimethoprim and discs containing both drugs in a ratio of 20 : 1 are widely used for sensitivity tests, such discs may be misleading. When both drugs are present it is impossible to know whether the organism is sensitive to both or only to one of them. Each drug should therefore be tested separately and if sulphonamide is already being tested one extra disc containing trimethoprim alone is needed. An organism should be reported sensitive to cotrimoxazole only when some sensitivity is shown to both drugs. See also page 239.

Polymyxin and colistin

Polymyxins diffuse very poorly in agar and the measurements above cannot be applied. The *Esch. coli* control gives a zone radius of 3 to 4 mm. and this can easily be reduced by too heavy an inoculum. *Proteus* species are normally resistant but almost all other Gram-negative bacilli are sensitive. (When resistance is seen on a primary test *Providence* or *Serratia* should be suspected.)

Ps. aeruginosa

The sensitivity of this species to gentamicin is much affected by the magnesium content of the medium ; when this is low the organism will appear unduly sensitive (Garrod and Waterworth, 1969).

A further difficulty arises because this organism is only moder-ately sensitive to both gentamicin and carbenicillin, usually the drugs of choice, and zones will be smaller than those of either the *Esch. coli* or the staphylococcal control. It will therefore be diffi-cult to detect moderate increases in resistance to these drugs.

Both these difficulties can be overcome by using a strain of the same species of known sensitivity (NCTC 10662) as a control for tests with this organism and when the zone is not significantly reduced, as judged by comparison with the control, the organism can be reported sensitive (Garrod and Waterworth, 1971).

Swarming organisms

These invade inhibition zones even when the organism is sensitive. The zone edge should be measured as for other organisms and swarming should be disregarded. In mixed cultures when no colonies forming a zone can be seen the test must be repeated, see also page 232.

Anaerobes

Streptomycin and the other aminoglycoside antibiotics are partly inactivated by anaerobiosis, therefore the control must also be incubated anaerobically. Disc diffusion tests are not as satisfactory for anaerobes as they are for aerobes (Garrod *et al.*, 1974). Nevertheless disc tests by the method described will often serve as a guide to treatment of infection by anaerobes which grow overnight. Primary tests of pus from pelvic sepsis, where anaerobes are very common, will give a quicker result if the culture is incubated anaerobically. Results for facultative anaerobes will not be invalidated and an additional aerobic test is not essential.

Technical faults (see also Inoculum)

Zones are sometimes found to be oval or eccentric. There are three common causes for this. If the disc is dropped on the medium and then shifted to a different position, sufficient antibiotic will be absorbed at the first site of contact to inhibit growth. If the medium " sweats " on incubation antibiotic may be carried in the surface moisture. If the discs are placed too near to the edge of the plate, i.e. less than 1 cm. from it, the zones may be eccentric, due to sideways diffusion of the antibiotic which is prevented from further radial diffusion when it reaches the edge of the plate.

If the culture is accidentally left on the bench for an hour or two after the discs have been placed on it, the zones will be large because the antibiotic will have had time to diffuse before growth began. Provided that the control and the test material were seeded before the discs were placed in position this does not greatly matter because when the result is read the test zone is compared with the control zone and both are similarly affected. Delay between seeding the test organism and the control may, however, lead to reduction in zone size of whichever was seeded first. This is most likely to happen in the paper strip method when all the cultures to be tested are not ready at the same time.

2. Determination of Minimal Inhibitory Concentration (M.I.C.)

Although the rapid test already described is a sufficient guide to treatment in the majority of infections, a more accurate estimate is sometimes desirable particularly when a patient needs prolonged treatment with an antibiotic which may be toxic. Numerous methods of measuring the M.I.C. have been devised (Florey *et al.*, 1951); they fall into two classes. Zones of inhibition can be measured by diffusion through agar under standard conditions, or the organism can be titrated in liquid or on solid medium containing falling dilutions of the antibiotic. Titration is usually preferred because zones are often ill-defined and difficult to read.

An organism of known sensitivity, usually the Oxford staphylococcus, should always be titrated in parallel with an unknown organism. When reporting, the M.I.C. for both test and control organisms should be stated so that a comparison can be made. A statement of M.I.C. without some standard of comparison and without details of the method employed has very little meaning.

Factors other than antibiotic concentration which affect the end point are as follows :

1. INOCULUM. Individual bacterial cells are not constant in their antibiotic sensitivity. Irregularity in response is much more marked with some drugs than with others. For example, sensitive staphylococci titrated against benzylpenicillin will show only a small rise in M.I.C. when the inoculum is increased one-thousandfold ; the end point with methicillin, however, varies widely with inoculum. A strain may appear fully sensitive when tested with a small inoculum, but when tested with a very heavy inoculum the M.I.C. may rise to a concentration unattainable when normal doses are given. Tests in liquid medium with very large inocula are misleading because when even a single bacterial cell out of hundreds of millions is uninhibited growth will be seen. Under these conditions in the patient the drug may succeed because very small numbers of uninhibited cocci would find it difficult to withstand the normal tissue defences and it is unlikely they would be able to flourish *in vivo*.

The M.I.C. for penicillinase producers titrated against penicillins which are inactivated by the enzyme can be varied between wide limits by altering the inoculum. When very heavy inocula are employed pre-formed penicillinase is introduced into each tube, thus reducing the activity of the drug enabling growth to begin.

Titration of sulphonamides is also very susceptible to changes of inoculum but such tests are rarely necessary. Inoculum effect is much less important in tests on solid medium because very small numbers of survivors of a heavy inoculum can be seen and ignored. Moreover, when large numbers of strains are to be tested a multiple inoculator can be employed. This method is therefore preferred.

2. COMPOSITION OF THE MEDIUM. It seems logical, when sensitivity testing as a guide to treatment, to employ a medium which mimics as closely as possible the milieu in which the drug will meet the bacterium in the patient. A standard sensitivity test medium with 5 per cent blood added, which must be lysed horse-blood when sulphonamide or trimethoprim are tested, is suitable. If titration is to be done in liquid medium 25 per cent serum broth should be employed because the effect of serum binding of some antibiotics should be taken into account. The difficulty of getting rid of inhibitors from liquid media makes testing against sulphonamide or trimethoprim unsuitable by this method unless special inhibitor-free medium can be obtained. In order to facilitate reading, which may not be clear-cut in a medium containing so much serum, a low concentration of glucose and an indicator are added so that growth will declare itself by fermentation and colour change ; the M.I.C. for non-fermenters must be judged by opacity. Although carbohydrates are known to affect zone readings in agar diffusion tests (page 213), a small concentration in a medium containing so much serum will not appreciably alter the M.I.C. in a tube titration test. Changes in pH, including that brought about by incubation in an atmosphere with added carbon dioxide, may affect the end point, particularly in streptomycin and tetracycline titrations, but the presence of serum minimizes the effect. For the effect of other constituents of medium see pages 213 and 224.

3. CULTURE CONDITIONS. Anaerobic incubation partly inactivates streptomycin and the other aminoglycoside antibiotics, neomycin, kanamycin, framycetin, paromomycin, gentamicin and tobramycin. The control must therefore be incubated under the same atmospheric conditions as the test organism even though it does not need special conditions. For incubation in air plus CO_2, see above.

4. PROTEIN BINDING. Estimation of M.I.C. in a high concentration of serum is desirable because the action of antibiotics, like that of antiseptics, is inhibited by the presence of protein. Protein binding is a variable property of antibiotics. For example, cloxacillin is more

highly protein bound than benzylpenicillin (Rolinson, 1967) and, moreover, the degree of binding is not the same for sera from all species ; therefore in sensitivity tests normal human serum is the best choice.

5. INCUBATION TIME.   Tests should be read as soon as possible, 12 to 18 hours is the usual time.   Further incubation will allow growth to appear in higher concentrations, partly because slightly inhibited bacteria will begin to grow and also because many antibiotics are unstable in low concentration at $37°$ C ; as they are destroyed, bacteria previously inhibited will flourish.   In the patient, doses will be repeated and the drug level maintained so that this effect will not be seen.

6. GROWTH PHASE.   Penicillin and many other antibiotics are most active against rapidly dividing bacteria.   Anything which reduces the rate of cell division will render the organisms less sensitive.   For example, penicillin is relatively harmless to sensitive bacteria at low temperatures.   Appropriate dilutions of an overnight liquid culture should therefore be employed in order to standardize inoculum and minimize differences due to testing during different phases of growth.

Titration methods

The range of dilutions to be tested can usually be limited in the light of the paper disc method result.   When no disc test has been made it should run from a concentration higher than can be maintained in the blood to a concentration lower than that which is known to inhibit the control strain.   When the organism to be tested was recovered from urine the range must be increased to include the high concentrations obtainable in urine.

Antibiotics used for making stock solutions are best obtained pure from the manufacturers for laboratory use only.   Capsules and tablets often contain material other than antibiotic which will affect the result.   Solutions prepared for parenteral treatment when available are satisfactory with the exception of chloramphenicol and clindamycin phosphate, see page 212.   For details of the solubility of different drugs see Garrod *et al.* (1973).

Tube titration method

MATERIALS

Sterile graduated 10 ml. pipette.
Sterile graduated 1 ml. pipettes.
Sterile 3 × 1·25 cm. tubes or small screw-capped bottles.
Pasteur pipettes.
Overnight broth culture of test organism.
Overnight broth culture of control staphylococcus.
Antibiotic solution of known concentration.
Indicator serum medium (page 371).

EXAMPLE

An organism shown by the disc method to have approximately the same sensitivity to chloramphenicol as the Oxford staphylococcus is tested for its ability to grow in the presence of varying dilutions of the drug. In each batch of tests the potency of the antibiotic solution and the suitability of the medium are controlled by titrating in parallel an organism of known sensitivity, the Oxford staphylococcus, and by including one tube containing medium without antibiotic both for the test organism and for the control. Since the control staphylococcus is sensitive to 5–10 μg/ml. (Table 30) dilutions of chloramphenicol ranging from 100 μg/ml. to 0·25 μg/ml. will include the end point for both test organism and control. There is no need to include all these dilutions in the control series. Pure chloramphenicol powder must first be dissolved in ethyl alcohol to give a stock solution containing 10 mg./ml. The steps between dilutions should be twofold but a long series of twofold dilutions leads to inaccuracies therefore the dilutions in the figure are preferred. The original tenfold dilutions are made each time with a fresh pipette.

PROCEDURE

Make master dilutions in sterile distilled water tenfold stronger than the amount required, i.e. from 1000 to 2·5 μg/ml. Then make the required dilutions by delivering 2·7 ml. of serum broth into one tube for each dilution required and adding 0·3 ml. of the appropriate master dilution to each, when more than 1 strain is to be titrated larger volumes will be needed. Start with the weakest dilution and as each final dilution in medium is made mix well and transfer 1 ml. to each appropriate tube for the test and control series, one pipette only is then required.   Do not make these ten-fold dilutions directly in medium in the tubes to be inoculated because it is essential that the dilutions for the control and the test series be identical.

Inoculate each tube with 1 drop 1/100 dilution of an overnight broth culture.   This gives an inoculum of about 10^5 organisms/ml. and is comparable with the inoculum recommended for the disc diffusion method.

All tubes are incubated and read after 18 to 24 hours' incubation, or as soon after this as growth is apparent.   The indicator facilitates reading in this serum broth which may be slightly cloudy.

Sensitivity is expressed as the highest dilution of the antibiotic which inhibited growth.   The sensitivity of the control is also reported.

After further incubation growth may be seen in tubes in which it was originally inhibited; it is disregarded (see page 228, No. 5).

Plate titration method

Making antibiotic containing medium with accurate dilutions for this method is a skilled process which cannot be undertaken by most media departments and must usually be done by the person making the titrations.   The possibility of using impregnated dried blotting paper as a measured source of antibiotic for the preparation of antibiotic containing nutrient agar has been suggested by Rolinson and Russell (1971).

MATERIALS

    Sterile graduated 1 ml. pipettes
    Sterile graduated 25 ml. pipette
    Screw-capped sterile bottles 30 ml. (1 oz.)
    Sensitivity test agar, sterile horse-blood
    Sterile Petri dishes
    Overnight broth cultures of test organisms and control.

PROCEDURE

Prepare molten sensitivity test agar with 5 per cent horse-blood added at 56°–60° C.

Make a range of dilutions of antibiotic in sterile distilled water twenty times stronger than the final dilution required in the medium. Label one sterile 8·5 cm. Petri dish for each dilution and one for the control plate.   Add 1 ml. diluent (distilled water) to the control plate and then, starting with the weakest dilution of antibiotic, transfer 1 ml. of each dilution to the appropriate labelled dishes.

Using the 25 ml. pipette (with its tip cut off to allow rapid delivery) transfer 19 ml. medium to each dish, mixing well. A mechanical rotator fitted with a large tray to take the dishes can be used to facilitate mixing.

When the medium has set, dry the plates in the incubator with the lids of the dishes tilted for from 30 minutes to 1 hour according to the humidity of the incubator. Well-dried plates are essential ; once dried they can be stored at 4° C. for at least 1 week (Ryan *et al.*, 1970).

Inoculate the plates using either a loop or a multiple inoculator with overnight broth culture diluted 1/100 approximately.

Include a standard sensitive control organism on each plate.

Read after overnight incubation disregarding abnormally minute colonies and growth of one or two colonies only.

When swarming bacteria are tested by this method cut ditches between each inoculated area. Reliable titrations cannot be made on media such as MacConkey agar which prevents swarming by chemical means. High concentration of agar in sensitivity test medium can be employed but the plates are very difficult to prepare.

Multiple inoculators used for staphylococcal 'phage typing can be employed for this test. An inoculator specially made will allow up to 32 strains to be tested per 8·5 cm. plate.

Relation between disc diffusion and titration

Under standard conditions rapidly growing organisms, showing the same zone of inhibition as the Oxford staphylococcus after overnight incubation, will prove to have the same order of sensitivity when tested by titration. It does not follow, however, that one showing half as large a zone will be half as sensitive when titrated because the relation between concentration of the drug and distance from the disc is not a simple linear one (see methods of assay, page 256).

The relation between zone size and minimal inhibitory concentration (MIC) is likely to be quite different if the organism grows slowly. The zone will tend to be large because the antibiotic will have had time to diffuse further from the disc. When the lag phase of growth is prolonged for several days this effect may be somewhat counteracted by destruction of the drug.

During titration the dose of antibiotic remains constant until it begins to degenerate. Therefore, when the drug is stable rates of growth will not affect the result.

Spreading organisms are capable of swimming on the surface of moist medium near discs even when they are sensitive. Therefore in the disc method, readings should be taken from the edge of growth of inhibited colonies. Parallel tests with *Proteus mirabilis* by tube titration and agar diffusion are consistent only when spreading is disregarded.

Sensitivity tests for *Myco. tuberculosis*

Testing the sensitivity of *Myco. tuberculosis* is potentially dangerous and most hospital laboratories in Britain isolate few strains. Experience shows that much better results are obtained when specialist laboratories undertake this work, moreover, the inevitable additional delay in transporting cultures and returning reports is not important because it is a small fraction of the time taken to test such slow-growing microbes. Methods of testing which can be undertaken in diagnostic laboratories are retained for the sake of completeness.

The behaviour of *Myco. tuberculosis* in culture is so different from other bacteria that the methods already described are not suitable for testing its drug sensitivity. The only reliable method for all types of specimen is isolation of the organism on solid medium followed by titration of a pure culture in medium containing serial dilutions of the drug. Sensitivity can be tested either in liquid or on solid medium. In laboratories where tests are seldom requested solid medium is to be preferred because in inexperienced hands it is safer and more reliable. The liquid medium has the advantage that drugs are added immediately before inoculation so that there is no danger that they will deteriorate on storage. It is easier with this method to test a wide range of dilutions and to vary the range tested when this is desirable. To ensure a satisfactory growth, however, the tubes must be inoculated with a pipette which is a hazardous procedure especially when handling resistant strains. As regards speed in reading the result there is very little difference between the two methods. Theoretically the liquid method is quicker, but in practice it may be difficult to obtain a smooth growth in the first liquid culture and the tubes are more liable to contamination, so that a reading in the shortest possible time (3 weeks) by this method is not always achieved. Tween 80 and other surface-active agents appear to affect the end-point particularly in testing streptomycin sensitivity (Michison *et al.*, 1958); resistant strains sometimes appear sensitive in this medium. Youman's liquid medium, which does not contain Tween 80, has proved satisfactory.

The solid medium method has the advantage that tubes can be inoculated with a loop and it is recommended for routine use. It is not, however, free from considerable risk of laboratory infection because the inoculum must be standardized and this involves either grinding living bacilli or shaking them vigorously.[1] A safer method in which lumps of bacilli are broken mechanically has been devised by Marks (1964). Technicians should not be allowed to handle liquid cultures or suspensions of living tubercle bacilli unless they have at least 2 years' bench-work experience and are tuberculin positive. A safety cabinet must be employed to protect the worker from the effects of accidental splashing and to prevent contamination of the laboratory. Laboratories without a well-maintained safety cabinet should not undertake sensitivity tests.

SOLID MEDIUM METHOD (Medical Research Council, 1953)

As soon as possible after growth is apparent on the primary culture scrape off a representative loopful from all parts of the growth, being careful not to remove particles of medium, and suspend it in 0·3–0·5 ml. of sterile distilled water. To obtain an opalescent suspension either grind it in a Griffith's tube (see page 59) or shake in a small screw-capped bottle, containing sterile glass beads, very vigorously. The opacity should correspond approximately to Brown's opacity tube no. 2, i.e., 1000×10^6 bacilli per ml.

Spread a 3 mm. loopful of the bacilliary suspension over the surface of each of a series of Lowenstein-Jensen slopes containing appropriate quantities of the drugs and over a slope containing no drug. Incubate at 37° C. and read at 14 and 28 days. Include a series of slopes of the *same batch* seeded with the control sensitive tubercle bacillus H37Rv. (For interpretation of results see page 236.) For satisfactory results even inoculation is extremely important.

---

[1] See Report to the Public Health Laboratory Service and to the Central Pathological Committee of a Special Working Party, *Month. Bull. Min. Health Lab. Service* (1958), **17,** 10.

Lumps of growth in the suspension must be avoided and the loop should be dipped edgewise into it to ensure the transfer of approximately constant volumes to the slopes.

It has so far proved impossible to devise a reliable test which can be made routinely on the original tuberculous material (unlike tests in some other infections where this is the rule). The main difficulty lies in the scanty and uneven distribution of tubercle bacilli in the specimen which makes standard inoculation of a primary series of cultures impossible. The two stage tests described above are very slow, at best the result is known four weeks after the specimen is received and if the strain grows slowly, or the cultures are contaminated at any stage, the procedure may take more than twice that time. The need for a rapid test is great and when the specimen is very heavily infected it can sometimes be met by culturing the original specimen on glass slides in liquid media containing varying concentrations of the drug. Sometimes after one week's incubation the bacilli will be found to flourish in the presence of high concentrations of streptomycin (100 μg./ml.). The patient can then be spared several weeks' treatment with a toxic drug. Strains which prove to be sensitive or moderately resistant should be re-tested after culture on solid medium.

Slide culture technique for rapid determination of the drug sensitivity of tubercle bacilli in sputum (Rubbo and Morris, 1951, modified)

Materials

Sterile microscope slides (standard size bisected longitudinally = 1 × 6 cm. approximately).

Test-tubes approximately 1·5 × 10 cm. each, containing 5 ml. of ox serum medium (page 372).

Three 18-cm. Petri dishes, two of them sterile.

Metal wire rack to hold slides in the dishes.

Two flat glass plates to cover the sterile dishes.

Streptomycin solution of known concentration.

Sputum produced on the day of the test containing numerous acid-fast bacilli.

Method

A small piece of sputum, heavily streaked with purulent or caseous material, is separated from the specimen and part of it is examined for acid-fast bacilli in the usual way. If numerous bacilli are seen, the remainder of it is spread on ten of the specially cut narrow glass slides and dried in the incubator for 15 minutes. Five of them are placed on the metal rack and submerged for 4 minutes in 6 per cent sulphuric acid (1·2 N) in the unsterile Petri dish. The sterile dishes are half filled with sterile distilled water and the acid is washed from the slides by immersion for two minutes in each of them. The treated slides are then transferred to five tubes of medium, the first containing 100 μg./ml. of streptomycin the second 10 μg./ml. and the third 2 μg./ml. The remaining two are controls. The procedure is repeated with the other five slides. Separate sets of Petri dishes and separate metal racks are required for each specimen to be tested.

The tubes are sealed by impregnating their plugs with paraffin wax and are incubated at 37° C. On the sixth day of incubation one of the control slides is removed from its tube with forceps, fixed and sterilized by heating on a hot plate at 90° C. for 15 to 30 minutes and is then stained and examined. Growth is seen as thick ropes of acid-fast bacilli which are short at first but increase in length until the smear is a network of them against a background of sputum debris. When growth is good on the control slide all the other slides are examined ; if it is poor or absent they are left to incubate and another control is examined on the tenth day. If growth is absent on the tenth day it is unlikely to occur later.

Limitations of the method

If slide cultures are made from positive specimens without streptomycin, i.e. ten slides per specimen with no inhibitory substance in the medium, the results are not always consistent. Very strongly positive specimens usually yield growth on all slides, but with specimens which fail to show bacilli in every field, when stained with auramine and examined by the 8 mm. objective, the chance of growth on every slide is less. If in the test growth is not seen in one of the control tubes, the fallacy is apparent, but chance distribution of the negative slides may lead to a false report of sensitivity. Individual bacilli showing no signs of division are often found on control slides and when seen in cultures containing streptomycin do not necessarily mean a mixture of sensitive and resistant organisms. The bacilli may have been dead originally, they may be in a resting phase or they may have become detached from neighbouring ropes of multiplying bacilli during manipulation.

The test is reliable and rapid for detecting resistant strains, but sensitivity must be reported with caution and should be checked by culture on solid medium with standardized inocula as previously described.

Interpretation of results

When the organism is as sensitive as H37Rv, and when it is many times more resistant, interpretation presents no problem, but unfortunately slight degrees of resistance are often encountered and it is then difficult to advise future treatment.

In the solid medium method occasional colonies of H37Rv are seen on slopes containing five or even ten times the amount of drug which normally inhibits, although growth never approaches that seen on the drug-free slope.   The presence of a few resistant bacilli in a predominantly sensitive strain does not prevent a good clinical response ;   therefore growth is considered to be insignificant when there are less than 20 colonies on a slope.   Resistance and sensitivity must always be judged by comparison with the result of the parallel test on the sensitive control, H37Rv.   When the control behaves normally, i.e. when the end-point of the titration is as expected (Table 25) and the unknown proves to be about 8 times less sensitive it can safely be reported as a resistant strain (M.R.C. 1953).

Treated patients often suffer from infection by a mixture of sensitive and resistant bacilli.   When the sensitive ones heavily predominate treatment with a drug which is only active against them can be expected to improve the clinical condition at least temporarily, but the resistant bacilli will increase and what is more important the drug will be powerless to prevent them becoming resistant to other drugs.   The problem is further complicated because tests on single specimens may vary, presumably because a different proportion of sensitive and resistant bacilli are present in different parts of the lung and in different samples of sputum (Mitchison and Monk, 1955). When slight resistance (less than 8 times more resistant than H37Rv) is observed the test should be repeated on a culture from a fresh specimen.

Catalase test.   Isoniazide resistant bacilli are almost always catalase negative, therefore the test can be employed as a guide in assessing the significance of minor degrees of isoniazide resistance. When the strain proves to be catalase negative the isoniazide resistance is likely to be significant.   The test is not helpful when positive because resistant strains do sometimes produce catalase. Catalase negative strains are often non-virulent to guinea-pigs, they may be virulent to mice and their virulence for man remains in doubt at present.

Opportunistic mycobacteria are often resistant to isoniazide ; they are, however, almost always strongly catalase positive (see Table 18). Method, see page 377.

Difficulties are encountered in testing " second line " drugs used in the treatment of patients infected with resistant strains. Considerable experience is needed for reliable titration and interpretation

TABLE 25.

Drug sensitivity of the standard strain *Myco. tuberculosis* (H37Rv) in microgrammes per millilitre of medium.

	Lowenstein-Jensen solid medium	Tarshis blood agar diffusion method
Streptomycin*	1	0·5–0·125
Isoniazide*	1–3	0·1–0·025
P.A.S.*	2	0·25

* Oxoid solid medium supplied with drugs incorporated in it by Oxoid Ltd., London, England.

of results. Since most hospital laboratories encounter few strains, staff have no opportunity of gaining sufficient experience. It is therefore better, when faced with a culture isolated from a patient in relapse after treatment, to test " first line " drugs only (see Table 25), and at the same time to send the strain to a reference laboratory for tests of " second line " drugs. There will then be no unnecessary delay in choosing active drugs for treatment.

Blood agar diffusion method

A technique employing pre-diffusion in blood agar, Tarshis medium (page 370), is satisfactory for " first line " drugs. Tarshis medium, without penicillin, is measured in 10 ml. aliquots into sterile plastic 30 ml. containers and sloped. The drugs to be tested are diluted in sterile water, or buffer, and 0·5 ml. of each dilution is delivered into each of as many tubes as are required for the batch of cultures to be tested, including a series for the control, H37Rv. The medium bottles are again sloped so that the liquid covers the surface of the medium and the drug is allowed to diffuse thus overnight at 4° C. Next day the tubes are inoculated with the strains to be tested, and the control strain as already described, and the tubes are incubated upright. Results are read after 2 and 4 weeks' incubation.

The advantages of this method are that the range of dilutions can easily be varied, there is no degeneration of drug during heating of the medium and the experienced worker doing the test is also responsible for the addition of drugs so that accidental errors are less likely than in a busy media department.

Antibiotics in Combination

Acute bacterial infections by single pathogens usually respond to treatment by an antibiotic which inhibits growth of the organism in laboratory tests, provided the drug is able to reach the site of infection in sufficient concentration. In certain circumstances, however, it may be necessary to give more than one antibiotic, either because two infecting microbes with different sensitivity patterns have to be dealt with, or because treatment must be prolonged and there is danger of resistance developing. In bacterial endocarditis caused by bacteria resistant to benzylpenicillin a combination of antibiotics is usually needed to ensure killing rather than inhibition. In this disease killing by the drugs is essential for a permanent cure; bacteriostasis will temporarily relieve symptoms but, unless the bacteria infecting the valves are killed, relapse is inevitable. Numerous courses of inadequate treatment finally result in death from damage to the heart or from some other vascular accident (Clinicopathological Conference, 1966).

The laboratory should therefore be prepared to test strains isolated from the blood by a method which shows bactericidal power as well as inhibition and which will demonstrate antagonism when inappropriate combinations of drugs are proposed for treatment. Once treatment has started, the patient's serum should be tested against his own organism to check that the strain is killed by the concentration of the antibiotic achieved.

The synergistic action of some drugs against some bacteria is well known. For example, penicillin and streptomycin are synergistic against *Strept. faecalis* (Jawetz and Gunnison, 1950; Robbins and

Tompsett, 1951). The addition of a tetracycline or chloramphenicol to the penicillins antagonizes their effect on sensitive organisms (Garrod *et al.*, 1973).

The other aminoglycosides, kanamycin and gentamicin, behave like streptomycin and whichever of them is most active against the organism in ordinary growth inhibition tests should be chosen for combined treatment when a penicillin or cephalosporin alone fails to kill.

In endocarditis results of therapy match well with tests of bactericidal effect of combined antibiotics, but in other situations, for example deep-seated chronic abscesses, the response is less certain. When two drugs are combined, the proportion of each at the site of the lesion is unknown and, moreover, the number of bacteria to be killed may be overwhelmingly large. With many unknown factors it is difficult to devise a test relevant to all conditions and it is too much to hope that an apparently good combination will always succeed in fibrosed and avascular lesions. One which is ineffective in the laboratory, however, is extremely unlikely to benefit the patient.

Synergism and antagonism in disc tests. Tests of mixtures of drugs on discs are misleading and do not necessarily indicate combined action. Two single discs or strips of blotting paper each impregnated with one of the proposed drugs can be placed side by side or at right angles on blood agar pre-seeded with the pathogen. After overnight incubation the shape of the inhibition zones may indicate synergism or antagonism but cannot be relied upon to do so. None of these tests indicates the effect of combining the drugs on killing power and they are therefore insufficient as a basis for recommending treatment in endocarditis.

The synergistic action between sulphonamide and trimethoprim which are normally given as the mixture, co-trimoxazole, is frequently seen when these drugs are tested separately and the discs are placed 3 cm. or less apart. This is an exception to the general rule that mixing drugs on discs is undesirable because cotrimoxazole discs will show a larger zone than either constituent separately when tested against an organism sensitive to both. Nevertheless the increase is not very great and an organism resistant either to sulphonamide or trimethoprim but not to both is likely to be reported sensitive to the combination. Bacteria are much more often resistant to sulphonamide than to trimethoprim and this can be recognized by testing separately for sulphonamide. Trimethoprim-resistant strains are sometimes sensitive to sulphonamide and they will not be discovered unless trimethoprim is also tested separately. Ideally three discs should be used, each drug separately and the combination. All essential information will however be revealed if sulphonamide and trimethoprim are tested separately and the combination is omitted. It can safely be assumed that when activity of whatever degree to either constituent is seen (when tested against the disc content recommended, see Table 22) the combination will be active, whether synergism can be easily demonstrated or not.

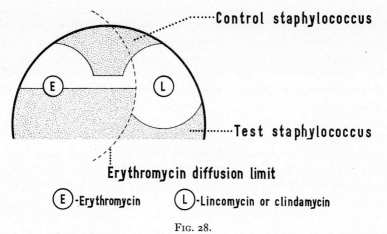

Erythromycin diffusion limit

Ⓔ-Erythromycin   Ⓛ-Lincomycin or clindamycin

Fig. 28.

Antagonism is often seen in tests on staphylococci when erythromycin and lincomycin discs are placed about 3 cm. apart. Strains which when tested alone are erythromycin-resistant but sensitive to lincomycin are seen to be resistant to lincomycin also when erythromycin is present (see Fig. 28). It is well known that lincomycin resembles the macrolide antibiotics too closely to be useful in combination with them, but it is not always appreciated that when lincomycin is used alone to treat staphylococcal infection by an erythromycin-resistant strain the organism will quickly become resistant to lincomycin also. When reporting the lincomycin result a warning should therefore be given.

Bactericidal tests. Two kinds of method are suitable for bactericidal tests in hospital laboratories. In one a series of dilutions of the drugs are tested both singly and in combination against a heavy and a light inoculum of a liquid culture of the pathogen. This method is straightforward and can be performed by any competent benchworker without special equipment. It has the disadvantage that only one dose of each drug can conveniently be tested and there is no guarantee that this will be the operative strength at the infection site. The other method is a diffusion method employing blood agar pre-diffused with drugs. The pathogen is seeded on a cellophane membrane tambour applied to the surface of the agar. Both nutrient from the medium and drug pass through the membrane, enabling the bacteria to grow or be inhibited. After growth has appeared the culture can be transferred on the membrane to drug-free blood agar and reincubated. Subsequent growth in the inhibition zones indicates bacteriostatic drug action. When the pathogen has been killed the inhibition zone will remain clear. The method needs a little practice in preparing and applying the tambours, and glass cylinders of appropriate size are required. When, as is usual, an urgent result is required the test should be set up in duplicate. It has the great advantage that the effect of a wide variety of concentrations of each drug separately and of a combination of the two can be observed. Sometimes the effect of a combination in different concentration is different. By the tube method such variation will go unrecognized since to test a large number of different concentrations of each drug in tubes is impracticable. It is therefore worthwhile to become adept at the cellophane transfer technique (see Fig. 29).

Combined tube dilution method (adapted from Chabbert, 1953)

A set of 10 tubes is arranged as in Table 26. Serum broth and antibiotic is added to each tube so that the volume of fluid per tube is the same and each contains the appropriate concentration of the drugs. The dilution of drugs in the tubes is that judged to be the average concentration of each at the infection site when full doses of each are given. If the volume per tube is about 2 ml., one drop of a 1 : 10 dilution of an overnight culture of the pathogen is seeded into each tube. A control tube is included which contains the same volume of medium but no antibiotic.

All the tubes are very thoroughly shaken and then subcultured immediately to sectors of two blood agar plates. The loop should be withdrawn edgewise to prevent over filling thus ensuring that each sector receives approximately the same volume. Duplicate

TABLE 26.

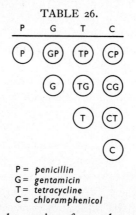

P = *penicillin*
G = *gentamicin*
T = *tetracycline*
C = *chloramphenicol*

seeding from each tube, using four plates, increases reliability. Tubes and plate are incubated overnight. Next day the tubes (which all appear clear except the control) are removed from the incubator, shaken thoroughly and again subcultured to blood agar. The first incubated plates are kept in the refrigerator so that overnight growth on them may be compared with that of the second subcultures which will be seen on the third day.

TABLE 27.

Result of combined antibiotic bactericidal test : $+ +$ = semiconfluent growth.

Tube	Inoculum	Blood agar culture	
		1 : 10	
		Before inc.	After inc.
1.	Ampicillin.	$++$	$\pm$
2.	Ampicillin and gentamicin .	$++$	—
3.	Ampicillin and tetracycline.	$++$	$+$
4.	Ampicillin and chloram. .	$++$	$+$
5.	Gentamicin	$++$	—
6.	Gentamicin and tetracycline	$++$	—
7.	Gentamicin and chloram .	$++$	—
8.	Tetracycline and chloram .	$++$	$+$
9.	Tetracycline	$++$	$+$
10.	Chloramphenicol	$++$	$+$

EXAMPLE : Meningitis in an infant, caused by *Esch. coli* sensitive to all four drugs, has responded poorly to ampicillin. Can improvement be expected if tetracycline is also given ?

Result : see Table 27.

Interpretation : A mixture of tetracycline and ampicillin or of chloramphenicol and ampicillin is less lethal to this *Esch. coli* than ampicillin alone. Tetracycline and chloramphenicol alone fail to kill large numbers of the microbe after overnight incubation. Gentamicin is the only one of these drugs which, when added to ampicillin, is likely to improve the patient's condition.

Cellophane transfer method (Chabbert, 1957, Garrod and Waterworth, 1962)

MATERIALS

Discs of cellophane about 15 cm. diameter.
Glass cylinders with flat ground rims, about 3 cm. deep and of a diameter to fit easily into a Petri dish.[1]
Sterile strips of good quality blotting paper.
Antibiotic solutions freshly prepared (Table 28).
Screw-capped jars sufficiently large to contain cylinders.
Blood sensitivity test agar plates.

Cellophane PT 300, commercially available as jam pot covers, is permeable to nutrients. When a culture is seeded on top of it there should be little difference between growth on its surface and growth on the medium without cellophane.

Cellophane stretches when it is moistened, it must therefore be kept moist from the time the tambour is made to the time of application to the surface of the medium ; if it is allowed to dry it will stretch and crinkle on application to the medium and a satisfactory transfer will be impossible.

METHOD

Wet the cellophane discs in distilled water. Stretch a wet disc over the flat ground end of a glass cylinder and fix it in position with a rubber band. Make sure the tambour thus made is flat and taut, and secure it with a second rubber band. Cut off any extraneous frill of cellophane. Place the tambour on moist blotting paper, cellophane downwards, in a screw-capped jar and autoclave at 15 lb. for 15 minutes.[2] On removal from the autoclave tighten the cap and the tambour will remain ready for use as long as the jar remains air-tight and the cellophane moist. Prepare solutions of the drugs to be tested (see Table 28).

---

[1] Obtainable from James A. Jobling and Co., Sunderland, England.
[2] A domestic pressure cooker is convenient and satisfactory.

Moisten a sterile strip of blotting paper in one of each of the antibiotic solutions to be tested, drain well, blot and apply them to the surface of a well-dried blood agar plate at right angles to each other (see Fig. 29). Mark the positions of the strips on the plate and incubate for 7 hours, or leave overnight in the refrigerator (about 4° C.). Remove the strips and apply a sterile tambour to the pre-diffused medium making sure there are no air bubbles between the

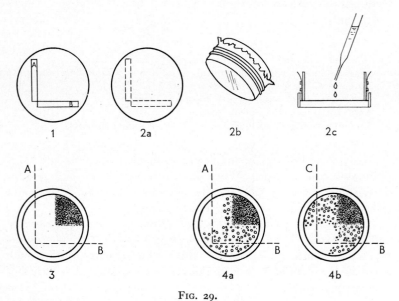

Fig. 29.

1. Prediffusion.
2. (a) Prediffused plate. (b) Tambour. (c) Tambour applied receiving inoculum.
3. Primary growth.
4. Secondary growth after transfer to antibiotic-free medium. (a) Antibiotics A and B combined = antagonism. (b) Antibiotics C and B combined = synergism. *Note.* A is a bactericidal antibiotic for this organism; B and C are incompletely bactericidal but kill the organism when combined.

membrane and the medium. Dry, medium downwards, with the lid of the Petri dish tilted on top of the glass cylinder for about 30 minutes in the incubator. Flood the tambour with a dilution of an overnight liquid culture of the test organism, remove excess fluid and dry again for about 30 minutes with the lid tilted as before. Then invert and incubate, medium upwards, with the glass cylinder standing on clean blotting paper on the incubator shelf without a lid.

The suspension or dilution of the test organism should be such that a heavy but not quite confluent growth is obtained. With staphylococci this is about 1 : 100 dilution of an overnight broth culture. When less robust organisms are tested, e.g. *Strept. viridans*, undiluted culture may be used. Drying after application of the culture and incubating inverted without a lid is important to prevent excessive moisture spreading over the cellophane surface carrying organisms with it, which would ruin the result.

After 7 to 18 hours' incubation, or when primary growth can be seen, mark the position of the antibiotic strips on the cylinder, transfer the tambour to a fresh, well-dried blood agar plate and reincubate inverted as before. Secondary growth, if any, within the inhibition zones indicates the effect of the drugs, both singly at the position of the tips of the paper strips, and combined near the angle at which they met (see Fig. 29).

Bactericidal activity of patient's serum

A test of the bactericidal activity of the patient's serum against his own organism is useful in endocarditis, particularly when this is due to a staphylococcus or *Strept. faecalis* which needs treatment with potentially toxic drugs. The dose must be sufficient to kill the organism but should be the minimum required for this.

Serum should be tested after at least 24 hours' treatment to ensure that blood levels are the maximum that can be expected from the dose schedule. It is worth while to test two samples of serum, one taken halfway between two doses, the other immediately before a dose, i.e. at the lowest level. Serum should be separated and tested as soon as possible to avoid inactivation of drug in the sample.

METHOD

Twofold dilutions of serum in indicator assay medium (page 371) are made. When there is sufficient serum it is better to test in duplicate so that the effect on two different inocula can be observed. Convenient volumes are of 0·5 ml. A control tube without serum is included.

Each tube is inoculated with one drop of diluted overnight liquid culture of the pathogen. For staphylococci suitable dilution for inoculation is 1 : 100. The dilutions are well mixed on a mechanical shaker or by shaking vigorously by hand in a screw-capped bottle to ensure even inoculation. Each dilution is well mixed after inoculation and two loopfuls from each and from the control are

explanted to blood agar.   The tubes are then incubated overnight.
Next day the inhibition titre is noted and each tube lacking growth
is again well mixed and explanted on blood agar, see page 241.

For a satisfactory result the undiluted serum of the sample taken
mid-way between doses should kill all organisms in one loopful from
the heavy inoculum.   The drug may be aided by the natural antibacterial mechanisms of the serum but this should also happen in
the patient.   When the infecting organism is a Gram-negative
bacillus, which are particularly susceptible to killing by fresh
normal serum, the patient's serum should first be inactivated by

TABLE 28.

Concentration of antibiotic solution employed on cellophane transfer technique
(from Garrod *et al.* (1973)†)
(Blotting paper strips on 30 ml. blood agar in 9 cm. Petri dishes)

	μg/ml.
Benzylpenicillin*	50
Methicillin	1,000
Cloxacillin	200
Ampicillin*	50
Streptomycin	1000
Kanamycin	1000
Gentamicin	100
Chloramphenicol	1,000
Tetracycline	200
Erythromycin	200
Fusidic acid	50
Vancomycin	500

* Use 250 μg/ml. when testing enterococci and Gram-negative bacilli.
† Consult for other antibiotics.

heating at 56° C. for 30 minutes.   The proportion killed can be
roughly calculated by the number of colonies seen on the explant
plate from the control tube before incubation which received the
same inoculum but no serum.   Degree of carry-over of drug to the
blood agar will be indicated by comparing the original explants
from the inoculated serum dilution tubes with those from the control.
Carry-over sufficient to invalidate the results is unlikely from the
tubes nearest the inhibition titre.   When in doubt it can be checked
by seeding loopfuls into 5 ml. broth so that any drug in the inoculum
will be diluted beyond the minimal inhibitory concentration.

When the heavy inoculum is not killed treatment may yet succeed
if the light inoculum has succumbed.   Advice about increased

dosage or a change of treatment must be assessed in the light of all other circumstances in each case (see Garrod, *et al.*, 1973).

ANTIBIOTIC ASSAY

Assay of antibiotics in body fluids may be required as follows:

1. *Blood*

(*a*) From patients with renal disease, so that the dose may be limited to avoid toxic symptoms due to poor excretion of the drug. For example, the aminoglycosides, particularly gentamicin, are often used alone or in combination with other antibiotics to treat serious infections including endocarditis. They are all potentially ototoxic and when renal function is impaired or treatment lasts longer than one week assay will be required.

(*b*) From patients with serious infection who show no clinical improvement when the pathogen is known to be sensitive *in vitro* and the dose has been calculated to be sufficient.

(*c*) From patients receiving a new preparation of a drug or receiving it via a new route.

2. *Cerebrospinal fluid*

(*a*) To find the concentration of drugs which cross the normal choroid plexus when meningeal infection is treated via an indirect route.

(*b*) To detect the presence of drugs held back by the normal choroid plexus as an early sign of meningitis (particularly streptomycin in miliary tuberculosis).

3. *Urine*

Assay is seldom helpful as a guide to treatment because levels vary with the dilution of the sample. If urinary excretion is to be checked an early morning sample should be taken. The standards must be prepared in buffer at the same pH as the urine. It may be desirable to alter pH to make the test more sensitive (see Garrod *et al.*, 1973).

4. It may also occasionally be desirable to check levels in serous and tissue fluids, in pus and in sputum.

METHODS

These are of two kinds. The drugs may be assayed either by diffusion through pre-seeded agar or by titration in liquid culture using a standard sensitive organism. The tests are similar to those already described for estimating sensitivity, but two difficulties arise which are not encountered in sensitivity tests. The fluid may contain inhibitors other than the antibiotic to be assayed and it may be contaminated with resistant bacteria.

Bacterial inhibitors. It is well known that serum and other body fluids often contain substances which inhibit bacterial growth. The amount of these substances varies from time to time in the same person. It is unwise therefore to assume that growth inhibition by body fluids is necessarily due to the presence of antibiotic in them. When fluids are assayed by agar diffusion the naturally occurring inhibitors are not likely to interfere with the result, because zones of inhibition caused by them are small in comparison with those produced by the antibiotics. Moreover, under these conditions the effect of natural bacterial inhibitor and antibiotic will not be additive. In liquid culture, however, these substances may invalidate the results unless the bactericidal power of serum is destroyed by heating at 56° C. for 30 minutes. Most antibiotics, even in high dilution, will withstand this amount of heat.

Bacterial contamination. Clearly the presence of resistant organisms will invalidate liquid culture methods. They do not, however, interfere with estimations by agar diffusion (unless they should themselves happen to be antibiotic producers or inactivators). It is very important that growth should not occur in the specimen before the test is made. Penicillinase producers will neutralize penicillin and other antibiotics may be destroyed by bacterial growth. Contaminated specimens must therefore always be refrigerated immediately they have been collected until they are tested.

It is difficult to sterilize the specimen without altering its antibiotic content. A proportion of the drugs is removed during Seitz filtration and even membrane filters absorb small quantities. Heating cannot be relied on to eliminate the organisms without damaging the specimen, therefore an agar diffusion method is preferred.

Controls. The unknown fluid is always tested in parallel with dilutions of known strength of antibiotic in a similar fluid. The pH of the test fluid and the controls should be the same. In streptomycin and tetracycline assay this is vital because slight changes in pH alter the end-point.

The standard solution for serum assay is diluted in normal human serum which has been heated at 56° C. for 30 minutes ; for cerebro-spinal fluid in buffered distilled water.

One further control should be included in fluid medium assays. An organism of the same species as the standard, but resistant to the antibiotic, should be tested for its ability to grow in the presence of the test fluid. When growth is inhibited bacteriostatic substances other than the antibiotic are present and the assay is invalid.

Assay of antimicrobial mixtures

It is often necessary to estimate one antibiotic in the presence of another. For example, if treatment is changed during the course of an infection in a patient with impaired renal function, traces of the first drug may remain in the serum when the blood level of the second is determined. When the first drug is a penicillin or cephalosporin it can be neutralized by adding beta-lactamases, page 16. Sulphonamides can be neutralized by the addition of 0·005 per cent para amino benzoic acid and trimethoprim by the addition of thymidine to the medium, with other mixtures an organism is used which is sensitive to the drug to be assayed but resistant to the other.

When penicillin is to be assayed in the presence of another drug (which is rarely necessary) difficulty may arise because penicillin-sensitive strains resistant to other drugs are uncommon. If one is not available assay may be possible by agar diffusion using the standard staphylococcus. In addition to the usual controls a parallel test is made in the presence of penicillinase. If the zones in the control penicillinase series are smaller than those in the test, the penicillin must have diffused further than any other inhibitors and the result can be considered reliable. If on the other hand the zones are equal in both series some other inhibitor must be responsible for them and the result does not indicate the penicillin level.

If a suitable resistant organism is not available when assay of one in a mixture of antibiotics is required, the predominant drug in the mixture can be assayed by the agar diffusion method using the Oxford staphylococcus. Since substances with different molecular weights diffuse differently, the zones of inhibition seen in the test will be caused by the antibiotic which diffuses furthest and the result will not be altered by the presence of small quantities of other antibiotics.

Standard strains for assay.   In the absence of strains specially
selected for assay (Garrod *et al.*, 1973) the standard control strains
recommended for sensitivity tests can often be used.   The Oxford
Staphylococcus (NCTC 6751) is suitable for most drugs but *Esch.
coli* (NCTC 10148) is more sensitive to some, e.g. chloramphenicol
(Table 30) and should be used to make the test more sensitive.
It should also be used to assay carbenicillin which contains traces
of benzylpenicillin thus making the use of the penicillin sensitive
staphylococcus impracticable.

Patients needing assay of aminoglycosides are often receiving
at least one other antibiotic and therefore a standard strain resistant
except to aminoglycosides is required.   A strain of *Klebsiella*
(NCTC 10896) is admirably suited to this purpose ; not only is
it resistant to a wide range of other drugs (penicillins, cephalosporins, macrolides, lincomycin, chloramphenicol, tetracycline,
fusidic acid, sulphonamide and trimethoprim) but it is a rapid
grower and, when inoculated heavily, zones can be seen after 4 hours'
incubation.   It can be maintained indefinitely by scraping growth
from nutrient agar to make a suspension in broth about equivalent
to an overnight broth culture, 5×10^7 bacilli/ml., which is frozen
at $-70°$ C. or colder.

Inoculum

Bacteria must be evenly distributed throughout the medium,
or over its surface, and the standards must be set up with the same
batch of medium having been inoculated at the same time as that
used for all the sera to be tested.   Zones will vary with inoculum,
as in sensitivity tests, therefore comparison must be made between
control and test sera using the same batch of inoculated medium.
Some variation in inoculum between batches does not matter.

Rapid assay of aminoglycoside antibiotics

This is an essential technique for laboratories in hospitals which
treat acute bacterial infections, endocarditis or renal tuberculosis.
Because of potential ototoxicity assay is needed whenever renal
function is impaired, or when treatment is prolonged, to prevent
accidental overdose.   Moreover, absorption of gentamicin particularly in acutely ill adults and in infants is sometimes less than
expected and the laboratory must recognize under-dosage (Garrod
et al., 1973).   There is some disagreement about the best time to

sample. For the prevention of toxicity there is good evidence that assay of trough levels, i.e. from a sample of blood taken immediately before the next dose is due, gives the most useful results (Line *et al.*, 1970, Mawer *et al.*, 1974). It is reasonable to assume that when the trough level is about 2 μg/ml. the peak level will be adequate and not likely to exceed 10 μg/ml. which is thought to be undesirable. If a peak level is to be measured blood should be taken one hour after intramuscular injection, although one cannot be certain that this indeed contains the peak level in all patients. In a patient having intravenous therapy blood should be taken immediately after the dose from a limb other than that receiving the injection.

Various techniques have been described for rapid assay of gentamicin, one of these requires only 2 hours' incubation (Noone *et al.*, 1971) but it is laborious to perform, needs special equipment and is liable to error except in highly skilled hands (Phillips *et al.*, 1974). In hospital laboratories rapid assay using the *Klebsiella*, see above, is preferred. Although 4 hours' incubation is needed it enables advice to be given before the next dose is due, the normal schedule being 8 hourly.

Agar plate assay (from Garrod *et al.*, 1973).

MATERIALS

Antibiotic standard solutions in sterile human serum (stable for several months if kept frozen).

Kanamycin 50, 10 and 2 μg/ml.

Gentamicin and tobramycin 10, 2 and 0.4 μg/ml.

Broth suspension of *Klebsiella* (NCTC 10896), see above.

Sensitivity Test agar.

Sterile cork borer, approx. 8 mm. diameter (No. 5).

Semi-logarithmic graph paper.

METHOD

i. Surface seeding. Thaw the standard dilutions. Thaw one aliquot of the *Klebsiella* suspension and dilute it 1 in 50 in sterile water. Flood well dried sensitivity test agar plates (which must be thin, 10 ml. per plate, and poured on a flat surface) with the diluted suspension and drain off excess fluid with a Pasteur pipette ; a minimum of three test plates is required. When the liquid has been absorbed, it should take only a few minutes, punch up to six holes per plate. Do not lift the surrounding agar from the plate when removing the agar cores. Fill three of the holes completely with one each of the standard dilutions and the other three with test sera. When serum levels are likely to be high the test serum must be diluted in normal human serum to come within the range of the standards. When in doubt also test the serum diluted ten-fold. Whichever of the two titrations comes within the standard range is then read and reported. The standards and each unknown serum, or serum dilution, must be tested in triplicate.

Read zone diameters either with calipers or with a millimetre rule placed across the plate. Plot the average of three zone readings of each of the three standards on semi-logarithmic graph paper as described for the vertical diffusion method (page 256). These should give a straight line. Then plot the average of three readings for each of the unknown sera and read the concentration in the serum from the graph. In plate tests there is usually no need to square the zone measurement to obtain a straight line.

ii. Agar inoculation. When many plates are needed this is less laborious than flooding.

Inoculate sensitivity test agar melted and cooled to 48° C. to give about 10^5 organisms per ml. Mix by inverting the bottle several times, then measure 10 ml. per plate. Get rid of any bubbles with a hot loop, not by flaming which will injure the inoculum. Dry with the lids tilted, then proceed as described above. Inoculated plates can be stored at 4° C. for use next day. Note that sufficient plates must be poured to test the standards and all sera to be tested in triplicate. No valid comparison can be made with medium inoculated on a different occasion.

Assay in liquid culture

It has been shown that agar diffusion gives more reliable results than titration in liquid medium (Reeves and Bywater, 1975), diffusion is therefore the method of choice. The liquid medium method however has the advantage for assay against the patient's own organism because tubes showing no growth can be subcultured to assess bactericidal activity as described above. Less preparation is needed for tube dilution assay and therefore it can be used in an emergency. Either the control staphylococcus or *Esch. coli* also used for sensitivity tests is suitable for assay provided it will not be inhibited by some other antibiotic which the patient may have received.

Sera without antibiotic are often inhibitory and bactericidal especially to Gram-negative bacilli. When *Esch. coli* is used the serum must be heated at 56° C. for 30 minutes but this is usually unnecessary for the staphylococcus. Non-specific inhibition can be recognized by inoculating a tube containing the strongest dilution with a staphylococcus resistant to the drugs assayed, if it fails to grow the assay is invalid.

The standard antibiotic solution to be titrated in parallel with the tests must be made in the appropriate diluent as for all assays, see page 248. The original dilution should be of a strength sufficient to include the minimal inhibitory concentration of the assay organism within a short range of two-fold dilutions, see Table 30. When very high levels are expected in the fluid to be tested it should be diluted tenfold in appropriate diluent, i.e. sterile human serum when serum is titrated. One or more dilutions are then titrated in parallel with the undiluted fluid and the standard to make sure that the end point will be seen. A long range of two-fold dilutions is likely to give an inaccurate result.

Assay of aminoglycosides is affected by the composition of the medium, some broths being unsuitable, serum glucose peptone water should therefore be used (Garrod *et al.*, 1973).

MATERIALS

    Indicator serum medium (page 371).
    Culture of sensitive organism (usually the Oxford staphylococcus).
    Culture of resistant organism.
    Standard solution of antibiotic.
    Sterile sample of fluid to be assayed.
    Sterile test-tubes (approximately 75mm. × 13mm. is a convenient size).
    Sterile graduated pipettes.
    Sterile diluent.

PROCEDURE. Make twofold dilutions of fluids to be tested and standard antibiotic solution in the indicator medium ; a convenient volume is 0·5 ml. per tube. Seed all tubes with a drop of diluted culture of the standard organism (1 drop of overnight broth culture in 5 ml. of broth, well shaken). Include three controls, one contains medium only, one contains the strongest dilution of the standard antibiotic solution and the third, the strongest dilution of the test fluid. Seed the first with the standard organism, the second and third control tubes are seeded with a comparable inoculum of an organism which is resistant to the antibiotic, they check the fluid for the presence of other bacterial inhibitors.

EXAMPLE. A patient suffering from severe burns followed by suppression of urine has been treated with gentamicin. Serum levels are required to avoid an overdose.

Immediately before the dose is due about 2 ml. of blood are withdrawn from a vein using aseptic technique; the blood is allowed to clot and the serum is separated from it. A standard solution of gentamicin is prepared in sterile normal human serum to contain 10 μg/ml. Dilutions are made as in the diagram. Each tube is seeded with one drop of diluted overnight culture of the Oxford staphylococcus except two of the control tubes which receive a comparable inoculum of a gentamicin-resistant organism.

The result is read after overnight incubation (see diagram).

RESULT. The serum contains 0·5 μg/ml. of gentamicin.

TABLE 29.

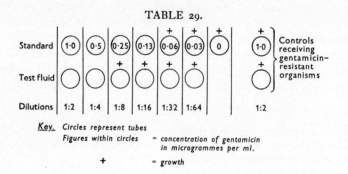

Key. Circles represent tubes
Figures within circles = concentration of gentamicin in microgrammes per ml.
+ = growth

Assay by vertical diffusion

(Streptomycin and aminoglycosides only)

The method described by Mitchison and Spicer (1949) using pre-seeded agar in 3 mm. glass tubing has proved simple and accurate for routine streptomycin assay.

APPARATUS

Sterile 8 cm. lengths of 3 to 4 mm. (internal diameter) glass tubing.
Assay agar 15 ml. in 1 ounce screw-capped bottle (page 372).
Overnight broth culture of the Oxford staphylococcus.
Sterile distilled water.
Water bath with thermometer.
One 1 ml. graduated sterile pipette.
Pasteur pipettes.

Plasticine.
Microscope with movable stage fitted with vernier millimetre scale.
Hair line to insert in eyepiece.
Standard solution of antibiotic. (Streptomycin 75, 15 and 3 μg/ml.,
for other drugs see page 251.)

PROCEDURE. The glass tubes are placed on end in rows in the plasticine, which is most conveniently held in a small shallow tin. Four tubes in a row are needed for each fluid to be tested and two rows of six tubes each for the standards. The assay agar is melted in a boiling water bath and then transferred to a 50° C. (agglutination) bath to cool. The overnight broth culture of the Oxford staphylococcus is diluted in distilled water (0·1 ml. in 10 ml. water) and *well shaken.* Water is taken from the 50° C. water bath and allowed to cool to 48° C., the assay agar is placed in it and is then inoculated with 1 ml. of the diluted culture. The seeded agar is very well shaken (about 20 times). Vigorous shaking is very important. A column of molten seeded agar about 3 cm. high is then delivered into each tube using a wide bore Pasteur pipette and working as rapidly as possible.

The agar is allowed to set for at least 5 minutes. The fluid for assay and the standards are then delivered undiluted on to the surface of the agar in the appropriate tubes. The depth of fluid must be at least 1 mm. The result is read after overnight incubation.

Seeded agar can be prepared one day previously and kept in the glass tubing overnight in the refrigerator. It is very important that the standards and the test should be set up from the same bottle of agar since slight variations in inoculum alter the zone depth. When the level expected is very high a ten-fold dilution should be tested in parallel.

READING. Each tube is in turn carefully removed from the plasticine and placed horizontally on a microscope slide which has a piece of plasticine at either end to hold it. It is then examined under the low power of the microscope and the zone of inhibition is measured to the nearest 0·1 mm., using the millimetre scale on the movable stage and a hair held between two rings of cardboard in the eyepiece, see Fig. 30.

One reading is first taken with the tube in position A and then another with it in position B, in relation to the hair line. Subtraction of the smaller from the larger reading gives the zone depth. All tubes in each row should ideally give the same reading. In practice the variation is remarkably small but occasionally one tube will show a grossly increased zone due to seepage of fluid between the glass and the agar; such a reading is disregarded. The zone depth for each fluid is the average reading of all the tubes, i.e. an average of four for the test fluids and of six for the standards.

Both the concentration and the zone depth are known for the standards so a graph can now be plotted from the results.

Only a straight line graph can be accurately made without multiple standards. Penicillin diffuses in such a way that the zone is lineally related to the logarithm of the concentration. Streptomycin and other aminoglycosides diffuse differently so that the

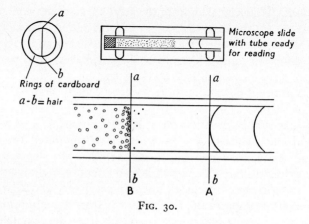

FIG. 30.

zone depth squared is lineally related to the logarithm of the concentration, within the limits likely to be encountered in body fluids. If graph paper with a logarithmic scale is used, mathematical calculations are reduced to a minimum. The figures for the two standards are plotted and a straight line is drawn through them. Finally the concentration of antibiotic in the unknowns is read from the graph (Fig. 31). In plate diffusion tests there is usually no need to square the zone measurements but this is worth trying when no straight line is obtained. A faulty test will not give a straight line whether zone measurements are squared or not.

It is necessary to make a fresh graph for each batch of tests but one sheet of graph paper may be used many times.

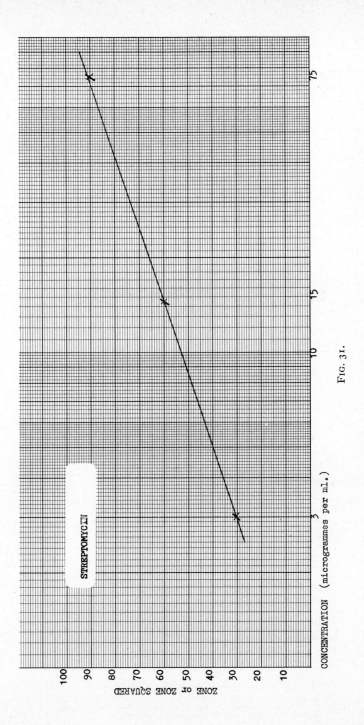

STREPTOMYCIN

ZONE or ZONE SQUARED

CONCENTRATION (microgrammes per ml.)

Fig. 31.

Anti-microbial drug	Drugs showing complete or partial cross-resistance	Approximate sensitivity of Oxford Staphylococcus (N.C.T.C. 6571)	Approximate sensitivity of *Esch. coli* (N.C.T.C. 10418)	Species sensitive (Britain, 1974)	Penetration	
					C.S.F.	Serous fluid
PENICILLINS Benzyl-penicillin† C	phenoxy-methyl penicillin	0·02–0·04 units/ml.	—	*Strept. pyogenes* *Strept. pneumoniae* *Act. israeli* *Clostridia*	—	±
Ampicillin C	amoxycillin carbenicillin		5 µg/ml.		—	±
Carbenicillin C	ampicillin amoxycillin		4 µg/ml. (32 µg/ml. *Ps. aeruginosa* NCTC 10662)		—	±
Methicillin† C	cloxacillin flucloxacillin	1 µg/ml.	—		—	±
Cepha-loridine† C	other cephalo-sporins (see also page 222)	0·025 µg/ml.	2·5 µg/ml.		—	+
Sulphonamide S	all sulphon-amides	—	—		+	+
AMINO-GLYCOSIDES Streptomycin† C		0·25 µg/ml.	0·5 µg/ml.		—	±
Kanamycin† C	neomycin* framycetin* paromo-mycin*	5 µg/ml.	10 µg/ml.			
Gentamicin† C	tobramycin†	0·12 µg/ml.	0·5 µg/ml.		—	
Tetracycline S	all tetra-cyclines	0·25–1 µg/ml.	2·5 µg/ml.	*Mycoplasma*	± or —	+
Chloram-phenicol S		5–10 µg/ml.	2·5 µg/ml.	*Haemophilus* *Mycoplasma*	+	+
Polymyxin† C	colistin†	—	0·25 – 1 unit/ml.	*Pseudomonas aeruginosa*	—	±
Lincomycin (S)	clindamycin	0·25 µg/ml.	—		—	+
Erythromycin (S)	oleondo-mycin spiramycin	0·05 µg/ml.	—		—	+
Novobiocin S Fusidic acid S Vancomycin† C Bacitracin*	ristocetin†	1 µg/ml. 0·06 µg/ml. 1 µg/ml. —	— — — —		— — —	+ + +
Nitro-furantoin Nalidixic acid		4 µg/ml. 500 µg/ml.	4 µg/ml. 1·5 µg/ml.			

* For local and intestinal use only, drugs not absorbed from the gut.
† Given parenterally to attain therapeutic serum levels.
‡ In patients with developing teeth and skeleton, in pregnancy and in renal failure.

| rmal Human body | | Potential toxicity with prolonged use | Main use |
Bile	Through placenta		
+	+	−	Drug of choice for sensitive Gram-positive bacteria and Neisseria
+	+	−	Penicillin with activity against some Gram-negative bacilli, inactivated by penicillinase
+	+	−	Penicillin active against *Pseudomonas* and indole positive *Proteus*. Inactivated by penicillinase
+	+	−	Penicillin active against penicillinase producing staphylococci
+	+	−	In some patients hypersensitive to penicillin. Prophylaxis in dentistry. Resistant urinary infections
+	+	−	Acute cystitis. Meningitis chest infections as an additional drug
+	±	+	Tuberculosis. Powerful bactericidal effect with penicillin
			Resistant urinary infections. Powerful bactericidal effect with penicillins
		+	Resistant Gram-negative bacilli, especially *Ps. aeruginosa*. Also staphylococci. Powerful bactericidal effect with penicillins
+	±	+‡	Wide spectrum antibiotics active against *Mycoplasma*, *Rickettsia* and *Chlamydia*
+	±	+	Haemophilus meningitis. Typhoid, when the bacillus is sensitive. Chest infections in the elderly
−		+	*Pseudomonas aeruginosa* infection. Resistant Gram-negative bacilli (except *Proteus*)
+	+	+	Sporing and non-sporing anaerobes, staphylococci, see page 240
+ +	−	−	Gram-positive bacteria and Neisseria in patients hypersensitive to penicillin and when bacteria are penicillin resistant Haemophilus, *Mycoplasma pneumoniae*
+ + + + +	− −	− − +	Additional drugs for resistant staphylococci
		± ±	Urinary infections only

§ Inoculum = 10⁵ cocci per ml. approx.
C = bactericidal in concentrations achieved in tissues. } brackets denote variation
S = bacteriostatic in   ,,   ,,   ,,   ,, . with different strains

Vertical blood-agar diffusion method (Fujii, Grossman and Ticknor, 1961, modified)

The vertical diffusion method employing nutrient agar and a staphylococcus has two disadvantages. The staphylococcus does not grow maximally throughout the medium because conditions become progressively more anaerobic from the surface to the bottom of the column of medium and staphylococci dislike anaerobiosis. The zone margin is well defined when caused by streptomycin and the other aminoglycoside antibiotics, but it is impossible to measure the ill-defined zones caused by other antibiotics sufficiently accurately. Both these disadvantages can be overcome by employing horse-blood agar and *Strept. pyogenes*. The zone measured is then the zone of lack of haemolysis, where the streptococcus has been inhibited, which is always well defined no matter what drug is employed, the colonies are disregarded. Moreover, the streptococcus grows as well anaerobically as aerobically.

The method is exactly as described above except that defibrinated horse blood, warmed to 37° C., is added to the molten agar in addition to the inoculum of streptococcus. The diluted culture must be well shaken before addition to the blood agar and the whole mixed well by inversion several times before distribution in tubes. Convenient volumes and concentration are as follows :

Molten nutrient agar at 50° C.   18 ml.
Defibrinated horse blood at 37° C.   2 ml.
1 : 100 dilution overnight glucose broth
culture of *Strept. pyogenes* well shaken   1 ml.

A series of standards is essential for each inoculated bottle of blood agar, therefore when many tests are to be made a larger volume of inoculated medium may be desirable. Unused inoculated tubes may be preserved in the refrigerator (about 4° C) overnight, in which case another set of standards must be set up in parallel with samples tested next day.

REFERENCES

ANNEAR, D. I. (1968). *Med. J. Aust.*, **1,** 444.
Bacteriology Committee A. C. P. Antibiotic Sensitivity Test Trial (1965). *J. clin. Path.*, **18,** 1.
BARBER, M. (1964). *J. gen. Microbiol.*, **35,** 183.

BLOWERS, R., STOKES, E. J. and ABBOTT, J. D. (1973). *Brit. med. J.*, **iii**, 46.

BOVALLIUS, A. and ZACHARIAS, B. (1971). *Appl. Microbiol.*, **22**, 260.

CHABBERT, Y. (1953). *Ann. Inst. Pasteur*, **84**, 545.

CHABBERT, Y. A. (1957). *ibid.*, **93**, 289.

CHABBERT, Y. A. and WATERWORTH, P. M., (1965). *J. clin. Path.*, **18**, 314.

Clinicopathological Conference (1966). *Brit. med. J.*, **i**, 93.

FELMINGHAM, D. and STOKES, E. J. (1972). *J. med. Lab. Technol.*, **29**, 198.

FLOREY, H. W., *et al.* (1951). *Antibiotics*, Oxford Univ. Press, London.

FUJII, R., GROSSMAN, M. and TICKNOR, W. (1961). *Pediatrics*, **28**, 662.

GARROD, L. P., LAMBERT, H. P. and O'GRADY, F. (1973). *Antibiotic and Chemotherapy*, Churchill-Livingstone, Edinburgh and London.

GARROD, L. P. and WATERWORTH, P. M. (1962). *J. clin. Path.*, **15**, 328.

GARROD, L. P. and WATERWORTH, P. M. (1969). *J. clin. Path.*, **22**, 534.

GARROD, L. P. and WATERWORTH, P. M. (1971). *J. clin. Path.*, **24**, 779.

GERONIMUS, L. H. and COHEN, S. (1957). *J. Bact.* **73**, 28.

HANSMAN, D., DEVITT, L. and RILEY, I. (1973). *Brit. med. J.*, **iii**, 405.

HEWITT, J. H., COE, A. W. and PARKER, M. T. (1969). *J. med. Microbiol.*, **2**, 443.

JAWETZ, E., and GUNNISON, J. B. (1950). *J. Lab. clin. Med.*, **35**, 488.

KEAY, A. J. and EDMOND, E. (1966). *Lancet*, **ii**, 1425.

LINE, D. H., POOLE, G. W. and WATERWORTH, P. M. (1970). *Tubercle*, **51**, 76.

MARKS, J. (1964). *Tubercle*, **45**, 388.

MAWER, G., AHMAD, R., DOBBS, S. M. and McGOUGH, J. G. (1974). *Brit. J. clin. Pharm.*, **1**, 45.

Medical Research Council Report, No. 3 (1953). *Lancet*, **ii**, 213.

MITCHISON, D. A., HOLT, H. D., and MOORE, S. H. (1949). *J. clin. Path.*, **2**, 213.

MITCHISON, D. A. and MONK, M. (1955). *J. clin. Path.*, **8**, 229.

MITCHISON, D. A., SELKON, J. B., HAY, D., STEWART, S. M. and WALLACE, A. T. (1958). *Tubercle, Lond.*, **39**, 226.

MITCHISON, D. A., and SPICER, C. C. (1949). *J. gen. Microbiol.*, **3**, 184.

NOONE, P., PATTISON, J. R. and SLACK, R. C. B. (1971). *Lancet*, **i**, 16.

PEARSON, C. H. and WHITEHEAD, J. E. M. (1974). *J. clin. Path.*, **27**, 430.

PHILLIPS, I., WARREN, C. and SMITH, S. E. (1974). *J. clin. Path.*, **27**, 447.

REEVES, D. S. and BYWATER, M. J. (1975). *J. antimic. Chem.*, **1**, 103.

ROBBINS, W. C., and TOMPSETT, R. (1951). *Amer. J. Med.*, **10**, 278.

ROBINSON, G. N. (1967). *Recent Advances in Medical Microbiology*, ed. A. P. WATERSON. Churchill, London.

ROLINSON, G. N. and RUSSELL, E. J. (1971). *Lancet*, **ii**, 745.

RUBBO, S. D., and MORRIS, D. M. (1951). *J. clin. Path.*, **4**, 173.

RYAN, K. J., NEEDHAM, G. M., DUNSMOOR, C. L. and SHERRIS, J. C. (1970). *Appl. Microbiol.*, **20**, 447.

SMITH W., and SMITH, M. M. (1941). *Lancet*, **ii**, 783.

THOMAS, M. and BALDWIN, G. (1971). *J. clin. Path.*, **24**, 320.

WATERWORTH, P. M. (1962). *J. med. Lab. Technol.*, **19**, 163.

8

Clinical immunology

Antigen-antibody reactions are employed both for the identification of microbes and to demonstrate antibodies in human serum to pathogenic organisms. Tests in routine use for the identification of pathogens have been described in Chapter 5. Those commonly made to demonstrate antibody in the patient's serum have now to be considered.

The value of a search for antibodies varies greatly in different clinical conditions. For example, in the diagnosis of syphilis a positive reagin test is very important because the reaction is usually specific, the antibody rapidly disappears in patients completely cured of the disease, and because culture of the spirochaete is impracticable. Even here, however, serological diagnosis is not really satisfactory because possession of treponemal antibody in latent syphilis does not protect a patient from other ills and its presence may lead to a mistaken diagnosis of syphilis as the cause of unexplained disease when in fact the patient has a new growth or a granuloma caused by some other infection.

Serology in pyrexia of unknown origin

The specific agglutinins formed in serum as a result of bacterial infection are of greater value in excluding the infection than in making a positive diagnosis. They are found in serum long after the patient has recovered and also in the serum of a certain proportion of healthy people with no history of severe febrile illness. Moreover, infection by other microbes or severe bleeding may be followed by a rise in titre of antibodies originally formed months or even years previously, the so-called anamnestic reaction. Lack of history of a previous infection does not exclude previous antibody formation because diseases such as typhoid fever, which are usually severe, can occur in a very mild and clinically unrecognizable form.

Agglutinins can almost always be found in the serum at the end of the second week in enteric fever. If a patient has suffered from a febrile illness for two weeks or more and his serum when adequately tested reveals no enteric agglutinins, this type of infection is very unlikely.

Undulant fever is often diagnosed by the presence of antibodies and a typical temperature chart. If the organism is not isolated the diagnosis remains essentially clinical because brucella agglutinins are sometimes found in serum from healthy people and their presence is therefore not diagnostic. Even if several samples of serum are tested and a rising titre is demonstrated, the evidence is still not good enough to stand alone and is of the same kind as the demonstration of acid-fast bacilli in sputum, which is discussed in Chapter 1. If the patient is seen in the acute stage of the disease blood culture should always be attempted. If performed by the method described on page 34 it is almost always successful.

With very few exceptions, which will be mentioned later, significant antibody titres are not found in the serum until infection has persisted for at least ten days. Even if antibodies are found in high titre in the first few days they are most unlikely to have been produced in response to the present infection and it is hard to assess the significance of low titres. Therefore serological tests will not as a rule aid early diagnosis. A positive result during the second week is, however, of much greater value if in the early stages of the illness no antibodies were found, but this does not mean that serum must be tested within the first few days of every undiagnosed fever. The following procedure is satisfactory and labour saving. As soon as possible after a generalized infection has been clinically diagnosed blood is collected for culture. About 5 ml. more than is needed for culture is withdrawn and placed in a sterile screw-capped bottle or plugged tube and allowed to clot. The serum is separated and kept in the refrigerator. If in ten days' time no diagnosis has been made a second sample of blood is taken and the two specimens are tested concurrently. In many cases by the tenth day the diagnosis will have been made, in which case the original sample of serum can be discarded. Apart from the time saved in avoiding unnecessary early tests, this method is more satisfactory because, when titres are to be compared in two samples of serum, the tests should always be made on the same day by the same person, using the same reagents.

When all appropriate tests described in this chapter are negative, the possibility of invasion by fungi, protozoa or helminths should be borne in mind. Enteroviruses can also cause febrile illness lasting for several months. The symptoms caused by them are extremely

variable and fever may be the only sign. There are so many serological types of these viruses that diagnosis by serology in the
absence of isolation is impracticable. Therefore specimens for
virus isolation should be taken at the outset when culture is most
likely to succeed. Faeces and a throat swab in virus transport
medium can be stored at −60° C. while investigations for bacterial
infection proceed. If at the end of a week the diagnosis is still
in doubt and fever persists virus culture is attempted. Isolation
of enterovirus alone is not sufficient evidence of infection because
healthy people often carry them. Diagnosis must therefore be
confirmed by demonstrating at least a four-fold rise in antibody
titre when the isolated virus is tested against the two samples of the
patient's serum already obtained for bacteriological tests.

Cytomegalovirus is another likely cause of fever without localiz-
ing signs, particularly in immunosuppression and in patients who
have received many blood transfusions. In this case the diagnosis
can be made when a marked rise in titre to complement fixing,
commercially obtainable, antigen is demonstrated ; isolation is not
essential.

Reactions of diagnostic value

There are three different types of antibody which are commonly
demonstrated in patients' serum as an aid to diagnosis. They are
agglutinins, complement fixing antibodies and precipitins. Antibody can also be demonstrated by specific binding with microbial
antigen in smears on glass slides. The attached antibody is then
recognized by staining with fluorescent antihuman globulin. In
addition to these a variety of different kinds of antibody such as
antitoxins, antihaemolysins, bacteriolysins, bacteriotropins and anti-
fibrinolysins can be found. These are not, however, sufficiently
constant, specific and easily demonstrable for aid in routine diagnosis ;
tests for them may occasionally be helpful.

There are also empirical antigen-antibody reactions in which the
antigens are not derived from the causal microbe. Examples of
these are reagin tests employing ox-heart extract as antigen, and
the Weil-Felix test for typhus employing three different Proteus
suspensions which react differently, not only with serum from
patients with different kinds of typhus but also with serum from
patients with other rickettsial diseases. Cold agglutinins and
streptococcus M.G. agglutination in mycoplasmal pneumonia is a

further example. In all these diseases the specific microbial antigen can now be successfully employed. Nevertheless, tests for reagin retain their usefulness because unlike more specific tests the titre falls rapidly during successful treatment. The Weil-Felix test can be employed as a screening test, especially in parts of the world where rickettsial diseases are uncommon ; the antigens are easy and inexpensive to prepare and the test is simple to carry out.

Although evidence that EB virus is the cause of glandular fever is now very strong the Paul Bunnell test which reveals the heterophile agglutinin normally present from the first week of the disease is still preferred for diagnosis. EB virus is also associated with other clinical syndromes and the presence of antibody to it is less easy to interpret. Tests for EB virus antibody in glandular fever-like disease with negative Paul Bunnell test are also negative (Joint Investigation, 1971).

Reliability of standard techniques

Although combination between antigen and specific antibody has been studied by many workers for a number of years many phenomena associated with the reactions are still not well understood. The tests which have been evolved for use in diagnosis are the result of much detailed investigation. It is impossible here to discuss fully the reason for each step as the tests are described, and in fact the only reason for many procedures is that they have been found to give reliable results in a large series of cases. It must be stressed, however, that if reliable results are to be obtained each step must be slavishly followed. Minor departures from the original technique may not apparently make any difference, but no difference is likely to be noticed unless a large number of sera (proved positive and negative by other means) are tested by the original and the modified method in parallel and this is necessary whenever a modification is introduced. The factors which are known to influence the reactions are : the proportion of antigen and antibody in the mixture ; the time and temperature of the reaction ; the rate at which the reagents are mixed (which includes the volume of each of the fluids to be mixed and the size and shape of the vessel in which they are mixed), the presence and concentration of electrolytes and the pH of the diluting fluids.

All glassware used for serology must be specially cleaned apart from the rest of the laboratory glassware. Saline solutions are made with pure ("Analar") sodium chloride in water twice distilled in a glass still. Buffer solutions are often recommended and can be obtained in concentrated form for easy preparation. The laboratory must ensure that the distilled water in which they are diluted is of very high quality. Bottles of distilled water or saline prepared in the hospital pharmacy for therapeutic use are sometimes employed. This is undesirable because pyrogen-free solutions for parenteral use in patients may be insufficiently pure for serological tests.

Instructions sent with commercially obtained antigens should be read with care and the technique recommended should be followed or there must be a good reason for any departure from it. The antigen may not perform reliably when the conditions of the test are altered.

Specific agglutination tests

The Widal test for typhoid and paratyphoid agglutinins is usually extended in Britain to cover antibodies to other *Salmonellae* and to the Brucella group, because these infections are common and can easily be mistaken for enteric fever. The typhoid and paratyphoid organisms possess both flagellar and somatic antigens and it is necessary to test for both types of corresponding antibodies. This means that in each test the serum is to be titrated against ten antigen suspensions, of which at most two are likely to agglutinate. For convenience a preliminary test is made. Two serum dilutions only are tested against each of the *Salmonella* suspensions and four against *Brucella*. One dilution is insufficient, even for a preliminary test, because of the prozone phenomenon which is shown by some sera containing a large amount of antibody ; in low dilution they fail to react, agglutination apparently being inhibited by excess of antibody, and in successively higher dilutions the reaction gradually becomes positive. If two dilutions are tested, the low one will be positive with low titre sera and the high one will be positive with sera showing the prozone phenomenon.

Br. abortus is the commonest cause of undulant fever in Britain. In the second week of an acute attack the patient's serum will agglutinate *Brucella*. It should be tested against both *Br. abortus* and *Br. melitensis* suspensions because these species have antigens in common. Moreover, some biotypes of *Br. abortus* give rise to antibody which, when tested against standard suspensions, agglutinate *Br. meliiensis* to a higher titre than the typical standard *Br. abortus* suspension (Kerr *et al.*, 1966). A prozone is often seen in brucella agglutination and may extend as far as a 1 : 400 dilution, hence the need for four dilutions (see below) in the preliminary test. Paratyphoid A and C are not endemic in Britain and tests for them may be omitted if the patient is known not to have been abroad.

METHOD

The separated serum is not inactivated since heating to 56° C. will destroy some agglutinating antibodies. It is diluted 1 : 10 and 1 : 100 in normal saline for the screening test for salmonella antibodies, and two further dilutions 1 : 200 and 1 : 400 are made for the brucella antigen tubes. No antigen controls are included in the screening test.

" H " *Agglutination.* Suspensions are kept refrigerated (4° C. approx.) diluted in saline ready for use. The serum dilutions are measured, one volume of each dilution for each antigen to be tested into Dreyer's tubes ; 0·2 ml. is convenient. An equal volume of the appropriate suspension is added to each and the tubes are placed in the 50° C. water bath. The level of water should extend to about two-thirds of the column of fluid in the tubes to ensure mixing by convection. Results are read at 2 hours and again after 4 hours or overnight incubation.

" O " *Salmonella and Brucella Agglutination.* The 1 : 10 and 1 : 100 dilutions are distributed into (50 × 12 mm.) tubes ; 0·25 ml. volumes are convenient. An equal volume of suspension in 0·4 per cent phenol in normal saline is added to each tube. The suspension is freshly diluted 1 : 15 from concentrated antigen. The tubes are shaken by hand to mix the reagents and are then incubated in a 37° C. water bath and read after 4 hours and again after overnight incubation (see Table 31).

Any positive sera are titrated by further twofold dilution, and an antigen control tube without serum is included. When the 1 : 20 final dilution tube only is positive the titration need only extend from 1 : 20 to 1 : 160. When the 1 : 200 dilution is positive titration extending from 1 : 200 to 1 : 6,400 is usually sufficient to include the titre of salmonella agglutination but brucella agglutination sometimes exceeds this.

It must be remembered that positive results in the screening test may only mean an unstable suspension. A report cannot be sent until results of the final titration including the control are known.

Technical note :   When delivering from a graduated pipette into narrow tubes, slope the pipette so that air is not prevented from escaping and the fluid runs in without bubbles.   When using a dropping method hold the pipette vertically to ensure a constant volume per drop.   Pipettes for delivering stock suspensions should be sterile to avoid contamination of the suspension by moulds.

TABLE 31

Extended Widal agglutination test.   (Diagram of tubes set up for screening test.)

	Salm. typhi	Salm. paratyphi A	Salm. paratyphi B	Salm. paratyphi C	Salmonella group	
'H' antigen 50°C. 2 hours	O O	O O	O O	O O	O O	
					Brucella	
'O' antigen 37°C. 4 hours	O O	O O	O O	O O	O O	O O *Abortus*
					O O	O O *melitensis*
Final serum dilution	1:20 1:200	1:20 1:200	1:20 1:200	1:20 1:200	1:20 1:200 1:400 1:800	

READING RESULTS.   Examine the tubes in a bright light against a dark background using a hand lens.   Note the degree of agglutination ;   it is judged not only by the deposit but also by clearing of the supernatant fluid.   The following degrees are recognized.

1. *Total.*   Agglutinated particles have sedimented to the bottom of the tube.   The supernatant fluid is clear.

2. *Standard.*   Clumps of agglutinated bacteria can be seen naked-eye but the supernatant fluid is still faintly cloudy.   Examination with a hand lens reveals fine agglutinated particles in the supernatant fluid.

3. *Trace.*   Agglutination is partial and can only be seen with the aid of a hand lens.

Note.   Clumps at the bottom of the tube without clearing or signs of agglutination in the supernatant fluid indicate either an unsatisfactory and possibly contaminated suspension or dirty tubes. The test must be repeated.

When no agglutination is found a negative report is sent.   In order that the work done shall be of value to future investigators the result is reported in detail thus :

Agglutination tests negative.
The serum gave the following titres :

	Antigen	
	" H "	" O "
Salm. typhi	<20	<20
Salm. paratyphi A	<20	<20
Salm. paratyphi B	<20	<20
Salm. paratyphi C	<20	<20
Salmonella group	<20	
Br. abortus	negative in all	
Br. melitensis	dilutions tested	

The laboratory record card will also be available for reference if necessary. The results are tabulated on the record card so that the range of dilutions examined and the result in each tube is recorded. This may be important for comparison of subsequent tests. A positive report might read as follows :

Agglutination test positive for *Salm. typhi.*
The serum gave the following titres :

	Antigen	
	" H "	" O "
Salm. typhi	1600	200
Salm. paratyphi A	<20	<20
Salm. paratyphi B	<20	<20
Salm. paratyphi C	<20	<20
Salmonella group	40	
Br. abortus	negative in all	
Br. melitensis	dilutions tested	

Antibodies in enteric fever

Salmonellae causing enteric fever stimulate the formation of three types of agglutinin which are of use in diagnosis; they are H, O and Vi. The H antibody, which is produced in response to stimulation by flagellar antigens, appears towards the end of the first week of the disease, and usually reaches the highest titre. It persists longer than the others after recovery, sometimes for many years, and its formation may again be stimulated non-specifically in subsequent febrile illnesses. Agglutination is rapid, it can usually be seen after two hours, incubation, and the agglutinated bacilli form large fluffy clumps. There is no evidence that H antibody is protective or helps to combat the disease.

Formation of O antibody is stimulated by O somatic antigens.

It also appears in the first week of illness and seldom rises above a titre of about 640. After recovery the titre falls and it is seldom demonstrable a year later. Production is not easily stimulated non-specifically in subsequent illness. Agglutination occurs more slowly, after four hours' incubation, and the clumps of bacilli are small and dense.

The Vi antigen, also somatic, is possessed by *Salm. typhi*, *Salm. paratyphi C* and other coliforms including certain strains of *Esch. coli*. Typhoid and paratyphoid bacilli which possess large quantities of this antigen often give rise to severe disease. Occasionally Vi antibody only can be demonstrated, such cases are rare, however, and there is no need to perform the Vi agglutination test in every case. The antibody is probably of value in combating infection by Vi strains.

In the majority of patients H and O agglutinins can be easily demonstrated and sometimes Vi agglutinins also. Occasionally one only of the three types of agglutinin can be found, O and Vi appear alone more commonly than H. Very occasionally in enteric fever diagnosed by blood culture no specific antibody of any kind can be demonstrated.

Significance of results

It is impossible to lay down definite rules for the significance of various titres. The agglutination results must be judged in the light of all other findings. Their significance depends on (*a*) the duration of the illness at the time the specimen was taken, (*b*) evidence of previous infection or prophylactic inoculation, (*c*) the level of titres found in the sera of healthy people in the population of which the patient is a member.

In Britain titres to the enteric group in healthy people are low, and at the end of the first week an H agglutinin titre of 1 : 80 in an uninoculated patient who has had no previous salmonella infection may be significant. An O agglutinin titre of 1 : 100 or more is good evidence of active infection, especially if it has been shown to rise to this level since the beginning of the disease. This finding has some significance even in inoculated and previously infected patients because O unlike H agglutinins are not maintained in the serum for long periods, neither are they often increased non-specifically. In general, diagnosis of enteric fever by agglutination tests alone is unsatisfactory and in previously infected patients and those who have received prophylactic T.A.B. inoculation, it is often impossible.

Agglutination tests for *Br. abortus* are more easily interpreted and therefore of more value in diagnosis. Although humans are not commonly inoculated against brucellosis, a kind of accidental inoculation often occurs in people whose way of life exposes them to the risk of infection, such as those who drink large quantities of raw milk and those in contact with infected animals or carcases. Provided such people are excluded, a titre of 1 : 80 or more in the second week of the disease can fairly safely be taken as evidence of infection. Later the titre rises much higher but at the end of three months or so no reaction may be seen. In chronic brucellosis the standard agglutination test is often negative, but antibody may be demonstrable by the antiglobulin agglutination (Coomb's) test (Wilson and Merrifield, 1951) or by a complement fixation test employing the agglutination suspension as antigen. Since symptoms are variable these additional tests should be done on serum from patients suffering from chronic undiagnosed ill health who have been exposed to brucella either by drinking raw milk or by the nature of their work (Kerr *et al.*, 1966).

Other causes of enteric-like fever

When the extended Widal test is negative in the second week of an enteric-like infection, other agglutination tests may indicate the cause. The Vi agglutination test may be positive, although by this time the infection will probably have been diagnosed by blood or faeces culture. The Weil-Felix test may indicate previously unsuspected typhus. Positive agglutination with leptospiral suspensions may enable a diagnosis of Weil's disease or canicola fever to be made. Infections which must be considered and which are diagnosed by other means, are meningococcal and other subacute bacteraemia, rickettsial diseases other than typhus, infective hepatitis, glandular fever and toxoplasmosis. In any undiagnosed febrile illness tuberculosis must be excluded ; it is sometimes very difficult to diagnose in the first few weeks. Finally, although very rare in Britain, generalized fungous infections such as histoplasmosis and coccidioidosis must be considered (see Chapter 6). Protozoa other than toxoplasma must be borne in mind, particularly in patients who have lived abroad within the last two years, but investigation for them is outside the scope of this book.

The Vi agglutination test is seldom required in Britain and is not easy to perform. Serum sent to about ninety laboratories to test reliability of technique revealed such variable results that it seems more reasonable to perform this test only in specialist laboratories. The technique has therefore been deleted from this edition.

The Weil–Felix Test (Felix, 1944)

METHOD

Using tubes of 13 mm. external diameter make twofold dilutions of serum in saline, each 1 ml. in volume, to cover a range between 1 : 20 and 1 : 640. Add to each tube 1 drop of concentrated Proteus XO suspension and also 1 drop of suspension to 1 ml. of saline which is the control. Incubate in a water bath for 2 hours at 37° C. and read. Stand at room temperature (in an ice chest in warm weather) and read again after 22 hours. The reading is made as described for the Widal test. Fresh sera sometimes show inhibition in low dilutions and should be heated at 45° C, for half an hour before testing. A titre of 1 : 80 (total agglutination) during the first week signifies infection in an uninoculated patient. A titre of 1 : 200 is evidence of infection even in a vaccinated patient. Agglutinins appear early, from the third or fourth day of the illness. When they are present it may be possible to demonstrate a rise of titre even after 24 hours.

Three series of dilutions are made to be tested against the three O suspensions. *Proteus* X19 agglutinates best with serum from cases of classical epidemic typhus and Brill's disease. *Proteus* X2 agglutinates with serum from cases of murine typhus and *Proteus* XK with serum from cases of scrub typhus. Non-specific titres up to 1 : 40 are common with all three suspensions but particularly with *Proteus* XK. Occasionally high titres are found after *Proteus* X infection, the antibody having been stimulated by the *Proteus* and not by a *Rickettsia*. Specific agglutination tests can now also be made with rickettsial suspensions. They are not, however, available for use in hospital laboratories in Britain.

A positive reaction with all three proteus suspensions is found in spotted fever caused by tick-borne rickettsiae which bears different names in different parts of the world, i.e. Rocky Mountain spotted fever, Kenya fever, etc.

Complement fixation tests

The main difference between complement fixation tests and other antigen-antibody reactions is that whereas the reaction between antigen and antibody in agglutination and precipitation tests is easily visible, the mixture of microbial antigen, antibody and complement in fixation tests appears to the naked eye no different after the reaction has taken place. (Bacteriolysis occurs when the antigen is composed of whole bacterial cells, but it is not easily demonstrated.) Therefore an easily visible indicator is necessary to reveal positive and negative reactions. The most convenient indicator is the lysis of sheep cells by complement-fixing antibody prepared against them in rabbit serum. Each complete test therefore involves two antigen-antibody systems. The first between microbial antigen, patients serum (with or without antibody) and complement,

TABLE 32.

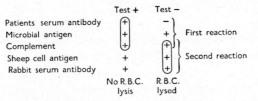

the second between sheep red cell antigen, rabbit serum antibody and complement, if still present.

The amount of complement used is carefully adjusted so that it is impossible in any one test for both reactions to occur completely. If antibody is present in the patients serum all the complement will be used in the first reaction leaving none available for red cell lysis by the rabbit serum haemolysin, and so the positive result in the first system will be demonstrated by lack of reaction in the second. Complement is present in varying amounts in all fresh serum. The source of complement for the test is guinea-pig serum which is particularly rich in it. The other sera, from patients and rabbit, are heated to destroy the unknown amounts of complement in them.

Titres of complement are expressed in terms of the minimal haemolytic dose (M.H.D.). One M.H.D. of complement is present in the highest dilution of guinea-pig serum which will lyse red cells under standard conditions. Serum haemolysin titres are similarly expressed.

Complement fixation tests are employed to aid the diagnosis of infections caused by all varieties of microbe, from viruses to protozoa and fungi ; they can also give evidence of helminth infestation.

Anticomplementary sera: The dose of complement employed with a particular antigen is standardized so that results obtained in different laboratories can be compared, but the amount for optimal results in different diagnostic tests varies. The less complement employed the more sensitive the test but the greater the risk of non-specific fixation. When specific complement fixing antibody titres are high the serum can be diluted and only large amounts of non-specific anticomplementary substances in the serum will then affect the test. When, in addition, anticomplementary activity of the antigen is low a small dose of complement can safely be employed. On the other hand, when antibody titres are low and the antigen alone destroys complement to some extent, a larger dose of complement is needed if false positive results are to be avoided.

When a serum is submitted to several different complement fixation tests it may be reported " anticomplementary " to only one of them. This happens in the test which needs the most concentrated serum or the lowest dose of complement and means that the serum contained a small amount of anticomplementary substance, insufficient to interfere with all the tests.

Preparation of haemolytic serum

Wash fresh sheep's red blood cells three times by adding saline to the deposited cells of centrifuged sheep's blood, mixing well and re-centrifuging. Centrifuge for the last time at 3000 r.p.m. for 15 minutes to obtain a very concentrated mass of cells. Inoculate an adult rabbit intravenously with the washed cells freshly prepared on each occasion, as follows : Inject 0·5 ml., then three days later give 1 ml. and repeat the 1 ml. dose on the seventh day. Ten days after the first dose, take a sample of blood from the rabbit and estimate the haemolysin titre. If it is satisfactory, that is 1 : 1000 or more, take a large sample of blood (40 to 60 ml. can be obtained from the ear) on the fourteenth day. If the titre is not satisfactory give another dose, 1 ml., on the same (tenth) day. Test the serum again on the fourteenth day and bleed immediately if the titre has risen. These injections of sheep red cells stimulate the formation of two antibodies in the rabbit serum which can be demonstrated by incubating mixtures of cells and serum at 37° C.

Sheep cells (washed) + rabbit serum (heated) ——→ agglutination
Sheep cells (washed) + rabbit serum + complement —→ haemolysis
(fresh guinea
pig serum)

The doses of sheep cells are injected within a short time to give a high haemolysin titre and a low agglutinin titre. In the complement fixation test the agglutinin can then be avoided by using the serum diluted beyond the agglutinating titre.

Haemolysin titration (see also page 282)

Guinea-pig serum possesses normal haemolysins for sheep cells which must be removed by absorption before the serum can be used as a source of complement in this test.

Method

Mix equal volumes of guinea-pig serum and packed washed sheep red cells. Refrigerate overnight and then centrifuge to get rid of the unlysed sheep cells. Make a series of dilutions in saline of the haemolytic serum to be tested 1 : 250, 1 : 500, 1 : 1000, 1 : 2000 and 1 : 4000. Deliver 0·5 ml. of each dilution into a small test-tube, add 0·5 ml. of about a 1 : 10 saline dilution of the treated guinea-pig serum. The dilution of guinea-pig serum does not matter as long as there is excess complement in each tube; serum already titrated for its complement content (see page 279) is used and allowance is made for a fall in titre during the treatment with red cells. Dilutions stronger than 1 : 10 tend to give equivocal results. The final dilution of the haemolytic serum in the tubes is now 1 : 500 to 1 : 8000. Add to each tube 1 ml. of a 2 per cent suspension of washed sheep cells in saline. The following controls are required:

Control 1. Saline (to replace haemolytic serum)⎫
             Complement   ⎬No lysis
             Red cell suspension   ⎭

Control 2. Saline (to replace complement)   ⎫
             Haemolytic serum (strongest dilution) ⎬No lysis
             Red cell suspension   ⎭

Control 3. Positive haemolytic serum (5 M.H.D.)⎫
             Complement   ⎬Lysis
             Red cell suspension   ⎭

Control 1 checks the efficiency of the removal of natural haemolysins from the guinea-pig serum used as a source of complement.

Control 2 checks the stability of the red cell suspension. If the 1 : 500 dilution of haemolytic serum contains agglutinins, the red cells will agglutinate. A serum with such a high agglutinin titre is useless unless the haemolysin titre proves to be about five times higher.

Control 3 checks that the complement is active.

Shake the tubes well and incubate in the 37° C. water bath for 1 hour. The titre is described as a minimal haemolytic dose (M.H.D.), which is the least amount of the serum which will lyse red cells under the standard conditions. If the titre is 1 : 1000 (1 M.H.D. = 0·001) or higher, the serum is suitable for use and as much blood as possible is obtained from the rabbit in the following manner :

To bleed a rabbit. Wrap the animal firmly in a cloth with its back flat and its head exposed, or place it in a special wooden box. Wash the margin of the ear which shows the largest marginal vein with soap and water and shave the fur from both sides of it in the neighbourhood of the vein. Dry the ear. Using a sterile swab apply sterile soft paraffin over the shaven area. Compress the vein at the base of the ear and with a sterile cutting needle make a longitudinal slit (about 3 mm.) in it through the paraffin at any convenient place about half-way between the tip and the base of the ear. Allow the blood to drop into a sterile bottle or tube. The flow can be increased by applying a little xylol to the middle of the outer surface of the ear. It causes slight irritation and dilatation of the blood vessels. If the ear is washed free of it after bleeding no damage to the skin ensues. In this way 40 to 60 ml. of blood can be obtained from a large rabbit. The animal appears none the worse and after a suitable lapse of time (about a month) it can be used for a test other than antibody preparation. The haemolysin titre of its serum will have fallen and it is unlikely to be of use for a second bleeding.

Complement titration

The performance of any complement fixation test is very simple once the dose of complement has been decided. When fresh guinea-pig serum is employed as a source of complement, titration on the day of the test is mandatory, it is also desirable when preserved complement is employed but may not be essential when small batches of sera are tested daily. Positive and negative sera must always be titrated in parallel with every batch of tests and should the dose of complement be insufficient (activity will not increase) this will be evident and the positive tests will have to be repeated. Such an event is not worth risking when large batches of serum are to be tested and the results of the tests are anxiously awaited.

Therefore complement titration on the day of the tests is the rule.

The method to be described is suitable for the majority of tests but note must be taken of any complement titration method recommended for use with a particular antigen. The dose of complement varies with the anticomplementary properties of the antigen and will have been calculated according to the recommended titration method.

MATERIALS

Guinea-pig serum (complement) fresh or preserved. Note that preserved complement must be diluted in high quality (double glass distilled) water which will be needed to make a 1 : 10 isotonic dilution.

Antigen, reconstituted according to the maker's instructions. Titration may be necessary, see page 282.

Barbiturate buffered saline prepared for complement fixation tests obtainable commercially.[1]

Previously tested positive and negative sera. They will have been heated at 56° C. for 30 minutes when originally tested and must be reheated for 10 minutes on the day of the test to destroy anticomplementary substances which may have developed during storage, see also page 284.

Sheep red cells thrice washed in buffered saline.

Anti-sheep-cell rabbit serum.

Graduated 1 ml. pipettes (to expel) labelled and kept exclusively for each reagent.

Graduated 10 ml. pipette (to expel) for buffered saline (additional to 1 ml. pipette).

Scrupulously clean glass ware, which must be detergent-free, to hold reagents. Small conical flasks are suitable for complement and antigen, beakers for buffered saline, 100 ml. for " clean saline " and 200 ml. for " dirty saline " which will be used to wash the serum pipette.

Tubes about 10 mm. × 75 mm., either specially cleaned glass or disposable plastic, or WHO plastic trays with 1·5 ml. wells or microtitre trays with loop diluters.

METHOD

Preparation of sensitized sheep cells

The percentage of red cells required may vary with different tests but 2 per cent usually gives results which are easy to read. To make 60 ml. of sensitized suspension proceed as follows : To 30 ml. saline in a 50-ml. measuring cylinder add 10 M.H.D. per ml. haemolytic rabbit serum (the final dilution will be 5 M.H.D.) and mix well. To 30 ml. saline in a 100-ml. measuring cylinder add about 1·3 ml. of thrice washed packed sheep cells, mix well and test the concentration by centrifuging a sample of the suspension

---

[1] From Mercia Diagnostics Ltd., Watford, Herts, England.

in a haematocrit tube for 10 minutes at 3000 r.p.m.   It should be 4 per cent (final dilution 2 per cent).   Adjust the suspension by adding more cells or saline if necessary.   Pour the 30 ml. of haemolytic serum dilution into the red cell suspension and mix well.   Close the top of the cylinder with a clean bung and stand in the 37° C. water bath for 30 minutes. During this time the hae-molysin is adsorbed on to the surface of the cells and they will then lyse rapidly in the presence of complement.   To ensure good sensitization it is essential to add cell suspension and haemolytic serum dilution to each other in equal volumes as described.

Complement titration method

Complement should be titrated in tubes or trays of the same kind that will be used in the test proper ; the standard volume should also be the same.

Set out tubes as in Table 33.   Four test rows and a row for dilution of complement are required.   To ensure accurate delivery to the bottom of the tubes dispense all reagents individually into each tube not by running in from a full pipette.   Accuracy from tube to tube is very important but the precise amount of the standard volume does not matter.   To ensure no loss of complement on the sides of tubes always dispense serum dilutions or saline first, then complement and finally antigen.   Follow this order also in the test proper.

Make the complement dilutions immediately before they are to be added from a 1 : 10 isotonic dilution and immediately return this master dilution to the refrigerator (4° C.) because complement is unstable at room temperature.   Mix the dilutions in row 5 very thoroughly by tapping each tube 10 times (when trays are employed the complement dilutions will be made in tubes).   Starting at the weakest dilution transfer 1 volume to each of the appropriate tubes ; continue thus until all tubes have received one volume of their appropriate dilution of complement. Add 1 volume antigen imme-diately and shake well.   Incubate at 37° C. for 1 hour.   Tubes should be placed in a water bath with the lid off, for fear of con-tamination with condensation water, and trays should be covered before they are placed in an incubator.

Add sensitized red cells, either 1 or 2 volumes according to the amount recommended in the test, shake well.   Incubate at 37° C. for a further 30 minutes and read.

When the activity of the complement is in doubt, for example fresh guinea-pig serum may have a titre less than 1 : 30 which is too low, set up row 1 in duplicate and add the sensitized cells to the tubes in the additional row before the first period of incubation. Lysis will be seen after 10 minutes' incubation in this row when

TABLE 33.

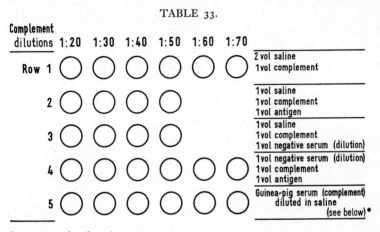

Complement dilutions	1:20	1:30	1:40	1:50	1:60	1:70	
Row 1	○	○	○	○	○	○	2 vol saline 1 vol complement
2	○	○	○	○			1 vol saline 1 vol complement 1 vol antigen
3	○	○	○	○			1 vol saline 1 vol complement 1 vol negative serum (dilution)
4	○	○	○	○	○	○	1 vol negative serum (dilution) 1 vol complement 1 vol antigen
5	○	○	○	○	○	○	Guinea-pig serum (complement) diluted in saline (see below)*

* Contents of tubes in row 5.

Tube 1. 3 volumes saline, 3 volumes 1 : 10 complement
     2. 6   ,,   ,,   3   ,,   ,,
     3. 6   ,,   ,,   2   ,,   ,,
     4. 4   ,,   ,,   1   ,,   ,,
     5. 5   ,,   ,,   1   ,,   ,,
     6. 6   ,,   ,,   1   ,,   ,,

Key
   Complement = guinea-pig serum diluted and used as a source of complement.

the complement is active and when it is inactive no unnecessary time will be lost in obtaining a fresh supply.

To read observe the tubes in a bright light. One minimal hae-molytic dose (1MHD) is contained in the highest dilution in row 4 which shows complete lysis.

The test proper

Arrangement of the work will depend on the number of sera to be tested, the urgency of the results, whether positive sera are to be titrated and how many are expected to be positive and also on the cost of the antigen. The three possibilities are shown overleaf.

(*a*) All sera can be titrated in full, see Table 34.   This is appropriate when the antigen is inexpensive, the number of sera to be tested few, the proportion of expected positive results is high and the results are urgently needed.

(*b*) Tubes 1 and 2 in the Table can be tested.   This has the advantage of being economical of antigen, a preliminary report can be sent without delay and positive sera only can be titrated later.

(*c*) Tube 2 in the table can be set up alone as a screening test. Tubes showing complete lysis can be reported negative.   Intact cells mean either that the serum contains specific antibody or it contains a complement-destroying substance.   Therefore the saline control, tube 1, must be tested before a positive report can be sent to confirm that no anticomplementary substance is present ; the full titration can be completed at the same time on these sera. When many samples are to be tested, few positives are expected and if the result is not urgent this is the most labour-saving method.

The dilution of patients' serum recommended for different tests is variable.   It may be necessary to test undiluted serum when the expected antibody titre is low as in the gonococcal complement fixation test.   A dilution of 1 : 4 or 1 : 5 is more usual, for example, in the Reiter protein test.   Small quantities of anticomplementary substances in the serum are less likely to interfere with the test when the serum can be diluted.

TABLE 34.

Tube		1	2	3	4	5	6
Patient's serum[*]		1:5	1:5	1:10	1:20	1:40	1:80
Dilution	1vol/tube	◯	◯	◯	◯	◯	◯
Complement	vol	1	1	1	1	1	1
Antigen	vol		1	1	1	1	1
Saline	vol	1					

[*] The initial dilution depends on the particular antibody investigated.

While the complement titration is incubating dilute and dispense each serum to be tested into tubes or wells (an electronic diluter described on page 287 greatly facilitates this procedure). Wash the serum pipette twice in distilled water and then twice in saline discharging the washings and then suck up and down thrice in saline. This must be done between each serum and a rubber teat is essential. Serum must not be mouth pipetted because of the danger of hepatitis (see Chapter 9).

When the complement titration has been read, dilute the original 1 : 10 dilution of complement further so that one dose will be contained in 1 volume. For example when the 1/60 dilution in the

TABLE 35.

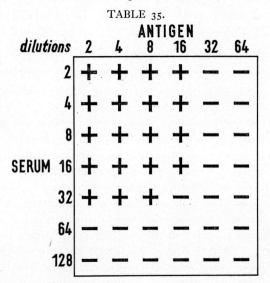

titration is the weakest to show complete lysis in row 4 of the titration 1MHD is contained in 1 volume of this dilution and if the dose recommended is 3MHD a 1 : 20 dilution of guinea-pig serum will be needed for the test.

Mix the diluted complement well and either add one volume to each tube immediately or return it to the refrigerator. Complement must not be left on the bench, its titre will fall.

When all tubes have received complement add immediately 1 volume of antigen to each appropriate tube and 1 volume saline to the saline controls. Shake well and incubate at 37° C. for 1 hour in the manner described for the complement titration.

Further standardization of reagents needed for some complement-fixation tests

i. Antigen titration

Commercially obtained antigens have usually been titrated by the firm supplying them but it is desirable to check that the stated titre is optimal under the local test conditions. This is done by testing serial dilutions of antigen against serial dilutions of a known positive serum in chess board fashion. The dilutions employed depend on the stated antigen titre and the titre of the positive serum and should be near the middle of the range of each respectively. The optimum dilution of antigen for use in the test is the weakest which shows the maximum titre of the serum. The example in Table 35 indicates that the antigen should be diluted 1 : 8.

ii. Chess board titration of haemolytic serum and complement

Complement fixation tests employing costly viral antigens are done in microtitre trays using very small volumes. Chess board titration of antigen described above should always be performed. The dose of haemolytic serum is also critical and its activity must be checked weekly. Complement dilutions are tested against haemolytic serum. An equal volume of 4 per cent sheep cell suspension is added to each dilution of haemolytic serum to make a series containing 2 per cent cells sensitized with increasing dilutions of haemolytic serum. After stabilization by incubating for 30 minutes at 37° C. these are added to the complement dilutions set out in a microtitre tray. The optimum dilution of haemolytic serum is the weakest dilution which shows complete lysis at the weakest dilution of complement and is expressed as 1 unit of haemolysis. Details of the method will be found in text books of virology.

Serological tests for the diagnosis of syphilis

Three kinds of antibody are present during the course of the disease. The first to appear and the last to disappear after successful treatment is the highly specific antibody to *Treponema pallidum* which is recognized in hospital laboratories by the fluorescent treponemal antibody (absorbed) test (FTA-ABS) and also in reference laboratories by the potentially dangerous and expensive treponemal immobilization test (TPI) in which living treponemes maintained by passage in rabbits are employed. A third test for this antibody has now been devised and is gaining popularity as a screening test in hospital laboratories ; this is the *Treponema pallidum* haemagglutination test (TPHA). Sheep, chick or turkey red blood cells are sensitized by attaching killed *Treponema pallidum* to them, they will then agglutinate in the presence of antibody. The test is much easier to perform and to read than the FTA-ABS test and no special microscopic equipment is required. It is less sensitive than the fluorescent test and therefore is not as reliable for the diagnosis of primary syphilis. Occasional false positive results are encountered and the diagnosis should always be confirmed by the FTA-ABS test.

Antibody detected by the Reiter protein complement fixation test (RPCFT) is a group treponemal antibody. It appears later than the specific antibody and some syphilitic patients do not develop it.

The detection of specific antibody is mandatory for the diagnosis of the disease but the titre does not alter very much during treatment. A positive test does not necessarily imply that further treatment is required neither does a continuing positive test imply that treatment already implemented is ineffective, indeed this is to be expected.

The Wassermann reaction, which has in the past been the most widely practised test for diagnosis, is a complement fixation test employing non-specific antigen derived from animal tissue. It recognizes reagin, an antibody produced in the active stages of the disease which is much more commonly found in syphilis than in other infections but is not specific for this disease. Sera from some patients suffering from leprosy, exanthemata or having recently undergone primary vaccination against small pox, sometimes give positive reactions as do occasional sera from pregnant women and some other healthy people. These biological false postives can be distinguished because they are usually transitory and also because tests for the specific antibody are negative. A quantitative test for reagin is needed to monitor progress since the titre of this antibody falls rapidly during successful treatment. Indeed in the past when the Wassermann reaction alone was relied on for the diagnosis there was a danger that treatment with penicillin for other infection might obliterate reagin before the diagnosis of syphilis could be made. This led to a late diagnosis when the disease relapsed after inadequate treatment and reagin reappeared.

There are now a large number of tests for reagin. The original Wassermann reaction and its modifications are less favoured than previously partly because it does not lend itself to automation but also because other more easily performed and more sensitive tests are now available, one of the most popular being the venereal diseases reference laboratory (VDRL) test. Readers are referred to the Public Health Laboratory Service monograph number 1 (1972) for an account of the performance of this test, the Wassermann reaction and the RPCFT.

Although there is an increasing tendency for all venereal disease serology to be performed in reference laboratories by automated methods, hospitals with large out-patient departments of genito-urinary medicine usually prefer the hospital diagnostic laboratory to undertake these tests and indeed discussion between clinician and laboratory is as rewarding here as in other fields. The automated reagin test (ART), which is a modification of the VDRL test, is ideal for screening blood for transfusion but even large diagnostic departments handle insufficient blood samples to justify the purchase of the costly equipment required ; mechanization using an electronic diluter which can easily be employed for other purposes is more satisfactory, much less expensive and no special maintenance is needed.

The tests which have so far proved most satisfactory for diagnosis and monitoring treatment are the rapid plasma reagin (RPR) card test, and the *Treponema pallidum* haemagglutination test (TPHA). As a check on positive results and for diagnosis of primary infection the absorbed fluorescent treponemal antibody test (FTA-ABS) is also needed. In each of these tests known positive and negative sera must be tested in parallel.

Cerebro-spinal fluid can be tested by the TPHA and FTA-ABS methods but not by RPR card test.

METHODS

Control sera. Sera, either positive or negative to all these tests, which are clear and non-fatty are pooled and then distributed in aliquots sufficient for one batch of tests. These are frozen and one aliquot of each is thawed for use on the day of the test when it is heated again at 56° C. for 10 minutes. Both a weak and a strong positive control serum is needed for the FTA-ABS test. Each new pool of positive serum is tested in parallel with the previous pool.

i. *Rapid plasma reagin (RPR) card test*

This test is a flocculation test employing a modified VDRL antigen containing carbon particles. It was originally developed in 1962 for work in the field and it can be done on either plasma or unheated serum without special laboratory equipment. It was modified by Portnoy, one of the original inventors, in 1963 for large-scale screening in the laboratory and a quantitative test was developed. The advantages of the test for laboratory use are that it can be done in small batches as the sera arrive, without inactivation, as a screening test. In the quantitative test the end point can be read without a microscope. No glassware is employed and if it seems desirable a permanent record of the result can be made by allowing the card to dry and spray varnishing it so that the flocculated carbon particles adhere. It is at least as sensitive and specific as other reagin tests (Scrimgeour and Roslin, 1973).

MATERIALS

Qualitative test kit no. 110 (catalogue number 75005).
Quantitative test kit no. 112 (catalogue number 75239) manufactured by Hyson, Westcott and Dunning Inc., U.S.A.; obtainable in Britain through Becton Dickinson UK Ltd., Empire Way, Mddx HAQ OP5.
Mechanical rotator to rotate at 100 r.p.m. 1·9 cm. circle.
Unheated patients' sera.
Positive and negative serum controls. These will be inactivated frozen aliquots, see above, the titre will be a little reduced by inactivation but will remain stable for several months.
All reagents must be allowed to reach room temperature, not be tested cold direct from the refrigerator.

METHOD

Instructions are included in each test kit. The serum is dispensed with the plastic tubes provided which resemble short plastic drinking straws with a flattened end. When squeezed between thumb and finger sufficient serum can be drawn up and transferred to the appropriate, pre-numbered circle on the card; it is distributed over the area of the circle using the flattened end of the tube which is then discarded. A drop of antigen is added from the container and delivery needle provided and the card is then placed on the rotator for 8 minutes which mixes antigen and serum. The serum and antigen must not be mixed before rotation. The tests are then read in the wet state. In practice about 40 tests can be done at a time—there are 10 circles on each card—attempts to do more than this will lead to drying during rotation.   While 4 cards are rotating another 40 tests can be prepared.   Covers for the cards during rotation can be obtained from the manufacturers but in Britain where the atmosphere is usually damp they are not needed. The majority of tests are easy to read.   All positive tests and any where the result is in doubt are tested quantitatively.   The speed of testing is more rapid than the present automated test (ART) which performs 100 tests per hour.

Control test cards can be separately obtained already having dried positive and negative serum on them which are reconstituted with distilled water, they are useful in laboratories dealing with few positive sera.

The kit for the quantitative test contains disposable pipettes with which two-fold dilutions can be made in saline on the cards. We have found the titre to be approximately two-fold higher than for the quantitative VDRL in parallel tests.   A prozone effect is occasionally seen, the reaction becoming more positive on dilution and then fading away but false negative results in the screening test due to this effect are very rare (Scrimgeour, 1973).

ii. Treponema pallidum *haemagglutination test (TPHA)*

The reagents for this test are commercially prepared and provided in kits containing an absorbing diluent, lyophilized sheep cells sensitized with *Trep. pallidum* and unsensitized control cells. A reactive (positive) serum is included.   The reagents cannot be obtained separately.   Control and sensitized red cells are standardized and batches are not interchangeable, excess reagents must therefore be discarded.   Two standard dropping pipettes for delivery of sensitized and control cells are included ; a leaflet describing methods for macro- and micro-assay is provided.   Since the reagents are costly laboratories will normally use the micro method.

Microtitre trays and diluters (Takatsy loops) are needed for the micro method described in the leaflet.   Alternatively, if the method is slightly modified, an electronic dispenser-diluter, Compu-pet 100[1], can be employed which is highly accurate and enables large numbers of tests to be done by one person without fatigue ; for a description of this machine see Cremer *et al.* (1975).

[1] Obtainable from General Diagnostics Warner and Co., Eastleigh, England SO5 3ZQ.

Since the TPHA is recommended for screening and all positive sera are further examined by other methods a quantitative test is not essential although this can be carried out if desired. No prozone in this reaction has been recorded therefore one dilution only need be tested. The method using the Compu-pet 100 and one dilution only is believed to be the most economical and least laborious way of performing this valuable test.

Mechanized micro method

MATERIALS

TPHA Test Antigen kit (FD101E).[1]
Microtitre trays with U wells and covers.
Compu-pet 100 with microtubing and pipette.
Distilled water (this must be of high quality to avoid spontaneous agglutination of the sensitized cells).
3 conical flasks (100 ml.) or beakers.
Wash out receptacle and clean absorbent paper wipes.

METHOD

Reconstitute the red cells with distilled water as directed.

Fill one flask with absorbing diluent and prime the machine by placing the non-delivering end of the microtubing in it and pressing the constant delivery button.

When the tubing and pipette are full of diluent and free from air bubbles make a 1 : 20 dilution of each serum to be tested and dispense them into wells of the microtitre trays as follows:

Pick up 5 microlitres serum, deliver 100 microlitres into well 1.

Pick up 10 microlitres of air (to separate diluent from serum dilution during dispensing).

Pick up 50 microlitres of the 1 : 20 dilution from well 1 and deliver 20 microlitres into each of the next two wells. Wash the micropipette by delivering 20 microlitres three times into the waste receptacle. Wipe the tip of the pipette on fresh absorbent paper and proceed to the next serum. Continue thus until all sera have been diluted and dispensed including a positive and negative control which should be dispensed in this order to check efficiency of washout.

[1] Manufactured in Japan but obtainable from Diamed Diagnostics Ltd., 24–36 Queensland Street, Liverpool, England, L7 3JQ.
Less expensive reagents using sensitized turkey red cells are now obtainable from Wellcome Reagents. The technique must be slightly adjusted to suit these different reagents.

Leave the diluted sera at room temperature for 30 minutes to complete absorption of non-specific reactive substances which they may contain.

Prime the microtubing in the manner described above, with sensitized cells and dispense 60 microlitres into each well 2.

Wash the whole of the microtubing and pipette through with distilled water, prime with control unsensitized cells and deliver 60 microlitres into each well 3.

Mix well by tapping the tray gently. Cover and leave undisturbed at room temperature for 3 to 4 hours when a preliminary reading is possible. Read finally after 18 hours at room temperature. Appearances of positive and negative tests are described in the manufacturer's leaflet.

The absorbed fluorescent treponemal antigen (FTA-ABS) test

This test has replaced the original indirect fluorescent test in which the patient's serum was diluted 1 : 200. The sorbent removes the group treponemal antibody enabling the test for specific antibody to be performed on serum diluted 1 : 5. This greatly enhances the sensitivity of the test.

It is an indirect fluorescent test in which absorbed patient's serum is applied to killed *T. pallidum* suspension on microscope slides. After incubation the slides are thoroughly washed in phosphate buffer to remove serum which has not bound specifically to the treponemes, and then treated with fluorescein-isothiocyanate-conjugated antihuman-gamma-globulin serum. When this is washed off only those treponemes which have been in contact with specific antibody in the patient's serum will retain the fluorescent antiserum.

The reagents are costly, therefore the test is limited to sera which are positive in at least one of the screening tests and to those from patients suspected of primary disease. Confirmation of a positive result, by repeating the test, is often desirable and is necessary in primary syphilis when the diagnosis rests on this test alone ; to retest repeatedly during treatment serves no useful purpose.

MATERIALS

Sorbent	Difco 3259–57 Bacto FTA sorbent
Antigen, *T. pallidum* suspension	BBL 40610 *T. pallidum* (Nichols) suspension for FTA test (from Becton Dickinson & Co. in U.K.)

Fluorescein-conjugated anti-human gamma globulin serum	Wellcome MFO1 Fluorescent Antibodies anti-human globulin (sheep)
Phosphate buffered saline (PBS)	Mercia Phosphate buffered saline pH 7·2 for immunofluorescent ANF test (100 ml. to make 2·5 litres)
Microscope slides 0·8–1 mm. thick	Chance
Cover slips to cover whole slide	Chance Propper Ltd. 22 × 70 mm.
Mounting fluid	Difco Bacto 2329–56 FA Mounting fluid
Pasteur pipettes of uniform size to drop 0·05 ml. approx.	Disposable Pasteur pipette, Bilbate Ltd.

Coplin jars.

Moist chamber, a plastic box with well-fitting lid containing moist blotting paper.

Monocular microscope with dark ground condenser and objective to magnify about × 500.

Bright light source either iodine quartz or mercury vapour discharge lamp.

Filters. A yellow barrier filter (OG1) for the eyepiece, a blue exciter filter (BG12 when mercury vapour lamp is used).

Control positive and negative sera, see page 284.

METHOD

i. Preparation of slides

Mark circles, using a diamond, about 10 mm. diameter on the thin microscope slides, not more than 3 per slide. Reconstitute the commercially obtained lyophilized antigen in the appropriate volume of distilled water. Using a wire loop about 5 mm. diameter transfer 1 loopful of antigen to each circle and spread over the marked area. Allow to dry in air, this can be speeded by using a hair dryer. When completely dry submerge in 10 per cent methanol for 5 minutes to fix the treponemes to the slide. Blot dry. At this stage the antigen can no longer be seen on the slide. Store in a closed slide box at 4° C., they will be stable for at least a month. Before use mark with a permanent marker (Pentel) pen so that each area of antigen can be easily seen and to prevent spread of serum from the test area. Pre-ringed teflon coated slides can be obtained.

ii. Titration of conjugate

The activity of commercially produced conjugate varies considerably and a good conjugate is essential, those supplied by Nordic Pharmaceuticals, Baltimore Biological Laboratories and Wellcome

Reagents Ltd. have been found satisfactory by the Reference Laboratory (Wilkinson *et al.*, 1972). The working titre of each batch should be tested as follows.

Make twofold dilutions of conjugate from 1 : 10 to 1 : 320 in phosphate buffered saline (PBS). Test each against a strong positive, a weak positive and a negative (PBS only) control. Use slides with three test areas each, i.e. one slide for each dilution of conjugate tested.

The working dilution is half the weakest dilution which gives brilliant fluorescence with the strongly positive, and weak fluorescence with the weakly positive control. Treponemes in the negative control area should not fluoresce with the strongest (1 : 10) dilution of conjugate.

Store undiluted conjugate frozen in small aliquots. Once thawed for use store at 4° C., preferably with 0·1 per cent azide as a preservative ; repeated freezing and thawing is harmful.

iii. *The test*

Place in a rack one small (75 × 13 mm.) tube for each serum to be tested and deliver into each of them 4 drops sorbent from a standard dropper. Then deliver 1 drop serum into each using a separate pipette for each serum, include strong positive, weak positive and negative serum controls, see page 284. The pipette must be held vertically and the drops allowed to form naturally, not prematurely shaken off. Mix by shaking the rack and stand at room temperature (22° C.) for 30 minutes. Then place 2 drops of the serum sorbent mixture from each tube on to a prepared area on a slide and spread gently over the area with the tip of the pipette taking care not to scrape the slide.

Incubate in a moist chamber at 37° C. for 30 minutes. Wash the slides well with PBS into a sink and then place them in PBS in a Coplin jar for 10 minutes changing the buffer three times during this time. Blot dry carefully.

Spread 2 drops conjugated antihuman globulin serum at the optumum titre on each test area and reincubate in the moist chamber as before. Repeat the washing procedure and blot dry.

Mount the slides using the fluid and coverslips listed above. Read.

When an ordinary cardioid dark ground condenser is used, non-fluorescent immersion oil (Gurr Ltd.) must be placed between the condenser and slide. An oil immersion objective is not required.

Inspect the controls first ; with the blue filter in position it will be impossible to see non-fluorescing treponemes, inspect negative areas without this filter to check they are present. Very faint fluorescence is occasionally seen with negative sera and should be disregarded.

The microscope and lamp described for viewing acid-fast bacilli stained with auramine (page 359) can be employed for immunofluorescence. The condenser and lenses, however, must give maximum fluorescence over a small field (one does not have to search for the treponemes) ; immunofluorescence is weaker than simple staining with a fluorescent dye. Laboratories using both techniques are strongly advised to acquire two sets of equipment because to change from one to the other technique is tedious and likely to result in less than optimum viewing conditions for either. For reliability in any fluorescent technique precise adjustment of the optical system is vital.

Specific immunoglobulins

An indirect fluorescent test, similar to the FTA-ABS, but using specific fluorescent serum conjugate which will react only with IgM human globulin can be employed to distinguish recently acquired antibody from old-standing antibody acquired as a consequence of past infection (Baublis and Brown, 1968). The IgM conjugate will react with antibody in recent infection whereas residual antibody after successful treatment being composed of IgG will give a negative result.

Significance of serological reactions in the diagnosis of syphilis

Much has been written on this subject and it is impossible here to do more than indicate briefly a few important points.

1. *Rapid plasma reagin (RPR) card test, VDRL, ART, WR, Kahn.* All these test for reagin. A positive result in a syphilitic patient (diagnosed by specific tests) indicates the need for treatment. The RPR is the most sensitive of these tests (Scrimgeour, 1973). In secondary and active tertiary syphilis it is almost invariably positive.

None of these tests alone or combined are satisfactory for the

292   *Clinical immunology*

diagnosis of syphilis. Non-specific results are comparatively common.

2. Reiter Protein complement fixation test: a group specific treponemal antibody test, positive in syphilis and yaws shortly after the onset of symptoms. It may be positive before tests for reagin and has been a useful screening test. The more specific TPHA test is now preferred for screening. A positive test does not necessarily indicate active disease. This antibody is not invariably found in syphilis but is absent in leprosy and other febrile illnesses. After cure it slowly disappears.

3. *T. pallidum* **haemagglutination (TPHA) test.** A positive TPHA test in Britain is good evidence of syphilis ; it will also be positive in yaws and related treponemal diseases but not in other infections. It is a sensitive screening test and becomes positive earlier than the Reiter protein CFT, which it is replacing, but is less sensitive than the FTA-ABS test. Occasional false positives are seen and the diagnosis should always be confirmed by the FTA-ABS test.

4. Absorbed fluorescent treponemal antibody test (FTA-ABS). The most specific and most sensitive of the common diagnostic tests. It is believed to recognize the same antibody as the TPHA. When both tests are positive on more than one occasion the diagnosis of treponemal disease is not in doubt but the infection may not be active.

5. Treponemal immobilization test (T.P.I.): the first highly specific test for treponemal antibody which is similar to, if not identical with, that detected by the fluorescent test. Living, virulent, *T. pallidum* and sterile antibiotic-free human serum are employed. Antibody immobilizes the actively motile organisms when incubated in the presence of complement. The test is costly and hazardous and is therefore carried out in reference laboratories only. When a reagin test, the TPHA and FTA-ABS are all positive this test is redundant. It may occasionally be worth doing in a patient with negative reagin tests but positive TPHA and FTA-ABS, since it recognizes the same antibody by a different method and is confirmatory evidence when the diagnosis seems clinically to be improbable. It is also positive in yaws and related treponemal infections which cannot be distinguished serologically from syphilis.

Gonorrhoea

The gonococcal complement fixation test (GCFT)

In the past this test was routinely performed on the serum of patients suspected of venereal disease, but this is no longer the case, partly because culture methods (page 101) are so much improved that even in chronic infection culture is a more reliable method of diagnosis than a serological test and partly because it has proved very difficult to maintain a supply of satisfactory antigen commercially. Antigens are marketed by reputable firms but they are often insufficiently active.

The test is occasionally needed for a patient with undiagnosed arthritis, clinically suggestive of gonorrhoea, when culture results are negative and also in suspected meningococcal or gonococcal septicaemia when antibiotics prevent isolation. (There is sufficient cross reaction between antibodies stimulated by both these pathogens to make the test useful in either infection.) Serum from these patients is best examined in a reference laboratory where freshly prepared antigen can be used.

Non-specific urethritis

This condition is much more commonly encountered than syphilis and gonorrhoea combined, its cause at present is unknown. *Mycoplasma hominis* and T-mycoplasmas have been thoroughly investigated as possible causes but convincing evidence is lacking. Evidence, both cultural and serological is accumulating that *Chlamydia* are responsible for at least a proportion of the infections (Oriel *et al.*, 1972). Antigen is not yet available for a routine diagnostic test.

Skin Tests

1. Hypersensitivity

In chronic infective disease when no likely pathogen and no specific antibodies can be found, it may be possible to demonstrate that the patient's tissues have been invaded by the microbe suspected of causing his disease by inoculating intradermally a small quantity of antigenic material prepared from cultures or from experimentally infected tissues. A positive reaction indicates that previous infection with the suspected microbe, or close contact with some closely related antigen, has rendered his tissues hypersensitive. Hypersensitivity usually develops about four to eight weeks after the original microbial invasion and lasts for many years, a positive reaction does not necessarily indicate present infection. The power to develop hypersensitivity is possessed by a very high proportion of people but not by everyone. Negative reactions to some antigens, for example tuberculin, make long-standing infection by the microbe very improbable, but steroid treatment often renders previously positive reactors negative. Very ill-nourished patients are also likely to give false negative reactions which become positive when nutrition improves (Lloyd, 1968 ; Harrison *et al.*, 1975).

The tuberculin test for evidence of tuberculous infection is one of the most useful and frequently performed. It has been described in Chapter 6. Similar tests are made for the diagnosis of many chronic infections ; examples are the Casoni test for hydatid disease, the Frei test for lymphogranuloma venereum and coccidioidin and histoplasmin tests. Brucellin is occasionally of use in the diagnosis of brucellosis and mallein in suspected glanders. Results are usually read 48 hours after inoculation. The Casoni test antigen stimulates an immediate or a delayed reaction and a comparatively large wheal is seen. Different types and preparations of antigen give different results and instructions for reading the results are usually issued with each batch. If its activity is unknown at least one person known to give a positive reaction and several negative controls should be inoculated for comparison with the patient's test inoculation.

2. Sensitivity tests

When small doses of diphtheria toxin or erythrogenic streptococcal toxin are injected intradermally they give rise to a local reaction, unless the person injected possesses specific neutralizing antitoxin. This is the basis of the Schick and Dick tests which show the patient's susceptibility or immunity to diphtheria and scarlet fever respectively.

Schick test. Inject intradermally 0·2 ml. of diphtheria toxin (containing one Schick dose) into one forearm and 0·2 ml. of toxin, inactivated by heating to 70° C. for 5 minutes, into the other, to serve as a control. The following reactions may be observed.

(*a*) Negative : No reaction in either arm.
(*b*) Positive : No reaction in the control arm. In the test arm a flush appears after 24 to 36 hours which increases to become maximal between the fourth and seventh day.

(c) Negative + pseudo reaction : A flush appears in both arms within 24 hours which rapidly fades. It has usually disappeared by the fourth day.

(d) Positive + pseudo reaction : A flush appears in both arms within 24 hours. It fades rapidly in the control arm and is replaced by the true positive reaction in the test arm which develops as described above.

A positive reaction indicates susceptibility to diphtheria. A very high proportion of negative reactors are immune.

Dick test. Inject intradermally 0·2 ml. of erythrogenic streptococcal toxin (containing one skin-test dose) into the skin of one forearm, and into the other inject 0·2 ml. of the toxin, inactivated by heating to 96° C. for 45 minutes, to serve as a control. The positive reaction appears within 6 to 12 hours as a bright red flush and reaches its maximum in 24 hours. If a pseudo reaction occurs it is smaller, reaches its maximum later and fades more slowly. As in the Schick test, positive reactions indicate susceptibility, negative reactions immunity.

Schultz–Charlton reaction. Human serum containing erythrogenic antitoxin, when injected into skin covered by a scarlatinal rash, will cause an area of blanching. The reaction is of use in identifying scarlatinal erythema.

Autogenous vaccines

At one time autogenous vaccines held pride of place in the treatment of chronic infections. Killed bacteria isolated from the patient's discharges were injected in the hope that they would stimulate the formation of specific antibody which would overcome the infection. The value of vaccine therapy has never been satisfactorily investigated. At present their reputation as therapeutic agents rests mainly on the quite inadequate evidence of doctors who have treated individual patients with good results when all previous forms of treatment have failed. (Those in whom autogenous vaccine therapy also failed tend to be forgotten.) The treatment is popular with the patient, and perhaps even with the doctor, because of the lurking superstition that the vaccine is the " hair of the dog that bites you ", and may as a result have some psychological value which needs to be taken into account when assessing the results of treatment. The physical value of autogenous vaccine therapy may yet be proved, in the meantime it is used as a last resort when antibiotics and other forms of treatment have failed.

METHOD OF PREPARATION

Make cultures of the infected material and examine stained films. Isolate and identify the bacteria present. Select those which seem to be the most likely pathogens as judged by their reputation as pathogens, their predominance in the film and in the plate culture. Seed pure cultures of each of them heavily on nutrient agar or blood agar plates. Wash the growth off the surface of the agar into saline, scraping the medium very gently with a Pasteur pipette bent to a right angle. (It is important not to include particles of medium in the suspension because they may give rise to unpleasant reactions if injected into the patient.) Adjust the concentration of the suspension by adding more saline until the required strength is reached as judged by comparing the opacity with Brown's standard opacity tubes. Heat at 60° C. in a water bath for one hour. Test for sterility by inoculating one ml. into broth and one ml. into cooked meat broth. Incubate the broth cultures for 4 days and then sub-culture to blood agar incubated aerobically from the broth culture and anaerobically from the cooked meat culture. If these cultures are sterile add 0·1 per cent chloro-cresol to the suspension as a preservative. Transfer the fluid with strict aseptic precautions to a vaccine bottle and label with the patient's name, the date and the number of each type of organism per millilitre of fluid.

For recurrent boils a vaccine containing 250 million staphylococci per ml. is suitable. A series of weekly injections is given starting with 0·1 ml. and increasing by 0·1 each week for 10 weeks. Strepto-coccal vaccines are usually prepared less concentrated, and contain about 50 million organisms per millilitre. Suitable doses for various kinds of bacteria are listed in Martindale's *Pharmacopoeia*.

Empirical antigen-antibody reactions

1. Paul-Bunnell test for glandular fever

Heterophile antibody which agglutinates washed sheep red corpuscles was found by Paul and Bunnell in the serum of a high proportion of patients suffering from glandular fever (infectious mononucleosis). The reason for the development of the antibody is unknown. Nevertheless the Paul-Bunnell test is of value in diagnosis, particularly if the serum is examined after absorption with ox blood and also after absorption with guinea-pig kidney antigens which remove similar antibodies found in serum sickness and occasionally in normal serum.

That Epstein-Barr (EB) virus causes glandular fever can no longer be seriously in doubt and patients' serum can be examined for specific viral antibody. Nevertheless there is much to be said for retaining the empirical test because EB virus is also associated with lymphoma and nasopharyngeal carcinoma (Epstein and Achong, 1973). Moreover, EB virus antibody is often acquired without overt illness whereas the Paul-Bunnell reaction is associated with active disease (Joint Investigation, 1971).

TABLE 36.

Tube	1	2	3	4	5	6	7
	ml.	ml.	ml.	ml.	ml.	ml.	ml.
Saline	0·4⎫	0·25	0·25	0·25	0·25	0·25	0·25
Serum	0·1⎭	0·25	0·25	0·25	0·25	0·25*	—
(inacti-vated)		{from tube 1}	{from tube 2}	{from tube 3}	{from tube 4}	{from tube 5}	
R.B.C. (2%)	0·1	0·1	0·1	0·1	0·1	0·1	0·1
Final dilution	1 : 7	1 : 14	1 : 28	1 : 56	1 : 112	1 : 224	control

* Discard 0·25 ml. of serum dilution from tube 6.

METHOD. (Davidsohn, 1937, 1938). Prepare a suspension of fresh thrice washed sheep red cells, 2 per cent by volume in physiological saline. Separate the serum from a clotted sample of the patient's blood and inactivate it by heating to 56° C. for 30 minutes. Make a series of twofold dilutions of the serum in saline, each 0·25 ml. in volume, from 1 in 5 to 1 in 160 (Table 36). To each tube add 0·1 ml. of sheep cell suspension. Incubate in the 37° C. water bath for an hour and read. If no agglutination is seen, refrigerate the tubes overnight at about 4° C., re-incubate for an hour on the following day and read again. The titre is the highest dilution of serum which shows agglutination when viewed in a bright light with the aid of a hand lens.

The three types of heterophile antibody which give positive reactions in this test can be distinguished by absorption.

TABLE 37.

Heterophile Antibody	Agglutination after Absorption with	
	Ox blood	Guinea-pig kidney
Normal serum	+	—
Serum sickness . . .	—	—
Glandular fever . . .	—	+

Absorption test (after Davidsohn)

PREPARATION OF ANTIGENS

(*a*) *Ox blood.* Wash the red cells three times in saline. After the final centrifugation resuspend them in four times their volume of saline. Boil the suspension in a water bath for one hour and replace the water of evaporation with distilled water. Add phenol to make a final concentration of 0·5 per cent and store in the refrigerator where it will remain stable for many months.

(*b*) *Guinea-pig kidney.* Remove the perinephric fat from guinea-pig kidneys and keep them frozen until required. Then thaw them and wash them in saline until they are free from blood. Mash them into a fine pulp and make a 20 per cent suspension of the pulp in saline. Boil in a water bath for one hour and replace the water of evaporation with distilled water. Add phenol and store as described above.

METHOD

Mix 0·1 ml. of inactivated serum and 0·5 ml. of absorbing antigen. Stand the mixtures for one hour at room temperature. Centrifuge and recover the supernatant fluid. Set up the test as already described but since the absorbed serum is already diluted approximately 1 : 5 (after removal of solid matter) deliver 0·25 ml. of it into each of the first two tubes omitting the saline in the first, then continue as before. When the red cells suspension has been added to all the tubes shake them well and allow them to stand at room temperature for 2 hours, then read.

SCREENING TESTS

When many tests are to be performed and a high proportion of negative results is expected, time may be saved by a screening test.

Absorb part of each sample of inactivated serum with the guinea-pig kidney antigen. Dilute the absorbed serum 1 : 2 and set up a single tube containing 0·25 ml. of the diluted absorbed serum and 0·1 ml. of 2 per cent sheep red-cell suspension. If after 2 hours at room temperature no agglutination is seen a negative report may be sent. If agglutination is present absorb the remainder of the original serum with ox blood and make the full titration as already described.

Sera giving a positive Paul Bunnell reaction will rapidly agglutinate formalinized horse-red-cells[1] in a slide test (Hoff and Bauer, 1965). Sera agglutinating sheep cells but negative to the Paul Bunnell absorbed test seldom give this reaction and, moreover, inactivation of the serum by heating to 56° C is unnecessary. This is therefore a useful screening test. Positive results should be confirmed by the standard absorption test already described.

False negative results are rare but in a patient with symptoms or with a blood-cell picture suggestive of glandular fever a negative screening test should be checked by the standard absorption test.

INTERPRETATION

The agglutinin is often demonstrable as early as the fourth day of the illness, but sometimes fails to appear until the second or third week or even later. Some cases which are clinically typical do not develop agglutinins and are also EB virus antibody negative (Joint Investigation, 1971). A titre of 1 : 56 with unabsorbed serum is usually considered to be significant if the patient has not recently received serum therapy. A titre of 1 : 14 or more with serum absorbed with guinea-pig kidney antigen combined with a negative result after ox blood absorption is good evidence that the patient is suffering from, or has had, glandular fever within the past year.

2. Empirical Tests in Mycoplasmal Pneumonia

Specific *Mycoplasma pneumoniae* antigen for a complement fixation test can now be obtained, but care is needed in the interpretation of positive results because high titres are sometimes seen in serum from symptom-free contacts with no history of infection (Dowdle *et al.*, 1967). As in other infections a rise in titre at least fourfold in samples tested in parallel is needed before the diagnosis can be made. Since isolation of *Mycoplasma pneumoniae* is uncertain even in the most experienced hands diagnosis by serology

[1] Obtainable from Reagents Ltd., Beckenham, England BR3 3BS.

has to be relied on.   The two empirical tests found to be of value before the cause of the disease was known are described since some laboratories without viral antigen may still wish to use them.

Cold agglutinins

An antibody which will agglutinate human group O red blood cells in the cold is found in the serum of about 50 per cent of patients suffering from mycoplasmal pneumonia.   The agglutinin appears about the fourth day of the disease and the titre rises to become maximal early in convalescence.   Cold agglutinins are also found in other conditions, for example haemolytic anaemia, Raynaud's disease, trypanosomiasis and paroxysmal haemoglobinuria. Provided these conditions can be excluded and particularly if a rising titre can be demonstrated, their presence is good evidence in favour of mycoplasmal pneumonia.   They are not found by the method to be described in influenza, the common cold, Q fever or psittacosis.   A test for them is therefore of value as a preliminary step in the serological investigation of the respiratory virus infections.

METHOD

Take about 5 to 10 ml. of blood by venepuncture into a sterile bottle.   Allow it to clot in the 37° C. incubator.   Separate the serum (do not inactivate it because heating to 56° C. destroys the agglutinin). Make a series of twofold dilutions in saline 1 : 10 to 1 : 160, each about 0·5 ml. in volume in small test tubes.   Include in the test a saline control.   Add to each dilution and the control an equal volume of a freshly prepared 2 per cent suspension of thrice washed group O red blood cells.   After shaking well, refrigerate the tubes for 2 hours or overnight and read.

The final dilutions of the series will be 1 : 20 to 1 : 320.   The titre of the serum is the highest dilution which shows agglutination when viewed with the aid of a hand lens.   The agglutination is reversible and will disappear on warming the fluid.

A titre of 1 : 40 or more is considered to be suggestive of mycoplasmal pneumonia provided the other conditions listed above can be excluded.   An attempt should always be made to demonstrate a rise in titre by taking at least two samples of blood, one towards the end of the first week of the illness and one early in convalescence.   If the patient is seen for the first time late in the disease it may be possible to demonstrate a fall in titre which is also of much greater significance than a single positive result.

Streptococcus M.G. agglutination (Thomas *et al.*, 1945)

Evidence for a diagnosis of mycoplasmal pneumonia is strengthened if agglutinins to Streptococcus M.G. can be demonstrated in the patient's serum.

Streptococcus M.G. can be recovered from the normal upper respiratory tract and has been isolated from the lungs of fatal cases of pneumonia. Its relation to the disease is still obscure. Serum in mycoplasmal pneumonia commonly shows both cold agglutinins and Streptococcus M.G. agglutinins. The two, however, do not necessarily appear together. The presence of either of them is in favour of the diagnosis, particularly if a rising titre can be shown as the disease progresses or a falling titre after convalescence. Streptococcus M.G. agglutinin usually appears in the second week of the disease; like cold haemagglutinin it is rarely encountered in influenza virus pneumonia, the rickettsial infections or other infective diseases. Two specimens of blood are taken by venepuncture, the first during the first week, the second between the fifteenth and twenty-fifth day of the disease. Serum from the first specimen is stored in the refrigerator at 2 to 4° C. to be tested later at the same time as the second specimen.

REAGENTS
    Patient's serum.
    Streptococcus M.G. suspension.
    Standard positive rabbit serum.
    0·85 per cent saline.

Note. The patient's serum must not be inactivated because heating to 56° C. lowers the titre.

METHOD (Standards Laboratory, 1952)

Make twofold dilutions of the patient's serum from 1 : 5 to 1 : 320 in saline using 75 × 13 mm. round bottom tubes and 0·5 ml. of fluid per tube. Make a similar series of dilutions of the rabbit serum to include the known titre of the serum. Deliver 0·5 ml. of saline into one tube to serve as the saline control. Add 0·5 ml. of the streptococcus suspension to all the tubes, shake them to mix the fluids well and incubate them in the 37° C. water bath for 2 hours; refrigerate at 2° to 4° C. overnight (22 hours) and re-incubate at 37° C. for 2 hours on the following day. Read the result in a bright light against a dark background.

The final dilution of patient's serum is 1 : 10 to 1 : 640 and the titre of the serum is the weakest dilution which shows agglutination by naked-eye examination. If a single specimen of serum only be available, a titre of 1 : 40 or more is usually considered to be significant.

Demonstration of antibodies as evidence of pathogenicity

When a microbe, usually considered to be harmless, appears to be causing either localized or generalized infection, part of the evidence which must be accumulated before its causal role can be established is the demonstration of specific antibodies to it in the patient's serum.

The type of test chosen in the first place depends on the identity of the organism and the way in which it grows. For example, a member of the family *Enterobacteriacae* which yields a smooth, easily emulsified growth will probably show satisfactory agglutination when tested in the manner described for the salmonella group. Antigen for an agglutination test must consist of a smooth stable suspension of bacteria in saline. Microbes which yield granular growth, autolyse easily or are autoagglutinable must be tested in some other way.

The complement fixation test has the advantage that it does not depend on the physical state of the bacteria in the antigen. A suitable antigen may be made by scraping growth from the surface of solid medium into saline, or by re-suspending in saline the centrifuged deposit from liquid cultures. The antigen is first tested for its action on complement by titrating it as described (page 278). If it proves anticomplementary, dilution or heating to 56° C. for 30 minutes, may improve it. In any antigen-antibody test excess of either antigen or antibody may inhibit the reaction. Therefore titration of a series of dilutions of antigen against dilutions of the presumed positive serum is required both to find the optimal proportions of each, when antibody is present, or to demonstrate its absence. Before investigating human sera it is advisable to check the reliability of the proposed antigen-antibody reaction. This can be done by injecting rabbits with killed antigen to stimulate antibody formation and then titrating the positive rabbit serum against the antigen. In a complement fixation test it is important to use the haemolytic serum optimally and a chess board titration of serum against complement may be required, see page 282.

If the patient's serum gives a positive reaction with the antigen and a number of presumably normal sera do not, and further, if a rise of titre during the disease followed by a fall after convalescence can be demonstrated, the result favours a diagnosis of infection. But it is only one small item in the mass of evidence which must be accumulated before pathogenicity is proved (see Chapter 1). At this stage the clinical bacteriologist must either hand over the investigation to an academic colleague or himself enter the field of medical research.

REFERENCES

BAUBLIS, J. V. and BROWN, G. C. (1968). *Proc. Soc. exp. Biol. (N.Y.)*, **128,** 206.
CREMER, A. W., MELLARS, B. and STOKES, E. J. (1975). *J. clin. Path.*, **28,** 37.
DAVIDSOHN, I. (1937). *J. Amer. med. Ass.*, **108,** 289.
—— (1938). *Amer. J. clin. Path., Tech. Suppl.*, **2,** 56.
DOWDLE, W. R., STEWART, J. A., HEYWARD, J. T. *et al.* (1967). *Amer. J. Epidem.*, **83,** 137.
EPSTEIN, M. A. and ACHONG, B. G. (1973). *Lancet*, **ii,** 836.
FELIX, A. (1944). *Trans. roy. Soc. trop. Med. Hyg.*, **37,** 321.
HARRISON, B. D. W., TUGWELL, P., and FAWCETT, I. W. (1975). *Lancet*, **i,** 421.
HOFF, G. and BAUER, S. (1965). *J. Amer. med. Ass.*, **194,** 351.
Joint Investigation by University Physicians and P.H.L.S. Laboratories (1971). *Brit. med. J.*, **iv,** 643.
LLOYD, A. V. C. (1968) *Brit. med. J.*, **iii,** 529.
KERR, W. R., COGHLAN, J. D., PAYNE, D. J. H. and ROBERTSON, L. (1966). *Lancet*, **ii,** 1181.
ORIEL, J. D., REEVE, P., POWIS, P., MILLER, A. and NICOL, C. S. (1972). *Brit. J. vener. Dis.*, **48,** 429.
PORTNOY, J. (1963). *Amer. J. clin. Path.*, **40,** 473.
Public Health Laboratory Service Monograph Series No. 1 (1972). H.M. Stationery Office, London.
SCRIMGEOUR, G. and ROSLIN, P. (1973). *Brit. J. vener. Dis.*, **49,** 342.
STANDARDS Laboratory for Serological Reagents (1951). Lond. Instructions issued with reagents.
—— —— (1952). Lond. Instructions issued with reagents.
THOMAS, L., MIRICK, G. S., CURWEN, E. C., ZIEGLER, J. E., Jr., HORSFALL, F. L., Jr. (1945). *J. clin. Invest.*, **24,** 227.
WILDE, C. E. (1973). *Lab. Equip. Dig.*, **11,** 130.
WILSON, M. M., and MERRIFIELD, E. V. O. (1951). *Lancet*, **ii,** 913.

9

Hospital epidemiology

The risk of infection

A large proportion of patients who attend a general hospital do so because they need treatment for some ailment caused by the invasion of their tissues by microbes. Many of them have infected discharging wounds or harbour virulent bacteria in their respiratory passages, on their skin and elsewhere. It follows that the bacterial population of the hospital is likely to be more dangerous than that of similar large institutions attended by a comparable number of healthy people. Comparison of air and dust samples from hospitals and other institutions confirms this view. Therefore to attend a hospital for treatment, or to work in a hospital, entails a risk of infection greater than that run in ordinary contact with friends and business associates. That there is increased risk to hospital staffs is shown by the high proportion among them of upper respiratory carriers of pathogenic bacteria when they are compared with members of the general population (Miles, Williams and Clayton Cooper, 1944). In spite of this, provided reasonable precautions are taken, they are unlikely to suffer because they possess the advantages of being healthy and of an age to combat infection successfully. The patients, however, are not so fortunate and are likely to suffer grave damage and even death as a result of hospital infection. When patients are particularly susceptible the benefits of hospital treatment may not be worth the risk of infection. Investigation in children's wards has led to a recommendation that no child should be admitted to hospital unless essential treatment cannot be performed at home or as an out-patient, because in childhood the risk of infection is so great.

As methods of investigation and treatment become more complex there is an increasing tendency to admit patients to hospital instead of treating them at home. It is therefore all the more important that measures for the control of infection in hospital should keep pace with advances in other branches of medicine.

Apart from natural lowering of resistance due to disease or age, some forms of treatment now undertaken, for leukaemia and organ transplantation for example, reduce natural resistance to infection to such an extent that they can only be legitimately carried out in special units. These are usually sited within hospitals but the bacterial population within them must be reduced far below that in surgical wards, and pathogens must be rigidly excluded if the patients are to survive (James *et al.*, 1967).

The limited value of antibiotics

Chemotherapy has greatly improved the prognosis of infective disease and it can be given prophylactically to cover short periods of special risk. Unfortunately it has sometimes led to the adoption of a complacent attitude towards prevention of infection. It is argued that infection can be treated successfully and therefore preventive measures are no longer of prime importance; this attitude cannot be too strongly condemned. In the first place it is quite unjustifiable to assume that the infecting microbe will be susceptible to treatment. Since antibiotics have been widely used, the frequency of hospital infection and the difficulty of preventing it has been more clearly demonstrated than ever before. Sensitive bacteria in wounds disappear only to be replaced by resistant ones, and although a proportion of these can be dealt with by the newer antibiotics there are several common species which resist all but those which give rise to unpleasant side effects. Moreover, if treatment is not immediately successful the lesion may become chronic and then application of the more active antibiotics may fail. In the second place, even if treatment is successful considerable harm is done. At best the patient suffers the inconvenience of a prolonged stay in hospital, and at worst he may succumb to iatrogenic illness. Moreover, the financial burden, often considerable, must be borne by someone.

Antibiotics can be used prophylactically to limit an epidemic in hospital but protection of the patient's tissues from exposure to microbes is likely to remain the most important factor in preventing infection.

Antibiotic prophylaxis

Prophylactic treatment aimed at eliminating hospital pathogens, see below, is doomed to failure. Indeed, stopping all antibiotic treatment is more likely to succeed (Price and Sleigh, 1970). Whenever antibiotics are given the patient's normal flora changes to microbes capable of existing in their presence. The microbial environment of the ward will inevitably change also, resistant microbes in large numbers, hospital pathogens among them, will be available to cause infection if there is a small breakdown in aseptic technique.

Prophylaxis aimed at particular pathogens known always to be sensitive to the drug given is worth while in some circumstances, for example, penicillin given to patients at risk of gas gangrene in mid-thigh amputation, see page 326.

Short-term prophylaxis aimed at the patient's own commensal bacteria is successfully given in patients with damaged heart valves undergoing dental surgery. If treatment is prolonged infection of the valves by a resistant strain is likely (Garrod *et al.*, 1973) because there is time for the commensal flora to change. This kind of prophylaxis can be given successfully in operations where colonized viscera such as the large intestine have to be handled or opened and the wound is at risk from contamination. For example two doses only of lincomycin plus gentamicin (or tobramycin) given intramuscularly immediately before operation and 8 hours later was shown to reduce infection and no resistance was encountered (Stokes *et al.*, 1974).

Hospital pathogens

The modern hospital environment favours bacteria with special properties. For success they need to be able to withstand and, if possible, grow in the presence of as many antibiotics as possible, especially those commonly employed such as the penicillins and tetracycline. If they can, in addition, withstand or actually multiply in dilute antiseptics commonly applied to wounds or skin so much the better for them.

Species which inhabit sites from which they are easily shed will constantly maintain their numbers in the hospital dust and air. If, in addition, they are virulent in the sense that small numbers only are needed to establish colonization or wound invasion their success as " hospital pathogens," i.e. species which often cause cross-infection, is assured.

The measures for the prevention of cross-infection originally recommended as the result of investigation of streptococcal wound infection in the second World War are not necessarily fully relevant to the changed conditions in modern hospitals. Other bacteria are now dominant and further measures aimed specifically at them are needed.

The widespread use of antibiotics has virtually abolished strepto-coccal wound infection although many strains, even of this organism, have within the last ten years become resistant to tetracycline and may in future acquire resistance to sufficient drugs to stage a come-back. Consideration of the properties of species notorious as hos-pital pathogens listed in Table 38 makes clear the reasons why *Staph. aureus* and *Ps. aeruginosa* are now especially troublesome.

TABLE 38

Comparison of properties of " hospital pathogens "

	Numbers in hospital environment*	Antibiotic resistance	Ability to become established in tissues	Growth at room temperature (18–25° C)	Growth in dilute antiseptics
Strept. pyogenes	±	±	+ + +	−	−
Staph. aureus	+ + +	+ +	+	±	−
Ps. aeruginosa	+ +	+ + +	+	+ +	+ +
Cl. welchi	+ +	±	±	−	−

* In the absence of an infected patient.

The recognition of carriers is important in controlling infection by *Strept. pyogenes*. Even a few organisms infect, so one carrier in a surgical ward is likely to start an epidemic unless aseptic tech-nique is rigidly applied. Isolation is required and, because the carrier rate in healthy people is low, it is also practicable.

In the case of staphylococci, however, there is no hope of isolating all carriers. Moreover, common carrier sites, the nose and skin, predispose to heavy dissemination and carrier rates of 60 per cent or higher are common among both patients and staff. Fortunately this organism is less adept at infecting wounds. Larger numbers are needed and tissue defences are better able to cope with them. Furthermore, all strains of *Staph. aureus* are not equally dangerous. Those which have caused major hospital epidemics are limited to a few 'phage types and are resistant to penicillin-tetracycline and often to other antibiotics. Attempts to isolate carriers and patients infected with these potentially dangerous strains have not proved successful (Williams *et al.*, 1966).   Isolation of all carriers of anti-biotic-resistant strains is not necessary because aseptic techniques are adequate to deal with all but the highly infective strains.   'Phage typing cannot be rapidly applied and, moreover, epidemic strains may prove to be non-typable by the standard 'phages (Temple and Blackburn, 1963).   Effective control by isolation is also impracticable because of the changing nasal flora of treated patients.   A patient who has acquired a resistant strain while in hospital tends to lose this strain once he has returned to an antibiotic-free existence in his home.   If his nose is swabbed on re-entry to hospital he will not be recognized as a carrier, but if he then receives antibiotics a few remaining cocci of his resistant strain will be favoured and he may revert to his previous antibiotic resistant, potentially infective carrier state.

Individual carriers of resistant strains are not especially at risk from their own strain when they have to undergo an operation.   In one series of 299 nasal carriers of resistant staphylococci who were also infected with staphylococci of similar antibiotic-resistance pattern, only 45 had the same type in nose and lesion (Stokes *et al.*, 1965).

Even though, in comparison with *Strept. pyogenes*, antistaphyl-ococcal measures in hospital are not much aided by antibiotics, the situation is not desperate because of their comparatively low virulence.   One nasal carrier of an antibiotic-resistant staphylo-coccus with epidemic propensity in a ward is seldom sufficient to overpower aseptic measures.   Such a patient will, however, pass on his resistant strain to other antibiotic-treated patients, whose sensitive staphylococci have succumbed, and in time this leads to a large number of carriers and resistant cocci in the environment and hence to wound infection (Stokes *et al.*, 1965).

It is important that all medical and nursing staff should have a clear idea of the constant risk of hospital infection and that they should know the common reservoirs of bacteria and how they gain access to the patient's tissues.   Once these principles are thoroughly understood methods of prevention are mainly a matter of common sense.

An important part of the bacteriologist's work is to advise his clinical colleagues of the best methods of safeguarding their patients against infection.   It is also his duty to help investigate cases of cross infection, so that in future they may be prevented.   Sufficient guidance will be given here to enable first steps to be taken in the investigation and prevention of infection; a full account of this subject is outside the scope of this book. The reader is strongly advised to study the two Medical Research Council War Memoranda numbers 6 and 11, which state basic principles of prevention of cross-infection, and *Hospital Infection* by Williams *et al.* (1966), which is a concise and practical account of the causes and prevention of infections in hospitals by four authors, each with wide experience in this field. As wound infection is by far the most important epidemiological problem in a general hospital it will be dealt with first. Methods for sampling the patient's environment will also be described.

Common Causes of Hospital Wound Infection and their Prevention

" Clean " operation wound infection

Investigation of sepsis arising in a " clean " operation wound is comparatively simple because the approximate time when the microbes entered the wound can be judged from the time of onset of symptoms and the nature of the lesion. If the wound was sutured without drainage and if no dressing was done in the ward before the sepsis was discovered the bacteria are almost certain to have entered during operation.

Theatre infection

The time of onset of symptoms varies with the species and virulence of the infecting microbe and also with the weight of infection. Symptoms usually appear between the first and the tenth day after operation. When local symptoms arise late the lesion is often a deep-seated abscess caused by an organism of low virulence. It is unusual in these cases for the patient to have been afebrile since the operation, but a normal temperature from, say, the second to sixth day, followed by fever and signs of wound infection later, does not preclude entry of the infecting microbe at operation.

The reservoirs of bacteria in any operation theatre can be listed as follows:

    1. The hair and skin of everyone present.
    2. The upper respiratory tracts of everyone present.

3. All unsterile textiles, particularly woollen materials.
4. Dust.
5. Bacteria-carrying particles suspended in the air.

The methods of entry into the wound of bacteria from these sources are well known and modern aseptic theatre technique is aimed at preventing them. It is likely to break down in the following ways.

Infection from the surgeon's hands

If the inside of the surgeon's rubber gloves are sampled after an operation bacteria can be recovered from them in spite of careful hand-washing and drying before they were put on. When the wearer is a skin carrier of *Staph. aureus* this organism may be found in profusion because it is carried in sweat from the deep layers of the skin to its surface. If the glove remains intact this does not matter, but unfortunately minute holes are often made unknowingly during operation and then the surgeon's skin bacteria will enter the wound. They may also travel from his skin to the wound if the sleeve of his gown just above the rubber glove becomes wet, because it then ceases to be an efficient barrier (Devenish and Miles, 1939).

Many surgeons, realizing these dangers, adopt as far as possible a "no touch" technique, but in many operations direct manual manipulation is essential at some stage.

Staphylococci on the skin can be reduced by washing in anti-bacterial detergents and drying without washing off the detergent, which forms a bactericidal film on the surface of the skin. Unfortunately it is not every surgeon whose skin will stand frequent application of these substances. One of the least harmful of them is hexachlorophane which is effective when incorporated in liquid or solid soap. Surgeons who find this irritant may be able to use "Betadine", an iodophor detergent which is also reliable and is more active against Gram-negative bacilli than hexachlorophane. Alcoholic chlorhexidine solution, commonly used on the patient's skin, is one of the best skin disinfectants for the surgeon's hands also (Lowbury, *et al.*, 1974).

Infection from equipment in the vicinity of the wound

Various pieces of non-sterile equipment are from time to time suspended above the wound and may be moved during the operation.

Bacteria-laden particles may then fall from them into the wound. For example, although the glass surface of the lamp over the table is usually spotlessly clean the screws round its rim and the chains from which it hangs may not be free from movable dust. In a busy theatre it needs damp-dusting at the end of each list.

X-ray apparatus is often manipulated close to the wound and it is difficult to clean. The wound should be protected by a sterile barrier, for example, a dry sterile towel, or if a clear view is required, a large sheet of sterile cellophane paper.

Ideally a separate machine should remain in the theatre so that it can be properly cleaned before use and so that contamination of the theatre by hospital dust brought in on cables and wheels is prevented.

Infection from bacteria-laden particles in the air

In spite of closed windows, prevention of draughts and other precautions, the air of operation theatres will be no more free from bacteria than the air of surgical wards or hospital corridors unless plenum ventilated. Theatres ventilated by exhaust fans which suck the air from the theatre to the exterior are hazardous because air enters the theatre through all available openings and, since the windows are firmly shut, it comes through open doors and the cracks under closed ones. Thus the air entering the theatre has passed through the corridors, lift shafts and wards and carries with it numerous bacteria, including a relatively high proportion of human pathogens. The stronger the fan, the more bacteria will be sucked in. When the theatre is on or above the first floor of the hospital the air outside the window is less contaminated than the air inside, even if the hospital is situated in the middle of a large city. If contamination from the air were the only consideration it would be safer to operate on a balcony outside the theatre.

Air contamination can be greatly reduced by plenum ventilation. A slight positive pressure is maintained by pumping outside air into the theatre. The air currents then travel from the theatre to the hospital corridors instead of in the reverse direction. The air supply is taken from the top of the building, through a duct when necessary. Contamination can be further reduced by filtering the incoming air. A less powerful exhaust fan can be placed near the sterilizers to make working conditions more comfortable.

Fortunately, even when the air is heavily contaminated, the

number of pathogens which fall into the wound is small. The chance of wound contamination can be judged by exposing blood agar plates near the wound during the operation. The known pathogens most commonly found are *Staph. aureus* and *Cl. welchi*. The staphylococcus is the more dangerous because it is able to initiate infection comparatively easily, whereas *Cl. welchi* needs special conditions before it can become established in the tissues.

When a small wound is exposed during operation for about an hour or less and no surgical complications arise, contamination from the air is not likely to cause infection, because the tissues can protect themselves against minimal doses, even of pathogenic bacteria. When, however, the wound is large and exposed for several hours the risk of infection from the air is correspondingly greater ; neurosurgical wounds and burns are especially susceptible. Surgical complications such as haematoma, gangrene of part of the skin flaps or subcutaneous collections of serous fluid, will allow a small number of contaminating bacteria to multiply unmolested, and it is well known that these conditions are associated with sepsis.

When no plenum-ventilated theatre is available, the following precautions will help to lessen the risk. If the corridor outside the theatre can be well aired with open windows, if the floor is oiled and the doors into the theatre are kept closed, fewer bacteria-laden dust particles will enter from the hospital. When the theatre is placed high in the building, i.e. on or above the first floor, it should be aired when not in use by opening the windows and closing all doors. The windows can be fitted with frames holding gauze of sufficiently fine mesh to prevent the entry of flies. Whenever the weather is warm and there is no danger of draughts the windows should be opened during operation.

Some of the reservoirs of bacteria in the theatre which have been listed above can be eliminated. Blankets and other woollens are particularly dangerous because, unlike other fabrics they are infrequently laundered and are heavily contaminated with bacteria. When shaken or moved, minute bacteria-laden particles become detached from them and remain suspended in the air ready to infect any exposed wound. Surgical blankets, indeed all hospital blankets, should be boilable ; cellular cotton is suitable. The patient should be brought to the theatre in clean linen and covered with freshly laundered cotton blankets. Changing rooms are best placed outside the theatre suite unless plenum ventilation is installed. Anaesthetists and spectators, as well as surgeons and nurses, should change into clean cotton garments.

Human hair is another fruitful source of bacteria, including *Staph. aureus.* Well fitting caps which completely cover the hair should be worn by all who enter the theatre. When the mask rubs against the surgeon's face minute flakes of skin may fall into the wound. Some surgeons apply cream to the face, eyebrows and eyelashes so that all particles will remain adherent.

Even if no known pathogens are found when the theatre air is sampled, a high total colony count indicates danger because pathogens will sooner or later make their appearance. Moreover, organisms usually considered to be harmless have caused fatal infection when introduced in this way (Cairns, 1939).

Air contamination measured by exposure of Petri dishes is expressed in terms of the number of colonies per minute of exposure. The counts are made after 24 to 48 hours' incubation. *Studies in Air Hygiene*, a report by the Medical Research Council, deals very fully with this subject. The highest count quoted for exhaust ventilated theatres is 4·7 colonies per minute and for plenum ventilated theatres 1·3 per minute. This method of sampling fails to take into account very small particles which remain suspended in the air, but they are more important in upper respiratory than in wound infection. A simple way of estimating air contamination is to compare plates exposed simultaneously in the theatre and at other sites in and outside the hospital (page 330).

Ward infection

Infection of " clean " operation wounds in the ward is usually the result of one or more of the following minor post-operative complications. 1. Blood or serum seeping through a drainage tube soaks the dressings which are then no longer an efficient barrier. 2. Some comparatively lengthy procedure, such as the evacuation of blood clot, is performed in the ward. 3. The dressing becomes loose and slips so that the wound is exposed.

Large wounds resulting from burns which are originally sterile are almost certain to become infected if dressed in the ward. High-pressure plenum ventilation in a special " burns unit " coupled with appropriate antibiotic treatment has improved the outlook for patients with extensive burns (Colebrook *et al.*, 1948, Lowbury, 1954). In a general hospital without such special conditions dressings are best done in the theatre, if it is plenum ventilated, if not, in a cubicle set aside for the purpose. The dressing-room should be damp-dusted and the floor mopped with disinfectant at the end of each session of dressings. The windows should then be opened wide for several hours. The room should be completely closed for at least an hour before the next dressing is done. No unsterile woollen textiles should be exposed in the room. After airing and closing the room no person should enter unless they are wearing cap, mask and gown. The patient's hair should be covered with a well-fitting cap, but he need not wear a mask unless he has a cough or cold or unless he wishes to talk while being dressed. If no cubicle be available the next safest place is a sheltered outside balcony near the top of the building not used by infected patients.

Infected traumatic and operation wounds

Hospital infection of wounds which are septic, or at least con-taminated, when first seen is much more difficult to investigate because, unless routine daily cultures are made, it is difficult to judge when the superadded infection occurred. The consequences to the patient may be just as serious as infection of a clean wound, but many cases are unrecognized and therefore are not investigated. An example of an actual case may serve to illustrate this point.

A young woman was admitted to hospital in 1940 with a depressed compound fracture of the skull, the result of an air-raid injury. The wound was full of dirt and pieces of felt from her hat ; these, and fragments of bone, were removed and for a few days she made good progress. The wound was swabbed within half an hour of admission and frequent routine cultures were made as part of a special investigation. In spite of original contamination, the wound was healing until *Strept. pyogenes* of the same type which had infected other patients in the ward was accidentally introduced into the wound. The result was that instead of leaving hospital healed within a few weeks she died a year later of osteomyelitis of the skull and extensive infection of the soft tissues of the head and face (McKissock, 1941). If the infection had not been specially investi-gated the osteomyelitis would have been thought to be the natural sequel to an injury heavily contaminated with dirt on admission and the real cause would have remained unrecognized.

Efficacy of preventive measures

During and since the 1939-45 war, the bacterial flora of wounds during treatment in hospital has been investigated by many workers and the chance of a wound becoming infected at the time of injury by any common pathogen is now well known. Although a series of cultures and typing of the pathogens may be necessary to prove hospital infection in individuals, a rough estimation of the efficacy of preventive measures can be made without multiple cultures by comparing the proportion of treated wounds found to be infected with the known chance that the bacteria gained entry at the time of injury.

The chance of original infection

An extensive investigation of small industrial wounds of the hand (Williams and Miles, 1949) showed a close relation between original wound contamination with *Staph. aureus* and the skin carrier rate of this organism ; a number of these wounds healed without clinical sepsis in spite of the presence of the pathogen. Original contamination with *Strept. pyogenes* (group A) was rare, 1 per cent in 432 wounds and by faecal bacteria very rare. It is probable that about 70 per cent of accident wounds which are not in communication with abdominal viscera or mucous membranes are originally contaminated with harmless organisms, if further infection is prevented an even larger proportion will heal rapidly without suppuration.

The presence of dirt in the wound is not evidence of dangerous bacterial infection. When swabs are taken to compare the bacterial flora of a road surface and a ward floor, the road swab looks very dirty but cultures on blood agar incubated at 37° C. usually yield a scanty growth of organisms harmless to man (Fig. 35). *Cl. welchi* is the only common pathogen and when it contaminates wounds which are treated under peace time conditions it usually behaves as a saprophyte and very rarely causes gas gangrene. In contrast, the ward swab looks comparatively clean but yields a greater total number of colonies and a much higher proportion of human pathogens. *Staph. aureus* is the most common of these, but *Strept. pyogenes*, other streptococci and organisms of faecal origin are sometimes found.

Staph. aureus is carried in the anterior nares, on the skin or in both sites in about 50 per cent of healthy people. Therefore, assuming that carriers are neither more nor less liable to injury than others, a wound infection rate of 50 per cent with *Staph.*

aureus can be attributed to self infection. The proportion actually infected in this way is probably much less. If the proportion of wounds undergoing treatment which are found to be infected with the organism exceeds this, as it commonly does, it indicates that hospital infection is not being adequately prevented. Barber and others have thrown more light on the frequency of cross-infection with this organism because they have shown that when *Staph. aureus* in a wound " becomes " antibiotic resistant it is usually due, not to the survival of resistant variants of the original strain, but to the introduction of a new penicillinase producing strain of a different 'phage type. The same resistant strain may haunt a hospital for months or even years, the main reservoirs of infection being the noses and skins of the patients and their infected wounds. Such " hospital staphylococci " are rarely found in patients newly admitted to hospital ; a high proportion of them isolated from wounds is therefore further evidence of poor aseptic technique.

Other resistant organisms such as *Esch. coli*, *Klebs. aerogenes*, *Proteus mirabilis*, *Ps. aeruginosa* and *Strept. faecalis* are more commonly found in wounds now than in pre-antibiotic days. When the wound is situated on the lower part of the trunk, the buttocks or the thighs, infection with the patient's faecal organisms is difficult to prevent because they may be carried from the anal area in sweat and thus gain entry to the wound. But infection of more distant sites with these organisms is again evidence of unsatisfactory technique.

Causes of breakdown in aseptic technique

Assuming that the " no touch " dressing technique, recommended by the Medical Research Council War Wounds Committee, is theoretically being employed there are four likely causes of breakdown in the technique. 1. Rapid inspection of wounds without adequate precautions. Senior members of the surgical and nursing staff may even touch the patient's skin near the wound with the unprotected hand when asked to view it. The temptation to do more than look seems almost irresistible. 2. An insufficient supply of forceps. 3. Unnecessary " cleaning " of operation wounds with antiseptic. 4. Too frequent dressings. 5. Hurried dressings due to staff shortage.

Cause 1. Bacteria will certainly be transferred in this way by unprotected fingers ; they may be harmless, but it is sometimes possible to grow " hospital staphylococci " in large numbers from swabs of ward furniture and equipment. In these circumstances the hands of the staff will inevitably be contaminated and unprotected fingers will be an important method of transferring them to the wound area. Indeed such a mechanism is much more likely than contamination by exposure to air. Culture plates which are larger in area than most surgical wounds usually have to be exposed for an hour or more (many times longer than a wound is exposed during dressing) at a busy time of day before hospital staphylococci

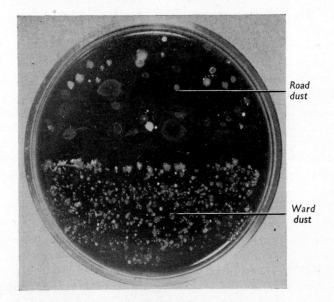

Road
dust

Ward
dust

FIG. 32.

Comparison of growth from road dust and ward dust.
(Blood agar incubated overnight at 37° C.)

can be grown from them. Because the person called to inspect a wound is not taking part in a dressings round he may not wash his hands before approaching the bedside ; even if he does so there is no certainty that he will completely remove the contaminating staphylococci.

The wound may be safely palpated if the fingers are protected with a sterile glove or by using a gauze swab. Sterile forceps, swabs and mask, minimum basic equipment, should automatically be provided for those asked to inspect wounds.

Cause 2 is usually easily remedied. Three pairs of forceps are desirable for doing each dressing, but only two pairs are really essential for most dressings. When two pairs only are available one of them is used to remove the inner dirty dressing and is then retained and used with the other pair to apply the clean dressing. The inner dirty dressing is contaminated only with bacteria from the patient's wound. The clean dressing will, in a few seconds, be similarly contaminated; it does not greatly matter if a few of the bacteria reach it a second or two earlier from the forceps which have removed the previous dressing. It is much safer to retain the contaminated forceps to help handle the clean dressing than to use only one pair of clean forceps and a hand. A shortage of forceps can often be overcome by using artery forceps discarded from the theatre or any other forceps which can be sterilized and used to hold dressings; the shape does not matter.

Cause 3. Accident wounds must be cleaned thoroughly. If particles of foreign matter and dead tissue remain infection is inevitable; *sterile* antiseptic lotions are employed for this purpose. An operation wound, however, should not be cleaned with antiseptic. Dried blood on the surface of the skin is not dangerous, indeed it is bactericidal, the proper time to wash it off is in the bath when the wound is healed. Purulent discharge may need to be wiped away with a dry swab, or local antibacterial treatment may be prescribed but antiseptic lotions should only be used at the specific request of the surgeon, never routinely. Antiseptics sufficiently harmless to be applied to wounds are not self-sterilizing. When treatment is necessary a small quantity of *sterilized* antiseptic is applied. When a dressing is stuck to the wound it should be wetted with the smallest possible amount of sterile saline. Moisture will carry bacteria from one area to another and flooding the wound area even with antiseptic may carry harmful bacteria into the wound, particularly *Pseudomonas* and *Klebsiella* which are comparatively resistant.

Causes 4 and 5. It is often insufficiently appreciated, that each time the dressings are removed the wound is exposed to the risk of infection, and further that if the wound is already infected its exposure involves dissemination of some of the pathogenic organisms into the air of the ward. Therefore dressings should not be done unless it is absolutely necessary. Many dressings combined with a shortage of staff means that they cannot be completed during the time that the ward is closed for dressings. If they are done at any other time the risk of infection is increased. Moreover, it may be necessary for one person only to do the dressing instead of having an assistant. This means that the wound is exposed to the ward air for longer than necessary. When a dressing is done in a hurry by one person the technique is likely to be imperfect. These conditions in a surgical ward lead to a vicious circle, because more wound infections occur which lead to more dressings and so on.

Investigation of hospital wound infection

When the bacteriologist is called upon to try and find the source of infection it is almost always because a " clean " operation wound has become infected. The first thing to be established is whether the infection occurred in the theatre or in the ward (see pages 309–313). When this point has been decided the procedure varies according to the nature of the infecting organism.

(a) Strept. pyogenes

The patient's nose and throat are swabbed. If the streptococcus is not recovered self-infection is most unlikely. When it is recovered self-infection is a possibility, but the upper respiratory tract may have become infected at the same time, or after the wound infection. When infection in the theatre is probable nose and throat swabs are taken from the surgeon and his assistants and from all the theatre staff, including porters and technicians, and from everyone else in the theatre during the operation. These swabs are seeded on blood agar which is incubated anaerobically to select haemolytic streptococci from the other bacteria present. Each strain isolated is grouped and all group A strains including the one from the wound are kept to be typed later. Although *Strept. pyogenes* is more commonly found in the throat than in the nose, swabs from both sites are essential. Nasal carriers are more likely to spread the organism than throat carriers, moreover some people harbour it only in their nose. When the swabs are taken each person is asked if he has any septic spots, discharge from the ears or any other lesion ; if so appropriate specimens are obtained.

When ward infection is probable nose and throat swabs are taken from all the staff who have dressed the wound or who have been present at the bedside during a dressing. All other wounds in the ward are sampled at the next dressing. If this preliminary swabbing reveals no carrier it will be necessary to take nose and throat swabs from the rest of the ward staff, including domestic staff and others such as radiographers and physiotherapists who enter the ward frequently, and from all the patients. Eczema infected with streptococci is particularly dangerous, very large numbers being shed from the lesion via bedding into the air and ward dust.

RECOMMENDATIONS. All patients who carry group A streptococci should be removed from the surgical ward. Symptomless throat carriers among the staff may continue work during treatment if special precautions are taken. As nasal carriers of streptococci are particularly likely to spread infection they should not be allowed to come into contact with surgical cases until cured. Fortunately they are rare and seldom carry *Strept. pyogenes* for longer than a few weeks. When the surgeon himself is a throat carrier he may operate comparatively safely while undergoing treatment if he takes special precautions. If he uses an antibacterial soap or detergent for washing and leaves a film of it on his hands and forearms, as already described, he is unlikely to transfer his streptococcus via the hands. Masks which are completely impervious to bacteria should always be worn in the theatre. A surgeon who knows he is a carrier should inspect his mask carefully and make sure that it is large enough and that the interleaved cellophane, when this type of mask is used, is in the correct position. He should use cream on his face, eyebrows and eyelashes to prevent particles from them falling into the wound and he will of course use a " no touch " technique whenever possible during the operation. After the operation his mask, which is particularly dangerous, should be carefully removed and, being handled by the fastening only, should be placed immediately into a bag which is closed and sent to autoclave or incinerator. Alternatively it can be immersed immediately in 5 per cent carbolic or equivalent antiseptic solution. On no account should it be thrown on the floor or left exposed to the theatre air.

Throat carriers among the nursing or lay staff can be employed in adult medical wards during treatment.

All carriers with symptoms should remain off duty until they are clinically cured. In an investigation of this kind carriers of group B, C and G streptococci are often found. It is therefore essential to group the strains as the investigation proceeds to avoid submitting these carriers to isolation and treatment unnecessarily. The bacitracin sensitivity test (page 121) is of value in reducing the number of strains which need to be grouped. Group A strains are always potentially dangerous, therefore typing is not urgent. Even if some carriers later prove to be harbouring a different type from the patients' strain no harm has been done by isolating and treating them.

(b) *Staphylococcus aureus*

Staph. aureus is ubiquitous in hospital wards and is the commonest cause of wound infection. It can be recovered from the noses of about 50 per cent of the patients and staff and from the skins of about 20 per cent of them. The air, dust and bedding in the ward is being constantly repopulated from these reservoirs. Although all strains are human pathogens the virulence of particular strains varies from time to time and an outbreak of staphylococcal infection is usually due to a single 'phage type, other types isolated from carriers in the same ward appearing for the present to be comparatively harmless.

Reference to Table 38 shows that the main reasons why *Staph. aureus* is a danger are the number of cocci inevitably present in the wards and the ability of this species to survive and remain virulent in an environment laden with antibiotics. In hospitals where antibiotics are widely employed, i.e. in all hospitals in countries which can afford them or are provided with them, "hospital staphylococci" which are resistant to the drugs locally most popular cause almost all hospital staphylococcal infection; it is on these that efforts of control must be concentrated. Antibiotic-sensitive staphylococci are at such a disadvantage that they are unable to cause epidemics in hospital.

Theatres and labour wards

It is neither reasonable nor necessary to attempt to exclude all staphylococcal carriers, but members of the staff suffering from staphylococcal infection should not be allowed on duty. Those who are heavy carriers of "hospital strains" should also be excluded and treated (see page 323). Exceptionally a senior, key member of the staff who is a carrier may continue to work while undergoing treatment. Swabs should then be cultured from all likely carrier sites, i.e. nose, axillae, perineum, conjunctivae and hair, and treatment must be aimed at controlling cocci from all these sites.

Maternity wards

Patients admitted to maternity wards are healthy young women with little previous hospital experience. They are very rarely carriers of hospital staphylococci on entry (Stokes *et al.*, 1965); the babies are sterile at birth. It is therefore worth-while to screen the staff for nasal carriage of hospital staphylococci on entry to avoid introducing such a strain into the hospital. The few carriers likely to be discovered should not work in theatres, labour wards or infant nurseries.

The nasal mucosa of newborn infants and their umbilical stumps are fertile soil for pyogenic cocci (Simpson *et al.*, 1960). As an added precaution, therefore, the babies should receive prophylactic nasal cream twice daily and be dusted with hexachlorophane powder instead of non-medicated talcum powder throughout their stay in hospital (Gillespie *et al.*, 1958). Hexachlorophane has become suspect since infants suffered toxic symptoms when bathed in liquid containing a high concentration. When used in powder form in low concentration as recommended it is harmless. No other anti-staphylococcal powder has so far been found equally effective and there have been serious staphylococcal infections as a result of stopping prophylaxis (Alder *et al.* and Ayliffe *et al.*, 1974). Mothers should be encouraged to do as much as possible for their babies, thus reducing handling by the staff. Prophylactic treatment, however, should either be done by the staff or by the mothers under supervision. When a baby or mother is infected both should be isolated. In a controlled trial these prophylactic measures proved highly successful (Stokes and Milne, 1962). Since they have become routine practice, staphylococcal infections on our maternity wards are much less common and those due to " hospital strains " are almost unknown.

General wards

Patients admitted to general wards are much more likely to carry hospital staphylococci because many of them have previous hospital experience and may have received chemotherapy for long periods. Those negative on arrival are likely to acquire hospital strains, particularly if they need antibiotic treatment (Knight, 1954). It has been shown in wards employing the non-touch dressing technique and having skilled medical and nursing staff that the number of wound infections due to hospital staphylococci is directly proportional to the numbers of carriers of these strains admitted to the wards (Stokes *et al.*, 1965). Very large numbers of cocci shed into the ward from carriers will allow some of them to find loopholes in the routine aseptic technique.

Since the introduction of many new anti-staphylococcal antibiotics the incidence of staphylococcal hospital infection has declined. So far no strain resistant to the penicillinase stable penicillins has managed to cause worldwide hospital infection on the scale of the notorious type 80/81 which caused havoc in the nineteen-fifties. Nevertheless localized outbreaks of infection by these strains have been reported and they are increasing in Britain (Parker and Hewitt, 1970). When more than a single infection is seen in a ward prophylactic measures, similar to those which have proved successful in maternity wards, offer the best hope of success in surgical wards also. Routine application twice daily of nasal cream, which should contain polynoxylin or gentamicin when the epidemic strain is neomycin resistant, to *all* patients, and routine use of hexachlorophane powder and hexachlorophane soap by both patients and staff are likely to reduce the load of dangerous cocci in the ward. Dissemination from carrier sites is very variable and a patient in the ward for investigation without an operation wound may well be potentially dangerous, especially if he has been transferred from another hospital department where he has acquired a hospital strain.

Nasal carriage of hospital staphylococci among patients is related to duration of stay as well as to antibiotic therapy. Different departments have different average carrier rates ; from low to high the order is : maternity, surgical, medical, geriatric. Radiotherapy patients also tend to carry these strains (Stokes *et al.*, 1965). Patients are much more often carriers than staff, and nursing staff who come into closer contact with them have a higher carrier rate than medical staff. Staff receiving systemic antibiotic treatment while on duty run an increased risk of acquiring hospital strains and therefore this should be avoided whenever possible.

Treatment of nasal carriers

At least 1 month's treatment is required and if conscientiously carried out it is usually successful in getting rid of hospital strains of staphylococci (Stokes *et al.*, 1965). These strains are not biologically dominant and in an environment free of antibiotic they tend to be replaced by sensitive wild types.

1. Apply nasal cream [1] to the anterior nares 4 times daily : on waking, on going to bed and twice during the day immediately before going on duty. Continue this for 4 weeks.

2. Use hexachlorophane " Cidal " soap both at home and at work.

3. During the first week after starting nasal treatment, wash hair in 1 per cent cetrimide and as soon as possible after this wash all woollen jumpers in 1 per cent cetrimide and do not rinse it out too thoroughly. Also send often-worn heavy garments and blankets to be cleaned. Have clean bed linen immediately after hair-washing. The cleaning programme should be complete by the end of the third week of nasal treatment.

4. After treatment has started boil face-flannels, loofahs, and powder puffs. Empty powder case, clean it with 1 per cent cetrimide and refill from a new box. Discard powder puffs which cannot be boiled.

All treated carriers should be followed up by culturing a nasal swab at least 3 days after the last application of cream and again 2 weeks or more, with no further treatment, after the first swab.

(*c*) Coliforms and enterococci

Esch. coli, Strept. faecalis, Klebsiella, Proteus and *Ps. aeruginosa* all commonly invade wounds particularly during antibiotic treatment. Although all these organisms are found in human faeces, carriers among the staff are unimportant as a source of infection. Their usual source in a ward is septic wounds, particularly faecal fistulae and colostomies. If dressings from such wounds are improperly handled they will contaminate the ward air and dust. Bedding is also liable to be contaminated from them. Primary infection in the theatre is rare and is usually the result of wound contamination by the patient's faecal flora.

Coliforms are becoming increasingly dominant in hospital infection and among them *Ps. aeruginosa* takes pride of place. Antibiotic-resistant *Esch. coli* and *Klebsiella sp.* are also troublesome. Although they are much less prevalent than staphylococci in most wards, they are able to multiply at room temperature and so increase

---

[1] Nasal cream is " Naseptin " I.C.I., containing neomycin and chlorhexidine; when the strain is neomycin-resistant, polynoxylin cream which liberates small quantities of formaldehyde can be substituted or gentamicin-containing cream can be used.

in contaminated fluids (see Table 38). *Ps. aeruginosa* is highly motile and when the trap of a bath or hand-basin drain is contaminated, the organism can be demonstrated in the basin itself, on soap and nail brush. A contaminated shaving brush has caused an epidemic (Ayliffe *et al.*, 1965). It will also grow in drainage bags containing urine and other body fluids and will contaminate moisturizers in ventilation systems (Anderson, 1959). Whenever cultures from wound or urine yield this organism, hospital infection should be suspected.

INVESTIGATIONS. *Ps. aeruginosa* is a normal commensal of the gut. It is found very frequently in chronic otitis externa, in tracheostomy wounds and in sputum from patients with chronic lung cavities. Often in these patients and sometimes in the hospital staff it can be cultivated from nose swabs. It is one of the organisms commonly responsible for delayed healing in burns (Lowbury, 1960). Its resistance to chemotherapy and antiseptics allows it to survive in these sites when other more virulent pathogens are killed. The presence of *Ps. aeruginosa* in a discharge does not necessarily imply infection. In tracheostomy wounds, for example, it may exist for long periods without apparently damaging the patient, and when the tube is removed the wound quickly heals. Nevertheless, in debilitated patients and especially in those whose body defences have been deliberately reduced, *Ps. aeruginosa* is difficult to control. Polymyxin (colistin) although always active against it in the laboratory has proved disappointing *in vivo*. Treatment with carbenicillin and an appropriate aminoglycoside should be given only when there is clinical and laboratory evidence of serious infection because of the danger of encouraging resistance. Hospital infection due to R factor carrying strains may be very difficult to eradicate (Lowbury *et al.*, 1972). The laboratory must be careful to distinguish between superficial contamination with this organism, which is common in tracheostomy wounds and in sputum, and true infection.

Because *Ps. aeruginosa* proclaims its presence in pus and in cultures by its striking green colour there is perhaps a tendency to over-emphasize its importance in individual patients. To the surgeon and hospital bacteriologist, however, its presence in a wound often indicates poor aseptic technique and it must not therefore be ignored.

In searching for the source of infection, contamination of medicated fluid applied to the lesion should first be checked. Medium containing cetrimide will select *Pseudomonas* from the majority of other contaminating organisms (p. 369). There are many harmless Pseudomonads that can be cultivated from dust, drains, etc., which grow poorly at 37° C. Strains of *Ps. aeruginosa* can be 'phage typed and pyocine typed to confirm or disprove their possible role in an epidemic. Pseudomonads other than *Ps. aeruginosa*, which are normally of low virulence, can sometimes infect. They are culturally different from *Ps. aeruginosa* and may be very antibiotic resistant, e.g. *Ps. multivorans* (Bassett *et al.*, 1970).

(d) Anaerobic sporebearers

The main reservoirs of the bacteria which cause gas-gangrene and tetanus are the faeces of both man and animals, and soil. *Cl. welchi* is almost constantly present in human faeces and can be recovered from soil and from street, house and hospital dust (Lowbury and Lilly, 1958). The other clostridia which cause gas gangrene—*Cl. oedematiens*, *Cl. septicum* and *Cl. histolyticum*, are mainly found in soil and infection by them is rare in peace time. *Cl. tetani* is found mainly in horse faeces and in soil contaminated with horse manure.

Virulent strains of all these bacteria have been recovered from wounds of patients who are not suffering from, and do not subsequently develop, gas-gangrene or tetanus. They are harmless in a wound as long as conditions are such that they are unable to multiply and produce their powerful exotoxins. A very low oxidation-reduction potential in the tissues is necessary for growth. It may be provided by the presence of necrotic tissue resulting from trauma, lack of blood supply, or the action of other bacteria. Suitable conditions are more easily produced in deep wounds.

INVESTIGATIONS AND RECOMMENDATIONS. These are the only infections which are likely, in a modern hospital, to be caused by faulty sterilization of dressings and ligatures. The infections already considered are caused by vegetative organisms which are killed by boiling water in 2 minutes. Only by some gross error of technique could they survive on dressings sterilized in the autoclave. Spores, however, may survive autoclaving if dressings are packed too tightly or if insufficient air is withdrawn from the chamber before sterilization. Such errors may pass unnoticed for long periods unless routine tests are made (see page 332).

When gas-gangrene occurs as a hospital infection the cause is almost always *Cl. welchi*. The patient can be self-infected from faeces. This is a hazard of mid-thigh amputation for gangrene because skin preparation will not necessarily remove all spores. If they enter the wound the patient's poor blood supply, plus unavoidable trauma during operation, is likely to result in favourable growth conditions and gas gangrene will result. Penicillin, which at present is always active against *Cl. welchi* should be given prophylactically. The placental site or gynaecological operation wounds may be infected from the vagina where a few *Cl. welchi* are normally found. It has already been noted that the organism can be recovered from theatre or ward air and dust. Small numbers from these sources probably enter wounds fairly often during operation but they seldom find conditions suitable for growth. The incubation period is very variable and therefore not of much value in estimating the time of infection. Symptoms may appear within a few hours of contamination, but the organism may remain quiescent for many days as a saprophyte and then if conditions become favourable it will multiply and symptoms will appear long after the original infection.

Hospital tetanus is a disaster; it has a high mortality. The incubation period is extremely variable. Spores have been known to remain in healed wounds for years and then to multiply and cause tetanus when conditions become favourable after a subsequent operation. The second wound need not be very near the original site because the spores can be carried to a distance by phagocytes. When there is no previous history of a soil-contaminated wound the cause is usually failure to sterilize something contaminated with tetanus spores which has been in close contact with the wound. Entry of spores from dust suspended in the air of theatre or ward is a possibility but is not very likely in most neighbourhoods. Catgut has been proved to be the source of some cases of hospital tetanus but manufacturers are now aware of this danger and no case from improperly sterilized catgut has been reported in Britain for many years.

It is difficult to sterilize rubber gloves at a sufficiently high temperature, or for sufficiently long, to be certain of killing all spores without destroying the rubber. The risk of tetanus from this source, however, is negligible because tetanus spores are not particularly heat resistant and gloves are extremely unlikely to be contaminated with them. Moreover, gloves are carefully washed free from external contamination before autoclaving. Small powder puffs are often sterilized with the gloves and the powder in them may be a source of danger. The puffs should therefore be sterilized first at a high temperature and then packed with the gloves to be autoclaved for the second time. Cotton wool or stockinette to be placed under a plaster should be autoclaved. Even if there is originally no wound the plaster may rub and damage the skin and allow spores to enter the tissues. All dressings and cotton wool should be autoclaved before being stored in the theatre, so that if they happen to be contaminated with dangerous spores these will not contaminate the theatre air or dust.

The introduction of building materials into or near the theatre is dangerous, particularly if the ventilation is such that dust raised by the builders is sucked into the theatre. If it is impossible to close the theatre when structural alterations are being made in the vicinity, the part under reconstruction should be sealed off from it in the manner used for sealing rooms during fumigation.

Unlike the vegetative hospital pathogens the spore-bearing *Clostridia* do not easily spread from patient to patient and isolation is not required. When more than one case is seen a common source of improperly sterilized equipment or of contamination after sterilization should be sought. Lack of appropriate prophylaxis will by chance cause an apparent epidemic in patients undergoing amputation.

Sampling the patient and his environment

1. *Nose and throat swabs.* When large numbers of contacts are swabbed it is difficult to avoid an occasional sterile specimen, which has to be repeated. If arrangements are made for the contacts to attend the laboratory for swabbing the number of unsatisfactory specimens can be reduced because there is minimal delay between sampling and culturing. Nasal swabs need to be moistened in peptone water, broth, or sterile saline before sampling.

2. *Skin.* A moistened swab is rubbed vigorously over the chosen site. The back of the wrist is a suitable area when looking for staphylococcal skin carriers. It is less likely than the hand to be contaminated with bacteria from extraneous sources. The swab is seeded over at least half a culture plate. It is unnecessary to spread the inoculum.

Since *Staph. aureus* produces phosphatase more rapidly than other staphylococci, phenolphthalein phosphate agar (Barber and Kuper, 1951) can be employed to detect them more easily in mixed culture. Phosphatase releases phenolphthalein and when exposed to alkaline vapour (by dropping about 5 drops of concentrated ammonia into the lid of an overnight culture plate) the colonies of phosphatase producers turn bright pink. Only pink colonies need therefore be tested for coagulase production. They can be detected among similar colonies and even when covered by spreading proteus ; for method of preparation, see page 369.

The skin of a surgeon or his assistants is best sampled at the end of a long operation by swabbing the hands or pressing the inner surface of carefully removed gloves on blood or phenolphthalein phosphate agar.

3. *Floors, furniture, baths, washing bowls, etc.* A moistened swab is rubbed over a wide area of the surface and then seeded on blood agar or appropriate selective medium. It is wise to streak out the pool of inoculum because even clean-looking surfaces often yield a heavy growth.

4. *Textiles.* Linen, cotton, and other smooth materials are most suitably sampled by the press plate method. The part of the material to be sampled is pressed against the surface of a culture plate. The remainder is folded thickly behind it to protect it from contamination by the hand.

Woollens are conveniently sampled either by shaking them over an exposed plate or by using the edge of the plate to brush the fabric so that particles from it will land on the surface of the medium.

5. *Unsterile equipment* such as plaster shears, bandage scissors, safety pins, sandbags and mackintosh sheets.

These are sampled as described for furniture but there is no need to streak out the inoculum from small articles.

6. *" Sterile " instruments and dressings.* The nurse, theatre sister or dresser is asked to prepare the instrument or dressing exactly as if it were to be used on the patient. Then the operating end of the instruments is dipped into bottles containing broth and cooked meat broth. Using scissors and forceps, dry-sterilized in the laboratory, portions of dressings are cut off and cultured in broth and cooked meat broth. After 4 days' incubation the cultures are seeded on blood agar plates which are incubated in the appropriate atmospheres.

This is a very sensitive method. If one viable organism enters the broth during sampling it will grow, so the method gives little indication of the weight of contamination. Growth of aerobic spore-bearers, micrococci and other microbes normally considered harmless is not necessarily a serious matter. The majority of cultures carefully made in this way should be sterile but growth of a single species in say 10 per cent of bottles, if no pathogen is recovered, is not evidence of faulty technique. Most surgical techniques are not sterile in the strictest sense ; they are aimed at excluding all known pathogens and reducing as far as possible the number of all other microbes which enter the wound.

7. " *Sterile* " *lotions.* If the solution is not antiseptic a large volume of it can be sampled by delivering about 10 ml. with a sterile pipette into an equal volume of double strength peptone water. Two such cultures are made, one for incubation at 37° C. and one at room temperature. Some antiseptic solutions, for example those containing mercuric salts, can be rendered harmless to bacteria before culture by the addition of a neutralizing chemical. When this is impossible the antiseptic properties are overcome as far as possible by dilution. The degree of dilution necessary can be ascertained by testing the growth of small inocula of the pathogen sought (one drop of a 10^{-4} dilution of an overnight broth culture) in serial broth dilutions of the antiseptic. Alternatively the fluid can be filtered through a membrane which after washing by filtering large volumes of sterile water can be placed on a culture plate. Any bacteria on it will yield visible colonies after incubation.

8. *Air.* Two kinds of bacteria-carrying particles are found in air—small ones which remain suspended for long periods and large ones which fall to the ground within about an hour in a still atmosphere. Infections caused by bacteria which gain entry via the respiratory system can be transferred by the inhalation of either large or small particles. If the risk of infection by this route is to be satisfactorily investigated a slit sampler is required (see Studies in Air Hygiene). The small particles, since they remain suspended, are unimportant in wound infection. The large ones can be collected by allowing them to fall on the surface of exposed blood agar.

To find the risk of aerial contamination of a wound in the theatre it is reasonable to expose blood agar plates in pairs at various sites for the duration of the longest operation likely to be performed.

At least one pair of plates should be prepared sterile on the outside so that they may be exposed near the operation site.

Petri dishes[1] are sterilized without covers in a flat tin in the oven. When cool the lid of the tin is lifted and blood agar is poured into them with minimal exposure.   The lid is then closed and time is allowed for the medium to set.   When the surgeon is ready to make the incision the tin is opened and he takes out the sterile blood agar plates and places them conveniently near the wound.   At the end of sampling they are handed to the bacteriologist who covers them with sterile lids in the usual way.

One plate of each pair is incubated in air, the other anaerobically. After 24 hours' incubation the number of colonies on the aerobic plates is counted.   Contamination of the air is expressed as the number of colonies per minute of exposure.   One colony of each type found on the anaerobic plates is streaked on blood agar to be incubated aerobically and then, without flaming the loop, on to blood agar for anaerobic culture.   Those which grow on the anaerobic but not on the aerobic subculture are anaerobes and need further investigation.   Known pathogens likely to appear are *Staph. aureus*, *Strept. haemolyticus*, coliforms and *Cl. welchi*.

To obtain a good idea of the risk of infection from the air it is necessary to repeat the sampling on several occasions because the result depends on a number of variable factors, the most important of them being : (*a*) the number of people in the theatre ; (*b*) the amount of movement, particularly of the patient and his coverings and of people entering and leaving the theatre ; (*c*) the number of operations previously done in the theatre since it was last well aired ; (*d*) the direction and strength of the wind and (*e*) the humidity of the atmosphere.

Autoclave sterilization

Sterilization of materials by steam under pressure is effective, quick and cheap, but its effectiveness depends on the proper handling and maintenance of the autoclave which needs frequent testing and supervision.   Although an old-fashioned boiler will not kill all spores and leaves instruments wet, all vegetative bacteria (i.e. all those which commonly cause hospital infection) and tubercle bacilli are killed after 2 minutes' boiling.   Disinfection of instruments by this simple method is safer than " sterilization " in a badly maintained and inefficient autoclave.

[1] Sterile plastic dishes can be transferred to a sterilized tin aseptically.

For preference instruments and dressings should be sterilized centrally pre-wrapped. They should emerge dry and should not be exposed to the risk of contamination after sterilization. In hospitals this is most readily achieved by two central, high-vacuum autoclaves operated or supervised by skilled staff. Two is the minimum number to allow for servicing. Theatre autoclaves for the sterilization of theatre instruments which need not be pre-wrapped can be of the cheaper downward displacement type. Wards should receive pre-sterilized dressings and instruments from a central sterile supply department (CSSD) because it is uneconomical to install ward autoclaves and usually impossible to maintain and operate them efficiently.

Tests of efficient sterilization

(*a*) *Downward displacement autoclaves.* Air is partly evacuated either by pump or by steam ejector and steam flows in through the top of the chamber, displacing the remaining air which flows out through the chamber drain at the bottom. In this type of autoclave packing is extremely important because the flow of steam must not be impeded either by overloading the chamber or packing too tightly. Waterproof materials, even when interleaved, are very difficult to sterilize by this method.

The cycle is prolonged because removal of air from the load is slow and subsequent drying is slow ; in a large machine it may take more than an hour. Articles most exposed to heat in the load may be damaged if those least exposed are to be properly sterilized. These machines are therefore suitable only for sterilization of instruments where no penetration of textiles is required. They are simple to use and efficient for this purpose. They are also employed for sterilizing bottled fluids.

TESTS. Browne's tubes [1], which change colour from red to green on exposure to heat for an appropriate time, are suitable for routine testing provided the type 1 tube (black spot) is used and that the tubes are stored in a cool place, not above 21° C. (70° F). Autoclave tape which becomes striped on heating will check that articles have been through the machine but not that they have been efficiently sterilized.

(*b*) *High-vacuum autoclaves.* These more costly machines over-

[1] Obtainable from Albert Browne Ltd., Leicester.

come the difficulty of penetration of steam into the load and of drying, by employing much more efficient evacuation of air using high-vacuum pumps before steam flows in, and by rapid evacuation of steam after sterilization. A higher temperature is reached but the articles escape damage as the cycle is much shorter, about 15 minutes with a large machine. These are the best machines so far made for the sterilization of dressings and wrapped instruments.

TESTS. The chamber may leak or the pumps be inefficient and this must be checked by daily inspection of a chart on which temperature and pressure during each cycle are automatically recorded. Browne's tubes [1] type 2 (yellow spot) can be employed within loads, but when temperatures are high (134–137° C) sterilization must not be more than 3 minutes. After this time the capsules may change colour in dry heat which is not a fair test as bacterial spores are not susceptible to dry heat for a short time at this temperature. A test of steam penetration within a load has been devised by Bowie and Dick employing autoclave tape ; it is reliable when properly carried out.

Bowie-Dick Test

MATERIALS. Huckaback towels 90 × 60 cm. to form a pile 25 × 27·5 cm. high when folded. Paper, firm but not waterproof. Autoclave tape, 3M brand No. 1222, sufficient to form a diagonal cross when stuck on the paper.

METHOD. The towels are washed when new or soiled, or are hung up for at least 1 hour before each test. The paper with tape on it is placed in the middle of the pile and the whole is held in whatever container is employed for dressings. It is then placed in the empty autoclave which is run for its normal cycle. When penetration is satisfactory the stripes will be uniformly darkened. The sterilizing time must not exceed $3\frac{1}{2}$ minutes at 134–137° C. because the tape will change colour in dry heat. It must be stored in a cool place and must not be left in the warm autoclave room before testing. The test should be done in an empty autoclave because conditions are least favourable in this type of autoclave when there is only one pack and a large volume of air has to be evacuated from the chamber (Bowie *et al.*, 1963).

LIKELY FAULTS. A common and obvious fault is wet dressings. This is often due to poor-quality steam supply, which is the province of the hospital engineer (Ministry of Health Memorandum, 1963).

[1] Obtainable from Albert Browne Ltd., Leicester.

It is also likely to happen if the jacket is not pre-warmed so that there is condensation on the inside of the chamber when steam flows into the cold machine.   It can also be caused by condensation on metal articles, jugs, bowls, etc., if they are sterilized with the dressings. This can often be avoided if they are packed inverted so that water cannot accumulate in them.   It may be necessary to pack them separately.   This is a very serious fault when dressings are packed in paper or linen because bacteria will be carried in moisture from the surroundings to the inside of the pack and the dressings will be contaminated after sterilization.   All autoclaves should be thoroughly tested when installed, and at regular intervals thereafter (see Ministry of Health Memorandum, 1963).   More detailed information on this complex subject is given in *Hospital Infection* by Williams *et al.* (1966).

Culture of blood for transfusion

Blood for transfusion is sometimes sent for culture when a patient has suffered an unexplained reaction.   Patients are not likely to suffer unless a bacterium, introduced into the blood at some stage, has had an opportunity to grow.   Symptoms are usually caused by toxins preformed during growth rather than by infection with the organism.   Toxin formation is likely when the contaminating organism grows well at $4°$ C. and may happen if the transfusion is given slowly and sufficient time elapses for growth to occur after the bottle is removed from the refrigerator.   Since many bacteria which multiply at $4°$ C. are incapable of growth at $37°$ C. routine methods have to be modified.

PROCEDURE

Make two smears for Gram and Leishman stains.

Seed three blood agar slopes (in screw-capped bottles to prevent evaporation), and incubate one at $37°$ C., one at room temperature in the dark and one at $4°$ C.   Deliver about 2 ml. of the blood into each of three bottles containing peptone water and incubate one with each of the blood agar slopes.

Examine the cultures daily.   If no growth is visible when they have been incubated for the same length of time that elapsed between the collection of blood from the donor and the transfusion, they may be discarded.   Growth may be difficult to see in the liquid cultures, they should therefore be subcultured on blood agar and Gram-stained before a negative report is sent.

The bacteria which have most often been found as the cause of this rare accident are coliform bacilli capable of growth at low temperatures (James and Stokes, 1957 ; see also A. C. P. Broadsheet 1974).

Dysentery and Food Poisoning

The patients and staff of hospitals are occasionally afflicted with epidemic diarrhoea and vomiting. Bacterial food poisoning is the most likely cause and the bacteriologist is called upon to find the source of infection and to recommend measures for its control.

In Britain the organisms which commonly give rise to these symptoms are salmonellae, dysentery bacilli, *Cl. welchi*, enterotoxin-producing strains of *Staph. aureus* and *B. cereus*.

PROCEDURE

The first step is to collect samples of faeces and vomit for culture. Then a very careful history is taken, noting especially the following points. 1. The number of people affected and the time of onset of symptoms. 2. The items of food and drink consumed by all the patients at the last meal attended by all of them.

The answers to the first question indicate the type of food poisoning. When it is caused by a bacterial toxin the onset will be abrupt, within a few hours of eating the food, and symptoms will appear in all those affected within about twelve hours. When it is caused by infection with *Shigella* or *Salmonella* the incubation period will be longer, from twelve hours to two days, and the distribution of the time of onset of symptoms will be much more scattered. When a few people only are affected it is as well to check carefully that they all have similar symptoms. When five or six people from the same department happen to fall ill at approximately the same time an epidemic may be wrongly suspected.

Answers to the second question usually indicate fairly clearly which dish was contaminated. It must be something consumed by all the people affected. When a number of people are symptom free, in spite of having eaten the suspected food, there are two possible explanations. Either the food was solid or semi-solid and the toxin or bacteria were not evenly distributed in it, or a part of it only was contaminated. For example, milk from an infected churn might be poured into one or two jugs only out of several consumed at the meal.

Any food having the following characteristics is a particularly likely source of infection.

1. Food which has been much handled in preparation.
2. Food composed of animal protein such as eggs (particularly dried egg), home cured ham, sausage meat, minced meat, and brawn.
3. Food, either cooked or uncooked, which has been prepared and allowed to stand for several hours before it is served.

It is important to realize that cooking often fails to kill food-poisoning bacteria. Penetration of heat through a large pie in a domestic oven is not rapid and although the temperature of the outside may be 100° C. the inside can be less than 60° C. *Staph. aureus*, *Shigella* and *Salmonella* are not particularly sensitive to heat. Egg dishes, for example, are seldom heated sufficiently to kill the *Salmonellae* which may be found in them.

Food is seldom *originally* contaminated with sufficient pathogens to cause infection. If the few pathogens originally present survive preparation and the food is not consumed immediately they may multiply and then, even if it is re-heated before serving, it is likely to prove infective. When toxigenic strains of *Staph. aureus* have grown in the food their subsequent death by heating will not prevent food poisoning, because the enterotoxin is relatively heat stable and is unlikely to be destroyed by temperatures reached in cooking.

Some strains of *Cl. welchi* also cause outbreaks of food poisoning (Hobbs *et al.*, 1953). Acute abdominal pain starts 8–24 hours after eating the food and is followed by diarrhoea. Vomiting, pyrexia and other symptoms are uncommon, and recovery within twenty-four hours is the rule. The food is almost always meat or poultry which has been boiled, allowed to cool and eaten perhaps some hours later either cold or reheated. It has a normal appearance and taste when eaten.

Direct smears of the food show Gram-positive bacilli resembling *Cl. welchi* and direct blood agar cultures yield a heavy growth of non-haemolytic or slightly haemolytic *Cl. welchi* colonies. The Nagler reaction is positive. Cultures of some strains recovered from food have been shown to withstand five, but not six hours' simmering.

Heat-resistant strains can be isolated from a high proportion of faeces from those affected. A suspension of the faeces in cooked meat broth which is steamed for one hour and then incubated can be expected to yield a pure growth. These strains are rarely found in normal faeces or in those from patients suffering from other intestinal infections. Some strains, however, are not particularly heat resistant so an unheated culture should also be examined.

Outbreaks are prevented by the method described for staphylococcal food poisoning, i.e. the food must be cooked and eaten immediately or cooked and refrigerated so that any *Cl. welchi* which may contaminate it are given no chance to multiply.

Bacillus cereus, contaminating cooked rice which is allowed to stand at room temperature, will grow in it and cause gastrointestinal symptoms, see page 162.

It is well known that freezing preserves rather than kills most bacteria. If pathogens have had a chance to multiply in food before it is frozen, food poisoning will result when it is consumed. Even if it is thoroughly heated toxins in it may give rise to symptoms.

Every effort should be made to culture the remains of the suspected food because a positive result may narrow the field considerably in the subsequent search for the source of infection.

Measures for control

(*a*) *In the kitchen.* Immediate measures are aimed at preventing multiplication of bacteria in the food. When several hours' delay between cooking and eating is unavoidable the food should be rapidly cooled and then refrigerated until its final preparation. In hospital kitchens, mixers, slicers, chopping boards and other equipment should be heat disinfected at least once daily. Facilities for handwashing after visiting the toilet are inspected and improved when necessary.

In outbreaks of food-poisoning caused by Salmonella and dysentery organisms, carriers in the kitchen staff must be sought. When *Staph. aureus* has been isolated from the food, and from the faeces or vomit of the patients, an attempt can be made to find a nasal carrier of the same 'phage type among the kitchen staff. If there are no facilities for 'phage typing, or if the staphylococcus was killed in cooking and the diagnosis remains clinical, no good purpose is served by swabbing the kitchen staff. About half of them will probably be found to be carrying *Staph. aureus* in the nose and without differentiation between enterotoxin-producers and others, isolation of carriers is impracticable. Further cases are prevented by keeping the food cool after preparation so that the staphylococci do not get a chance to grow and produce their toxin. Cooks with colds should be encouraged to report sick and remain off duty until cured, because frequent use of the handkerchief in the kitchen by a carrier is dangerous. Instruction of the staff in the need for frequent hand-washing and the provision of antibacterial soap and plenty of clean towels will also help to lessen the risk. Cooks with septic hand lesions should not be allowed on duty.

Two methods are available for searching for carriers of the intestinal pathogens—serological tests and faeces culture. In an outbreak of salmonella food-poisoning, when the kitchen staff have not received prophylactic T.A.B. inoculation, a search for salmonella antibodies in their serum may lead to the discovery of a chronic intermittent carrier who would otherwise have escaped notice. The sera can be tested against the epidemic strain. A positive agglutination test does not of course prove that an individual is a carrier, it merely shows the need for repeated stool cultures.

Dysentery carriers do not necessarily show antibodies to the bacilli and the method of choice for investigation is rectal swab or faeces cultures.

Kitchen staff are usually co-operative if the reason for the tests is explained to them and if they are warned from the outset that more than one specimen from each person will probably be necessary.

(*b*) *In the wards.* Staphylococcal food-poisoning is an intoxication, not an infection and therefore no precautions against the spread of infection are necessary.

The principles of barrier nursing patients with intestinal infections are described in the Medical Research Councils War Memorandum No. 11 and in *Hospital Infection* by Williams *et al.* (1966). A few practical difficulties may be worth mentioning here.

When there are only one or two infected patients in a general medical ward a special nurse is often appointed to look after them. This appears to be very desirable but in practice when the patients need frequent attention and when the nursing staff work in three shifts it means that six nurses are necessary if there is to be no overlap of duties, because at any time one of the infected patients may need attention while his nurse is having a meal. It is clearly impracticable to appoint six nurses to look after one or two patients. A nurse is unlikely to spread infection unless she contaminates her hands from the infected patients' excreta and then touches something which enters another patient's mouth. This can be prevented as follows. The nurses on each shift are divided into two groups. One group has nothing to do with the infected patients and as far as possible avoids handling bed-pans altogether. The other group is carefully instructed in isolation nursing technique and attends to the infected patients. These nurses also perform certain general duties such as giving bed-pans, helping with bed-making and doing dressings; they are absolutely forbidden to handle the uninfected patients' food or crockery. It is better if they do not enter the ward kitchen. They will not take temperatures of non-infected patients nor give them medicines except by injection. After attending the infected patients they will of course wash their hands very thoroughly. If the hands are heavily contaminated washing cannot be guaranteed to remove all the intestinal pathogens. Therefore when incontinent or helpless patients are given bed-pans the nurses' hands should be protected with rubber gloves. The gloves are either disposable or they are removed carefully and boiled for two minutes each time they have been used.

Hepatitis

Until the discovery of hepatitis B virus and the outbreaks of hepatitis in dialysis units, blood was regarded as non-infective and no special precautions were taken by those handling it. The degree of risk is uncertain but is not great outside special units.

It has not yet been established which of the particles seen by electron microscopy are infective. Since blood contains very numerous particles before hepatitis can be diagnosed clinically all blood samples must be regarded as potentially infective. The virus is believed to be resistant to boiling and some disinfectants but it cannot easily be studied at present. Decontamination is carried out either by autoclaving or by application of disinfectant containing free chlorine, 3000–5000 parts per million.

Patients who have received repeated blood transfusions, especially those on haemodialysis and haemophiliacs, and those having injections with inadequately sterilized equipment, such as drug addicts and the extensively tattooed, are likely to be carriers.   Their blood should be sent to the laboratory in additional sealed plastic containers with a warning biohazard label in case of spillage en route.   The infection cannot be transmitted by carriers except by contact with their blood, other patients are not therefore at risk except through faulty sterilization of needles, syringes or surgical instruments. Infected patients should be isolated for the first week of the illness.

Small quantities of human blood, plasma or serum needed in the laboratory, for example for coagulase tests, should be taken from unused donor blood which has been tested for hepatitis B (Australia) antigen.

Respiratory infection

Acute infections spread by droplets from the upper respiratory tract are not nursed in general medical wards unless, as in lobar pneumonia, factors other than the spread of the bacteria are needed to establish infection.   The most important infective respiratory disease which is nursed in general hospitals is pulmonary tuberculosis.   Little is known about the way in which the tubercle bacillus travels from " open " cases and initiates infection in contacts.   Probably the most dangerous infective material is minute particles of dried sputum which contaminate the patients' bedding and clothing. Tubercle bacilli are resistant to killing by disinfectants but are as sensitive as other vegetative organisms to heat.   The method of steeping linen in disinfectant for 12 hours cannot be relied on to kill them and prolonged treatment is usually inconvenient.   In hospitals which have their own laundries, linen from tuberculous patients can be washed in very hot water before sorting, in a machine set aside for the purpose.   If there is no hospital laundry, it is placed in bags and autoclaved at 5 lb. pressure for 30 minutes before sorting.   This has the disadvantage that heating before washing tends to fix stains in the linen which cannot be removed.

Termination of isolation

Fumigation by sealing the room and releasing formalin vapour is no longer considered to be necessary except in very dangerous infections such as smallpox when it is best carried out by the Public Health Department who have the training and equipment to do it properly.

Guidance on the need for isolation and barrier nursing will be found in " Control of Communicable Disease in Man ", (1970) an official report of the American Public Health Association approved by Public Health Authorities in Britain. After either of these procedures terminal cleaning is carried out. This is normally done by an orderly wearing a gown who cleans the whole room and furniture very thoroughly, first with disinfectant, e.g. 1 per cent Sudol, and then he removes the disinfectant in the same order of cleaning so that each area treated is exposed for an adequate time. Vigorous cleaning during application is very important, the disinfectant will not penetrate layers of dust. The room can be reoccupied without delay. Plastic-covered pillows and mattresses can be treated similarly otherwise they should be heat treated with bedding and clothing. A short disinfecting cycle is sufficient since in general spores can be ignored ; *B. anthracis* is the only sporebearer likely to cause transmissible infection. Heat sensitive articles such as shoes and other leather goods can be disinfected in a small formalin cabinet.

References

ALDER, V. G., BURMAN, D., CORNER, B. D. and GILLESPIE, W. A. (1972). *Lancet*, **ii**, 384.

ANDERSON, K. (1959). *Med. J. Aust.*, **1**, 529.

ASSOCIATION OF CLINICAL PATHOLOGISTS, Broadsheet No. 54 (1974) by Tovey, G. H. and Gillespie, W. A.

AYLIFFE, G. A. J., BRIGHTWELL, K. M., BALL, P. M. and DERRINGTON, M. M. (1972). *Lancet*, **ii**, 479.

AYLIFFE, G. A. J., LOWBURY, E. J. L., HAMILTON, J. G., SMALL, J. M., ASHESHOR, E. A. and PARKER, M. T. (1965). *Lancet*, **ii**, 365.

BARBER, M. and KUPER, S. W. A. (1951). *J. Path. Bact.*, **63**, 65.

BASSETT, D. C. J., STOKES, K. J. and THOMAS, W. R. G. (1970). *Lancet*, **i**, 1188.

BOWIE, J. H., KELSEY, J. C. and THOMPSON, G. R. (1963). *Lancet*, **i**, 586.

CAIRNS, H. (1939). *Lancet*, **i**, 1193.

COLEBROOK, L., DUNCAN, J. M., and ROSS, W. P. D. (1948). *Lancet*, **i**, 893.

DEVENISH, E. A. and MILES, A. A. (1939). *Lancet*, **i**, 1088.

GARROD, L. P., LAMBERT, H. P. and O'GRADY, F. (1973). *Antibiotic and Chemotherapy*. 4th edition. Churchill-Livingstone, Edinburgh.

GILLESPIE, W. A., SIMPSON, K. and TOZER, R. C. (1958). *Lancet*, **ii**, 1075.

HOBBS, B. C., SMITH, M.E., OAKLEY, C. L., WARRACK, G. H. and CRUICK-SHANK, J. C. (1953). *J. Hyg. Camb.*, **51**, 75.

JAMES, J. D. and STOKES, E. J. (1957). *Brit. med. J.*, **ii**, 1389.

JAMES, K. W., JAMESON, B., KAY, H. E. M. and LYNCH, J. (1967). *Lancet*, **i**, 1045.
KNIGHT, V. (1954). *J. clin. Invest.*, **33**, 949.
LOWBURY, E. J. L. (1954). *Lancet*, **i**, 292.
—— (1960). *Brit. med. J.*, **i**, 994.
LOWBURY, E. J. L., BABB, J. R. and ROE, E. (1972). *Lancet*, **ii**, 941.
LOWBURY, E. J. L., LILLY, H. A. and AYLIFFE, G. A. J. (1974). *Brit. med J.*, **iv**, 369.
LOWBURY, E. J. L. and LILLY, H. A. (1958). *J. Hyg.*, **56**, 169.
McKISSOCK, W. (1941). Personal communication.
MILES, A. A., WILLIAMS, R. E. O., and CLAYTON COOPER, B. (1944). *J. Path. Bact.*, **56**, 513.
Ministry of Health (1963). *Hospital Technical Memorandum No. 10.* H.M.S.O., London.
PARKER, M. T. and HEWITT, J. H. (1970). *Lancet*, **i**, 800.
PRICE, J. D. E. and SLEIGH, J. D. (1970). *Lancet*, **ii**, 1213.
REPORT (1942). Med. Res. Coun. Lond. War Mem. No. 6, " The Prevention of ' Hospital Infection of Wounds ' ".
REPORT (1948). Med. Res. Coun. London. Special Report Series, No. 262, " Studies in Air Hygiene".
REPORT (1951). Med. Res. Coun. Lond. War Mem. No. 11, " The Control of Cross Infection in Hospital ".
SIMPSON, K., TOZER, R. C. and GILLESPIE, W. A. (1960). *Brit. med. J.*, **i**, 315.
STOKES, E. J., HALL, B. M., RICHARDS, J. D. M. and RILEY, D. J. (1965). *Lancet*, **i**, 197.
STOKES, E. J. and MILNE, S. E. (1962). *J. Hyg. Camb.*, **60**, 209.
STOKES, E. J., WATERWORTH, P. M., FRANKS, V., WASTON, B. and CLARK, C. G. (1974). *Brit. J. Surg.*, **61**, 739.
TEMPLE, N. E. I. and BLACKBURN, E. A. (1963). *Lancet*, **i**, 581.
WILLIAMS, R. E. O. and MILES, A. A. (1949). Med. Res. Coun. Lond. Report No. 266.
WILLIAMS, R. E. O. *et al.* (1966). *Hospital Infection: Causes and Prevention*, 2nd edn. Lloyd-Luke. London.

Media-testing and other techniques[*]

Media-testing

Bacteria are extremely sensitive to slight changes in the medium on which they are grown. In experimental work with known stock strains, differences between different batches of media are easily recognized, but in diagnostic work little is known of the variety of bacterial species in each specimen and therefore only gross errors in media making are easily noticed.

Provided the ingredients used in the preparation of blood agar come from a constant source the medium should remain fairly constant. Faults in its preparation may be shown by damage to the red cells. Defibrinated blood is a particularly sensitive indicator as it lyses more easily than oxalated blood. Different batches of selective media, however, may appear identical and yet differ profoundly in their effect on bacterial growth (Stokes, 1968). Materials for the preparation of media should be purchased in large quantities, provided they are chemically stable, and the first batch of medium made from fresh material should always be tested in parallel with the previous batch. In hospital practice in Britain many of the important pathogens for which selective media are prepared are rarely encountered. Unless routine media tests are made, mild infections caused by these pathogens and carriers of them are likely to be missed because the medium may only be able to support the growth of massive inocula.

To compare solid media by seeding them with loopfuls of undiluted liquid culture will only reveal gross differences. Inoculation of a series of six tenfold dilutions using the same loop makes the test more sensitive, but a more accurate and no more laborious method is to make surface viable counts by the method of Miles and Misra (1938). To test liquid media satisfactorily some kind of viable count is essential because the approximate number of viable organisms in the inoculum must be known. The surface viable count is simple, quick to perform and is not extravagant in the use

---

* See also quality control, page 30.

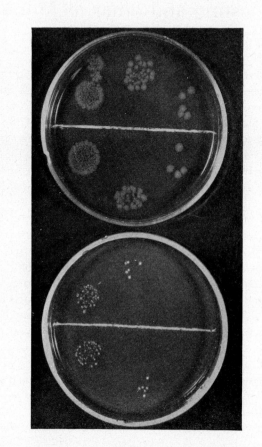

FIG. 33.

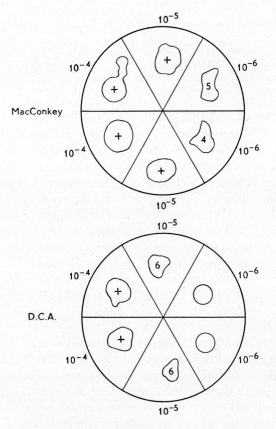

FIG. 34.

Comparison of solid media. Sample plates from a viable count made to compare the growth of a strain of *Shigella sonnei*.

On MacConkey's bile salt agar
and desoxycholate citrate agar.

Growth is 10-fold less on the more highly selective medium and the colonies are smaller in diameter.

of medium.   It can be applied to a variety of problems in addition to media-testing.

8 sterile constant dropping pipettes.   7 (75mm × 13mm) sterile test tubes.
Smooth liquid culture or suspension.
Dilution fluid, usually peptone water.
Solid medium with a well dried surface.
Metronome (not essential).

Preparation of constant dropping pipettes [1]

Make a Pasteur pipette, taking care that the fine-drawn part of the glass is circular. Pass its fine end through a circular hole (0·98 mm. in diameter) cut in a thin piece of metal, until the glass just sticks in the hole ; withdraw it a little to loosen it again and then cut the glass close to the metal with a sharp thin file (an ampoule file is suitable). Withdraw the tip of the pipette from the hole and test its diameter with a micrometer gauge ; measure in two positions at right angles to each other to check that the tip is circular.   Such a pipette will deliver drops of aqueous solutions each 0·02 ml. in volume provided that the glass is grease free, that the pipette is held vertically and rigid and that the fluid drops from it at a rate of 60 drops per minute. These " 50 droppers " can now be obtained commercially.[2]

When the number of viable organisms in a culture is to be estimated it is essential to use pipettes tested with a micrometer and to drop the fluids as strictly as possible in time to a metronome.   For media-testing, however, such a degree of accuracy is not required because drops will fall from the same pipette on the media to be compared.   If the tips of the pipettes when examined with a hand lens appear circular and evenly cut they need not be checked with the micrometer.   Provided they are held vertically and the drops are allowed to form slowly and regularly and are not shaken off prematurely they will be of approximately constant size and the metronome can be dispensed with.

Method for comparison of nutrient properties of solid media

Dry the plates open for one hour in the incubator.   Mark them in sectors, one for each drop.   (Sufficient medium of each kind,

[1] See M.R.C. " System of Bacteriology."
[2] From Bilbate Ltd., Daventry, Northants, England.

A and B in Fig. 35, is required to receive 6 to 8 drops of each dilution to be tested.) Set out the apparatus as in Fig. 35, the plates lid upwards. Inspect the pipettes to see that their tips have not been damaged during sterilization and flame them lightly to remove any trace of grease. Then deliver 45 drops of peptone water, or other diluent, from the first pipette into each of the seven tubes. Wash the pipette in the culture by sucking the fluid up and down ten times ; then deliver five drops of culture into the first tube to make

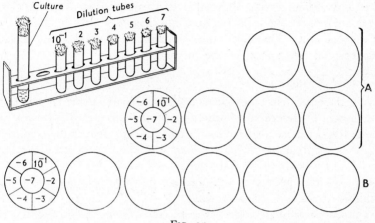

FIG. 35.

a tenfold dilution ; discard the pipette. Wash a fresh pipette ten times in the first dilution and at the same time wash the inner side of the tube so that small splashes from the five drops of culture are included in the dilution. Deliver five drops of this dilution into the next tube and six or eight drops on to the solid medium in the appropriate areas ; discard the pipette. Continue thus until all the drops are in their places on the plate cultures. Allow them to seep into the surface of the dried medium then invert the plates and incubate.

INTERPRETATION

When the plates are inspected after incubation it will be seen that growth decreases with dilution until each drop yields 5 to 20 discrete colonies. An average is taken of the number counted in each of the six or eight drops from this dilution. In addition to a good yield, as judged by the colony count, a highly nutrient medium will show large colonies approximately equal in size. When, therefore, the count from two batches of media is nearly equivalent their relative merits may be further demonstrated by measuring the diameter of each colony counted and calculating the average colony size and the scatter of the diameter measurements from each set of plates.

The results are usually clear cut and the counts need no statistical analysis. If, however, a variation less than tenfold in the count is the only demonstrable difference between two media, statistical evaluation of the figures is required and may show that the difference is insignificant. Details of suitable statistical methods are given in the original description of the method.

SOURCES OF ERROR

1. Bacteria, like other small particles, have a tendency to cling to glass. Therefore the lower part of the pipette, which will be wetted when it is filled with sufficient fluid to deliver the required number of drops, should be washed and the rest of it left dry to avoid unnecessary loss of bacteria.

2. Insufficient mixing and washing down of splashes inside the dilution tubes is a common fault.

3. Drops tend to skid on the surface of some solid media. If the pipette is held rigidly with its tip about 4 centimetres above the medium this can usually be prevented.

4. If, in an effort to keep time with the metronome, some drops are prematurely shaken off they will be unequal in size. It is more important to let them form naturally than to keep strict time.

5. A dilution fluid harmless to bacteria is essential.

6. Sometimes considerable plate to plate variation is seen, therefore all the plates must receive at least one drop of each dilution tested.

7. If accuracy is to be achieved in any dropping method the number of drops per measured volume should be large. To make tenfold dilutions 45 plus 5 drops is the minimal number required.[1]

8. Bacteria which cling together such as staphylococci and streptococci should be grown in a medium which discourages

[1] Fluids with different surface tensions yield different sized drops. This does not matter in media testing, since growth from drops of the same fluid is to be compared.

adhesion, and should be well shaken before dilution. One colony of these species represents one viable particle, not necessarily one coccus.

Nutrient properties of liquid media

The ability of liquid media to yield a heavy growth of bacteria and the time taken to do so can be judged by examining the opacity of growth at intervals after inoculation. In medical bacteriology, however, a most important property of liquid media is that they should be able to support the growth of *small inocula*, and these two properties are not necessarily related. For example, Dubos' bovine albumin tween 80 liquid medium for *Myco. tuberculosis* yields a heavy growth of many strains within three to five days, but it is incapable of supporting the growth of small inocula of freshly isolated

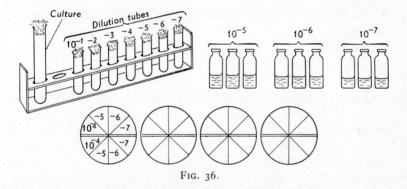

FIG. 36.

strains and is therefore useless as a medium for their primary isolation from blood or cerebrospinal fluid.

The viable counting method described above can be used to check the number of viable organisms inoculated into test broth cultures. A preliminary count is made of a culture of the test organism to determine which dilution contains only a few viable organisms per drop. If, for example, the 10^{-6} dilution of an overnight broth culture of the test organism yields an average of 2 viable bacteria per drop when counted on blood agar, one drop of this dilution seeded into the test medium should yield good growth and is a suitable inoculum for the test. It cannot be assumed that the count will be exactly the same on another occasion. Therefore the inoculum must be checked by another count performed at the time of inoculation.

METHOD

Set out the apparatus as in Fig. 36. Make tenfold dilutions of the culture as previously described limiting the count to those dilutions expected to yield isolated colonies. When the appropriate dilutions are reached deliver one drop into each of the test bottles of liquid medium. At least three are inoculated for each dilution because with small inocula one bottle might receive no viable bacteria.

Examine plate and liquid cultures after overnight incubation. From the count on blood agar the approximate number of viable organisms seeded into each liquid medium culture is known. If growth is visible in the liquid cultures which have received small inocula the nutrient value of the medium for this particular strain is satisfactory.

EXAMPLE

To test the ability of tetrathionate broth to support the growth of *Salm. paratyphi B.*

Preliminary count : One drop (0·02 ml.) of a 10^{-6} dilution of an overnight peptone water culture yields an average of 2 colonies on blood agar.

Test : A second count is made of an overnight peptone-water culture of the same strain and drops of the appropriate dilutions are seeded into bottles containing tetrathionate broth as well as on to blood agar for the count. Nine bottles receive one drop each of the following dilutions.

The first three bottles receive 10^{-5} dilution
The second three bottles receive 10^{-6} dilution
The third three bottles receive 10^{-7} dilution

Results read after overnight incubation : The count on blood agar on this occasion, yields 4 colonies per drop of 10^{-6} dilution, therefore the first three bottles received approximately 40 viable bacteria each, the second three bottles received about 4 viable bacteria each and some of the third three bottles may have received one or two viable bacilli. If the medium is satisfactory there should be heavy growth in all of the first six bottles. Lack of growth in any which received the 10^{-7} dilution does not mean that the medium is poor because they may have been seeded with drops of fluid containing no viable bacteria.

When a new batch of medium is compared with an old one known to be satisfactory, a single comparison using a freshly isolated strain is usually all that is required. Since, however, different strains of the same species vary greatly in their ability to grow in artificial media a number of freshly isolated strains from different sources should be tested before a new diagnostic medium is adopted.

Efficiency of selective media

The nutrient properties of selective media for the species selected can be tested as already described. When compared with blood agar or blood broth they usually prove less nutrient. Nevertheless if their selective power is good they may be invaluable in diagnostic work. Their efficiency can be tested by seeding known numbers of a pathogen into uninfected but contaminated material, e.g. a suspension of normal fresh faeces. The method described for testing liquid media can be adapted so that bottles of contaminated material are seeded with graded doses of pathogen. These artificial specimens are then shaken vigorously to distribute the pathogen and are immediately inoculated on to the test medium. The recovery of the pathogen on any one medium will be found to vary enormously according to all the other bacteria present and a number of experiments are usually required using different strains and different specimens before the medium can be either recommended or condemned (see Report of the Public Health Laboratory Service, 1966).

Anaerobic culture technique

There are a number of satisfactory methods of maintaining anaerobic conditions in liquid cultures. Reducing agents in common use are strips of sheet-iron, minced meat (which contains reducing fatty acids), and thioglycollic acid. If primary anaerobic cultures of material from mixed infections are made in liquid media only, the proportion of the different species present will not be revealed. A few organisms of a quickly growing species such as *Cl. welchi* may assume a predominance in the culture quite out of proportion to its numbers in the specimen. In order to judge which, if any, of the anaerobes found is likely to be the causal organism, primary seeding on solid medium is essential. This necessitates incubation in a jar from which all free oxygen has been removed (Stokes, 1958).

A modification of McIntosh and Fildes' jar fitted with a capsule which reacts in the cold is now obtainable[1] and has the following advantages. It can be set up within five minutes, which is particularly important in hospital work when specimens tend to arrive late in the day ; it has an external indicator which checks that all free oxygen has been removed ; it is sealed with a plastic gasket which makes it clean to handle.

The capsules are inactivated by moisture, sulphur and arsenic. Hydrogen sulphide given off from cultures will gradually inactivate them. Arsenic inactivation is caused by traces of arsine in the hydrogen. If the hydrogen has been generated from the action of hydrochloric acid on zinc it should be passed through a solution of 10 per cent lead acetate to remove hydrogen sulphide and then through 10 per cent silver nitrate to remove arsine.

Apparatus

Air tight jar, capsule, indicators.
Vacuum pump which may be driven by hand, by water or electrically.
Manometer.
Hydrogen.

Place the cultures in the jar and screw on the lid. Then attach the jar to the pump which is fitted with a mercury manometer. Remove air until there is a negative pressure in the jar of 60 to 70 cm. of mercury. Close the tap and attach the jar to a Kipp's apparatus for generating carbon dioxide, or a bladder filled from a CO_2 cylinder. Run in a small amount, about 10 per cent, because its presence improves the growth of non-sporing anaerobes (Watt, 1973). Attach the jar to a source of hydrogen. A convenient source is a balloon or football bladder filled from a hydrogen cylinder and attached to a wash bottle to show the flow of gas (see Fig. 37). When the tap is turned on there is first a rush of gas into the jar, then a pause and then a regular flow of bubbles as the palladium catalyses the reaction between the gases and the resulting decrease in volume allows more hydrogen to enter the jar. If the capsule is inactive this second flow of gas will not occur or will be very sluggish. Spare capsules are kept so that one may be exchanged with minimal delay. When the second flow of hydrogen into the jar is seen to be satisfactory, that is not less than about one bubble per second, leave it to flow for about 3 minutes then close the tap and place the jar in the incubator. The reaction will

[1] From Baird and Tatlock (London) Ltd. 2 capsules per jar give better results than one.

continue in the incubator and there will be more than sufficient hydrogen to combine with the oxygen present.

Explosions in anaerobic jars are due to ignition of the hydrogen and oxygen when they are present in the proportion of two to one. If at least two-fifths of the air in the jar is evacuated and replaced by hydrogen this proportion occurs during the first inrush of gas. Diffusion of gases through a jar filled with plate and tube cultures is probably slightly delayed and pockets of the explosive mixture may be present after the first flow of hydrogen. The most likely cause of an explosion is the escape of a fragment of glowing catalyst from

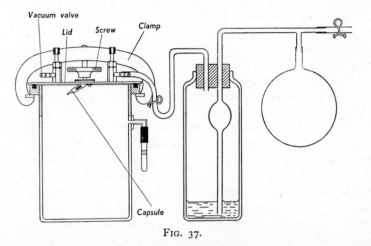

FIG. 37.

the gauze envelope of the capsule when the proportion of gases in the jar is optimal. It is therefore very important that the gauze, which protects against explosions on the Davy lamp principle, should have a fine mesh and should be intact. When the jars are removed from the hydrogen source and placed in the incubator they should be handled gently. If these precautions are observed explosions are exceedingly rare. In an explosion the base of the jar, which is slightly concave, blows out and the indicator tube blows off but the jar itself remains intact and there is no danger of flying glass.

An indicator which shows the presence of oxygen is essential for anaerobic work. It is not reasonable to report anaerobic blood agar cultures sterile unless they have been incubated in a jar with an indicator to check efficient anaerobiosis. A hiss of inrushing air on opening the jar combined with slight discoloration of the blood agar is evidence of an air-tight jar and some change in atmosphere, but both these signs can be present when some oxygen remains and must not therefore be taken as evidence of anaerobiosis.

PREPARATION OF INDICATOR

Lucas' semi-solid indicator is reliable and long lasting; it is prepared as follows.

Add 12 drops of 9 per cent thioglycollic acid and 2 drops of phenol red to 5 ml. of a 2 per cent borax solution in a boiling tube. The colour should be pale pink, it is *essential* to adjust the pH, when necessary, to achieve a pale pink colour. Then add 10 ml. methylene blue solution (one 19 mg. B.D.H. reductase tablet to 200 ml. of water) and 10 ml. of previously melted watery agar. Boil the mixture, which is blue, in a water bath until it is colourless. Transfer it with a warmed Pasteur pipette into a series of warm ampoules which are sealed off immediately. When required the ampoule is opened and the tube is held to the side arm of the jar with a short piece of thick-walled rubber tubing. The proportion of watery agar may need alteration with different batches of agar. When cool the indicator should just set, if it is too firm the colour does not change rapidly. In the presence of oxygen the indicator becomes blue to a depth of about 5 mm. When it is used every day exposure to the moist atmosphere in the jar prevents drying and the indicator will last for several months without rejuvenation. When the surface begins to dry, which is shown by cracks and by a deeper penetration of the blue colour, it can easily be rejuvenated by heating the top of the indicator in a Bunsen flame and pouring it off leaving the moist part below exposed. Rejuvenation can be repeated until no more material remains in the tube. Commercially prepared indicators are less satisfactory in our experience.

The indicators are inspected each morning before the jars are opened. If any of them shows a trace of blue colour the cultures removed from that jar are marked and repeated if necessary. The jar, capsule and indicator are tested before further use.

Gas pak[1] anaerobic jars can now be obtained (Collee *et al.*, 1972). The anaerobic atmosphere is achieved by a reaction between chemicals in a sachet provided, an indicator is also included. The method is expensive but since no hydrogen cylinder is required

[1] Obtainable from Becton Dickinson UK Ltd, Wembley, Middlesex HA9 OP5.

it is safer and very convenient for laboratories using only one or two jars daily.

Incubation in air plus 5 to 10 per cent CO_2

An old McIntosh and Fildes' jar without a capsule, which need not be quite air-tight, is suitable. The jar is evacuated to −40 cm. mercury and CO_2 is allowed to run in until the evacuation, as read on the manometer, is reduced by 7·6 cm. approximately. Assuming atmospheric pressure to be 76 cm. this is equivalent to 10 per cent.

A less accurate method is to use a biscuit tin and measured quantities of marble chips and hydrochloric acid. The following formula will give an atmosphere of approximately 5 per cent. If V = volume of the container in litres add 0·25 × V grams of chips to 2·5 × V ml. of 25 per cent hydrochloric acid.

A lighted candle placed in a jar which is then closed will reduce the oxygen and increase the carbon dioxide content of the jar. An unknown amount of fumes possibly toxic, will also be released from the candle and there will be a small amount of carbon deposits. Undoubtedly CO_2 dependant bacteria will grow in a candle jar which must be used for their culture when there is no better alternative.

Microaerophilic atmosphere

The oxygen requirements of bacteria can be studied in shake agar cultures, enriched with Fildes' extract and serum when necessary. The molten medium in a test tube is cooled to 45° C. seeded with liquid culture and well shaken; after incubation in air microaerophilic organisms usually grow most profusely 1 to 2 cm. below the surface. The amount of oxygen in the jar for plate cultures can be adjusted to suit the requirements of any particular organism. Most microaerophilic organisms grow well in a McIntosh and Fildes' jar without a capsule if two-thirds of the air is evacuated, CO_2 is run in to give a concentration of 5 per cent and the remaining space is filled with hydrogen.

The Preservation of bacteria

It is often impossible in a diagnostic laboratory to pursue detailed investigation of interesting bacteria immediately they are isolated. Strains temporarily put aside may be lost because they need frequent subculture, which is neglected, or because after many generations they become contaminated or lose their original characteristics. In most laboratories a small collection of stock cultures is kept comprising standard strains for antibiotic sensitivity tests and assay and a selection of the less common pathogens for use in media testing when freshly isolated strains are not available. Some species will remain viable up to a year without subculture under suitable conditions. For delicate species and longer preservation drying is required.

(a) Preservation without drying

Bacteria such as coliforms and staphylococci which grow readily will survive for several months in semi-solid (0·5 to 1 per cent) nutrient agar. A loopful of young culture is stabbed into the agar which is incubated overnight and then stored at room temperature (18°–21° C.) in the dark. Members of the salmonella group survive well, for a year or more, on Dorset's egg slope cultures in screwcapped bottles kept in the dark at room temperature. *Shigellae* keep better on nutrient agar of rather poor nutritive quality. Organisms which grow less well need more frequent attention. Streptococci will survive for at least a month in blood broth seeded with a young culture incubated overnight and then refrigerated. They survive nearly as well on blood agar slopes which are kept in the dark at room temperature and can more easily be checked for contamination. *C. diphtheriae* is difficult to maintain, most strains need weekly subculture on Loeffler's medium ; after incubation for 48 hours they are kept in the dark at room temperature. *N. meningitidis* and *N. gonorrhoeae* are notoriously difficult to preserve. They need continuous incubation and subculture on alternate days in an atmosphere containing 5 per cent CO_2. When they are well established subcultures can be reduced to once in four days and additional CO_2 may be discontinued. Anaerobes survive well in sealed cooked meat broth cultures. *Clostridia* will remain viable thus for many months if the tubes are kept at room temperature in the dark. Anaerobic streptococci and *Bacteroides* survive for several months in similar cultures if they are kept in the incubator. *Myco. tuberculosis* will survive on Dorset's egg medium for about 6 months if the medium remains moist and the culture remains in the incubator.

When an unknown organism is isolated which grows poorly on blood agar it should be subcultured on alternate days and incubated in the atmosphere of original growth to maintain viability during investigation ; when each subculture is made the remaining growth is reincubated. The characteristics of the strain will remain more constant if subcultures are made from the pool of inoculum rather than from single colonies.

Most bacteria and viruses survive well suspended in broth and frozen at a temperature less than $-60°$ C. A special low temperature refrigerator or a supply of solid CO_2 or liquid nitrogen to maintain this temperature is required.

(b) Preservation by drying

Even the most delicate bacteria can be preserved for long periods by rapidly drying serum broth cultures or suspensions in a vacuum. The apparatus required is expensive and not available in all diagnostic laboratories. A simple drying method in a gelatin ascorbic acid medium (Stamp, 1947) is successful for all but the most delicate bacteria. It is suitable for the stock strains and for preserving other organisms which after identification by the usual methods may be required later for detailed investigation.

For example, if bacteria from the blood of a patient with subacute bacterial endocarditis is preserved, a new strain can be compared with the old one when after apparent cure the infection recurs. If they are the same it is probable that previous treatment was inadequate; if they are different the reinfection may well be overcome by another course of the treatment previously given, provided that the new pathogen is equally sensitive to the drug. Bacteria preserved by this method retain their original characteristics well.

APPARATUS
    Desiccator containing phosphorous pentoxide.
    Suspending medium.
    Sterile waxed filter paper.

Suspending medium. This consists of three separately prepared solutions which are subsequently mixed.

A. Peptone 1 per cent.
      NaCl 0·5 per cent.
      Lemco Beef Extract 0·5 per cent.
B. Gelatine Coignet Gold Label (use 10 per cent).
            Nelson's No. 1 photographic (use 10 per cent).
            Fallowfield's photographic (use 5 per cent).
            (Any of these three varieties is suitable).
C. Ascorbic acid 0·25 per cent.

Solution A is made up double the required strength. It is adjusted to pH 9·3 to precipitate phosphates, then filtered and readjusted, to pH 7·6. An equal volume of the gelatin, also double strength, is then added. The mixture is cleared with egg albumin, steamed for 30 minutes and then shaken well. It is steamed for a further 2 hours then filtered and adjusted to pH 7·6 and finally sterilized in the autoclave at 115° C. for 15 minutes.

A 2·5 per cent solution of ascorbic acid is sterilized by heating to 100° C. for a few seconds, or sterile 5 per cent "Redoxon" solution in ampoules is used. It is added to the sterile medium, immediately before use, to make a final concentration of 0·25 per cent.

The organism to be preserved is seeded heavily on at least two blood agar plates. A thick suspension of about 5×10^9 per ml. is desirable. (This is obtained when the overnight growth of the standard Oxford staphylococcus is washed from two 8·5 cm. blood agar plates into 20 ml. of medium.) The growth must be young. For comparatively poorly growing organisms such as *Strept. viridans* three or four blood agar plates are seeded heavily and the growth is washed off with peptone water which is discarded after centrifugation ; the deposit is suspended in about 2 ml. of medium. Liquid cultures can be used but large volumes are necessary to equal the growth obtained from solid medium. (Further incubation to produce greater opacity is useless because a large number of the bacteria will then be dead.)

METHOD

Cut a disc of filter paper to fit into a Petri dish and immerse it in hot paraffin wax for two minutes to sterilize. Place it in a sterile Petri dish when dry. Drop about 20 to 30 drops of the suspension from a sterile pipette on to the waxed paper, transfer the covered Petri dish to the desiccator, evacuate to 10–30 cm. of mercury and leave on the bench for 4 to 5 days. When the desiccator is opened the drops will be found to be dry brittle discs. Dislodge them from the waxed paper with a short stiff loop and store them in a small sterile plugged tube in the desiccator. When the culture is required remove one disc with a moistened loop and seed it into about 2 ml. of a suitable liquid medium. When the disc has dissolved shake the fluid well, plate out a loopful to test purity and incubate the remainder.

Maintenance of control strains for antibiotic sensitivity tests and assay (see also page 215)

The standard Oxford H. staphylococcus (NCTC 6571) will retain its sensitivity to antibiotics for many years if maintained in the following manner. Three preparations are made, one dried by Stamp's method, one in a nutrient agar stab, which after 48 hours incubation is stored in the dark at room temperature (18°–21° C.), and one broth culture which is subcultured daily. The agar stab is subcultured monthly.

At the beginning of each week a fresh broth culture is made from the stab for daily subculture during that week. If the stab is contaminated the strain is recovered from the dried preparation. On no account is it plated out and recovered from a single colony, because this entails the risk of picking a variant and so altering the properties of the strain. *Esch. coli* (NCTC 10418) and *Ps. aeruginosa* (NCTC 10662) are similarly maintained. *Klebs. edwardsi* (NCTC 10896) kept for rapid assay of aminoglycosides is maintained as a suspension frozen at −70° C., see page 250.

Fluorescence microscopy

(a) Examination of acid-fast bacilli

Films of sputum, or other material, made on glass slides are stained with auramine phenol, a yellow dye which fluoresces in short-wave light. Acid-fast bacilli retain the dye after decolorization with acid-alcohol. The advantage of fluorescence is that the stained bacilli appear larger and are more easily seen, particularly in specimens such as pus, urine deposits and cerebrospinal fluid deposits where background fluorescence is slight. In sputum examination by this method has been shown to yield a higher proportion of positives than by conventional methods (Wilson, 1952). A low power objective can be employed to cover a wide field so that the whole film can be scanned within 5 to 10 minutes. Acid-fast bacilli appear as bright points of light which, when examined with 4 mm. objective, are seen to be typically shaped bacilli. Confidence can be gained initially by marking the position on the slide and re-examining after overstaining with the conventional Ziehl-Neelsen method. The only disadvantage is that a positive smear must be examined with each batch of slides to ensure that the microscope is set up correctly.   Preferably an unstained positive slide should be included so that both stain and optics are checked.

Originally a mercury vapour lamp was thought to be essential for optimal fluorescence but an iodine-quartz tungsten lamp[1] has been found to be equally good for most purposes, cheaper and easier to use.   It also has a longer life and does not give off so much heat ; when turned on it is instantly ready for use.

MATERIALS

Microscope with lamp built into the stand.
Iodine-quartz tungsten bulb (Tomlinson, 1967) 100 watt.
Aplanat condenser (preferably oiled).
5 mm. thick, blue PY filter to cut out all but short-wave light.
Yellow filter in eye-piece.
Eye-piece, either ×8 or ×10.
Par-focal objectives 16 mm. and 4 mm.

METHOD.   The positive slide is first examined.   A grease pencil mark is made on its upper surface to encircle the area of the stained smear.   A drop of oil is placed on the condenser and the slide is lowered into position so that there is an oil seal between slide and condenser.   The low power objective is then focused until the grease pencil mark can be seen.   The field inside it is then examined for acid-fast bacilli.   The unknown specimens are then focused in the same way and examined by scanning the field in strips until the whole area within the grease pencil mark has been viewed or until acid-fast bacilli have been seen.   When experience has been gained most fluorescent debris seen in sputum can be recognized as such under the low power so that time is not wasted by continually taking a closer look with a high power lens.   An oil immersion objective is not required.

(b) Examination for immunofluorescence

The same microscope and lamp can be employed but since fluorescence is feebler and some of the objects viewed (for example, spirochaetes) are small, a darkground condenser and a high-power oil immersion objective may be required.   A monocular microscope loses less light from the object because it travels through fewer prisms.   Thin microscope slides also allow better illumination.

[1] 100 W, Atlas A1/209.

MATERIALS

Monocular microscope and lamp as described above.
Darkground condenser, oiled.
2 mm. blue PY filter between lamp and condenser.
Par-focal objectives as above, also 2 mm. oil immersion objective.
Yellow filter in eye-piece.
Microscope slides 0·6–0·8 mm. thick.

METHOD. A positive slide is first examined as before, focusing on a grease pencil mark and using a lower power objective. Satisfactory fluorescence is checked by examination with the high-power oil immersion objective. Fluorescence and contrast can sometimes be increased by employing different filters. The microscope manufacturers[1] are knowledgeable about different combinations of filters for different purposes. Those described here are satisfactory for immunofluorescence of cultures of *N. gonorrhoeae* and for *Treponema pallidum* in the fluorescent treponemal antibody test.

For the staining method and examination of films see Chapter 6.

Staphylococcal toxin and antitoxin

1. *Toxin.* Although coagulase and DNase production are closely related to pathogenicity they may occasionally be absent when it would appear from clinical evidence that the strain is pathogenic. A test for alpha toxin (rabbit red cell haemolysin) production is then made.

Method. Incubate a subculture of the staphylococcus in 0·5 ml. nutrient broth in air containing 30 per cent CO_2 for 48 hours and then centrifuge. Deliver into a test-tube three reagents ; the supernatant fluid, saline (0·1 ml.), and 2 per cent suspension in saline of thrice washed rabbit red cells (0·2 ml.), and into a control tube the same volumes of supernatant fluid and red cells but 0·1 ml. of staphylococcal antitoxin (antihaemolysin) instead of the saline. Incubate both tubes in the 37° C. water bath for one hour and leave them for 18 hours on the bench. If alpha toxin is present it will lyse the cells in the first tube but will be neutralized in the control tube in which the cells will remain intact.

2. *Antitoxin.* *Staph. aureus* is commonly found on the skin and is the most likely pathogen to contaminate specimens. If in a

---

[1] Gillett and Sibert Ltd, London, S.W.11.

patient suspected of bacteraemia one blood culture yields this organism and subsequent cultures prove sterile the diagnosis of staphylococcal bacteraemia remains doubtful. It may then be of value to examine the serum for staphylococcal antibodies. If the titre is high and there is no recent history of acne or furunculosis, it is evidence in favour of the pathogenic rôle of the staphylococcus. A negative result is of little significance because the sera of heavily infected patients do not always show high antitoxin titres. The test is also valuable in the diagnosis of deep-seated osteomyelitis when no specimen for culture can be obtained. The reagents required are patient's serum, a 2 per cent suspension of thrice washed rabbit cells, staphylococcal α-toxin of known titre and antitoxin of known titre. Dilutions of the patient's serum are tested for their neutralizing power on a constant dose of α-toxin (haemolysin) and compared with a parallel titration of known anti-toxin on the same dose of toxin. For details of the method see Parish *et al.* (1934). The antileucocidin test will also aid diagnosis of deep-seated staphylococcal infection. High titres are found in only 2 per cent of normals and in 85 per cent or more of those with osteomyelitis ; for evaluation of these tests, see Lack and Towers (1962).

Transport medium for gonococci and oxygen sensitive anaerobes (Moffet *et al.*, 1948)

Gonococci are rapidly killed at room temperature by oxidation. If they are protected from oxygen by immersion in a non-nutrient reducing medium they regularly survive 24 hours at room temperature and may remain viable for several days. Protection from oxygen is also desirable for cervical swabs from cases of puerperal sepsis for the isolation of anaerobic streptococci, some of which are oxygen sensitive and are likely to die before cultures are made.

MEDIUM. Melt 190 ml. of previously autoclaved 0·3 per cent fibre agar.[1] Add 0·2 ml. of thioglycollic acid (B.D.H.) and sufficient N/1 NaOH to bring the medium to approximately pH 7·2. Then add 10 ml. of 20 per cent sodium glycerophosphate in distilled water and 2 ml. of 1 per cent $CaCl_2$ in distilled water. Mix thoroughly and while still hot titrate to pH 7·4 with N/1 NaOH. Add 0·4 ml. of methylene blue (0·1 per cent in water). Replace the medium in

[1] See Note below.

the steamer for a few minutes and then distribute it to screw-capped bottles, about 7 ml. in each. Bottles with well-fitting caps and sound washers are essential. Sterilize for 1 hour in the steamer. Use medium from one or two bottles to fill completely the remainder, so that as much air as possible is excluded. Screw the caps down firmly and store for 24 hours before use. Incomplete removal of oxygen is indicated by blue coloration (oxidized methylene blue) and bottles showing it are discarded.

SWAB. Boil absorbent cotton wool and wooden applicators in Sorensen's phosphate buffer solution pH 7·4 to remove acid and dry them in the oven. Make the swabs in the usual way and then dip them in 1 per cent watery suspension of BDH blood charcoal or " Norrit " before sterilization. The charcoal neutralizes inhibitory substances which may be present in the agar.

USE. Sample the cervix or urethra in the usual way. Plunge the swab into the medium so that it is completely immersed, break off the stick and screw down the cap firmly. For culture remove the swab from the bottle with forceps and seed it on to solid medium in the usual way. Smears are best made on sterile slides at the time of sampling but can be made later immediately before culture. The medium compares well with other transport methods (Cooper, 1957). Although non-nutrient to most bacteria some coliforms will multiply slowly in the medium at room temperature. The swabs should, therefore, be cultured as soon as possible.

MEDIA FOR PRIMARY ISOLATION

Note. The supply of agar for media making is variable and particular kinds are not always available. The quantities given in the formulae gave a satisfactory result originally but are not necessarily optimal for newly bought agar, particularly of a different kind. Some indication is usually given by the manufacturer of the concentration needed, which may lie between one to three per cent for the preparation of culture plates. Appropriate adjustment will be needed for semi-solid media.

Blood agar (U.C.H., 1953)

Nutrient broth. Infuse 400 g. of minced fresh ox heart in 1 litre of tap water for 3 hours, then boil it for 20 minutes in a container with a lid and leave it overnight. Skim off the fat and surface scum and filter through muslin and cotton wool. Make up to the original volume (1 litre) with tap water. Add 10 g. of Difco Proteose peptone (or Evans' peptone) and 5 g. sodium chloride and steam for 1 hour to dissolve them. Adjust the pH to 8·4–9·0 using 40 per cent sodium hydroxide, and phenol red as indicator. Steam for 2 hours to precipitate the phosphates. Filter through paper pulp and adjust the pH to 7·8 using concentrated hydrochloric acid and phenol red. Bottle and add 10 ml. of a 0·5 per cent solution of para-aminobenzoic acid per litre of broth. Sterilize in the autoclave at 15 lb. for 15 minutes.

Agar base. To 900 ml. distilled water add: 100 ml. of nutrient broth, without para-amino benzoic acid 20 g. of Davis' New Zealand agar [1] and 8 g. of sodium chloride. Add 1 ml. N/1 sodium hydroxide, shake well and dissolve in the steamer. Filter through paper pulp. Adjust the pH to 7·6 using hydrochloric acid and phenol red. Bottle and then sterilize in the autoclave at 15 lb. for 15 minutes.

Nutrient agar. To 1 litre of nutrient broth add 14 g. of Davis' New Zealand agar [1] and 4 ml. of N/1 sodium hydroxide. Dissolve by heating in the autoclave at 15 lb. for 15 minutes. Filter through paper pulp, adjust the pH to 7·8 using phenol red and hydrochloric acid. Bottle in 250 ml. quantities and autoclave at 15 lb. for 15 minutes.

Blood agar. Remove defibrinated blood from the refrigerator and stand it on the bench to warm. Melt 250 ml. of nutrient agar and 250 ml. of agar base in the steamer; when it is liquid stand the bottles for 15 minutes on the bench, then place them in the 56° C. water bath. Pour a layer of the base into sterilized Petri dishes and allow it to set. Stand the blood for not more than 1 minute in the water bath. Add 25 ml. of blood to 250 ml. of nutrient agar, at 56° C., mix well and pour on to the agar base immediately. The medium must be at least 4 mm deep, the blood agar layer being about 2 mm so that haemolysis is easily seen.

Kanamycin blood agar for selection of anaerobes

Prepare a solution in distilled water containing 15,000 μg/ml. kanamycin. Add 5 ml. of this solution to 1 litre blood agar immediately before pouring. Mix very well and pour into sterile Petri dishes without a base layer.

[1] See Note, page 363.

"Liquoid" (sodium polyanethyl sulphonate) broth for blood culture (von Haebler and Miles, 1938)

Prepare a 5 per cent solution of "Liquoid" (Hoffmann la Roche) in 0·85 per cent saline and sterilize it in the autoclave at 15 lb. for 15 minutes.

Add 10 ml. of this solution to 1 litre of nutrient broth, mix well and check that the pH is 7·6. Distribute in 10 ml. quantities in narrow-necked 30 ml. screw-capped bottles, and sterilize as before in the autoclave. Screw the caps on tightly to prevent evaporation ; they must be free from traces of inhibitory substances.

(About 5 ml. of patient's blood will be added to each bottle to give a final concentration of "Liquoid" of 0·03–0·05 per cent.

Brewer's thioglycollate broth (Southern Group Laboratory formula)

Hartley's digest broth	1000 ml.
Sodium thioglycollate	1 g.
Glucose	2·5 g.
Agar (New Zealand)	0·6 g.
Methylene blue	2 mg.
Adjust pH to 7·5	

Distribute in 80 ml. volumes in 100 ml. (3 oz) screw-capped bottles and sterilize in the autoclave at 121° C. for 15 minutes.

Diphasic medium for blood culture

Deliver 10 ml. Oxoid blood agar base no. 2 liquefied into a 50-ml. flat bottle. Autoclave at 121° C. for 15 minutes and slope. When cold add 10 ml. Liquoid broth aseptically. Incubate for three days. Examine in a bright light and discard any bottles showing turbidity.

Charcoal agar for the prevention of swarming (Alwen and Smith, 1967, slightly modified)

Acid-water washed charcoal[1]	1 g.
Nutrient agar (Oxoid blood agar base No. 2)[2]	4 g.
Horse-red-cells washed and packed	2 ml.
Distilled water	100 ml.

---

[1] Darco, G. 60 Atlas Chemical Co. from British Drug Houses Ltd., Poole, Dorset, England.

[2] Some but not all other nutrient agar bases are satisfactory.

METHOD. Boil the charcoal in 2N hydrochloric acid (Analar, mercuric chloride-free) in a reflux condenser for 3 hours. Allow to sediment, remove the supernatant acid, replace it with fresh acid and boil for a further 3 hours. Allow to sediment again, remove the acid and wash in distilled water until the washings are the same pH as the water. Dry the charcoal thoroughly in a hot air oven (150° C. for 3 hours), powder it in a pestle and mortar and store in a well stoppered jar at room temperature.

For use mix charcoal, nutrient agar base and water and autoclave at 15 lb. for 15 minutes. Cool to 50° C. and pour into cold, previously refrigerated, plates. The charcoal must remain suspended in the medium. This is easily achieved by placing each plate to set on a Luckham " Rotatest " machine set at 80 r.p.m. Incubate overnight and dry open in the incubator for 30 minutes before use. *The plates must not be refrigerated.*

Whole blood cannot be added but when the medium is employed for recovering fastidious organisms, or for primary culture of specimens, thrice-washed, packed, horse-red-cells warmed to 37° C. should be added before pouring.

Downie's tellurite blood agar. To unmelted nutrient agar add 0·5 per cent NaCl and 1 per cent Bacto-peptone. Steam for 2 hours to sterilize; for use melt and cool to 56° C. Add sterile potassium tellurite, to make a final concentration of 0·04 per cent, and 5 per cent horse blood. Pour into plates containing a layer of agar base as described for blood agar plates.

Note. The potassium tellurite can be conveniently stored as a 4 per cent solution sterilized by steaming for 30 minutes.

VCNT blood agar for *N. gonorrhoeae* (Phillips *et al.*, 1972).
Composition

Vancomycin	3 μg/ml.	⎫ In lysed
Colistin	100 units/ml.	⎬ horse-blood
Nystatin	12·5 units/ml.	⎪ agar.
Trimethoprim	5 μg/ml.	⎭

METHOD. Prepare stock solutions in sterile distilled water as follows: vancomycin 3000 μg/ml. colistin methane sulphonate 20,000 units/ml., trimethoprim lactate 5000 μg/ml. Sterilize a 10 per cent solution of saponin by autoclaving at 15 lb. for 15 minutes. Mix the solutions aseptically as follows: vancomycin 10 ml., colistin 50 ml., trimethoprim 10 ml. and saponin 30 ml. Distribute in 10 ml. amounts and store at −20° C. On the day of preparation make 1 litre Oxoid Blood Agar Base No. 2 and cool to 50° C. Reconstitute nystatin to make a suspension containing 12,500 units/ml. Add to the base 100 ml. horse-blood, 10 ml. VCT saponin mixture and 10 ml. nystatin suspension. Mix very well and pour 15–20 ml. per 8·5-cm. Petri dish. Use within 48 hours.

Lacey's selective medium (for *Bord. pertussis* (Lacey, 1951–1954)). The medium is prepared from an agar base to which four different fluids are added.

(*a*) Base

New Zealand agar (Davis)	about * 14·5 g.
Potato Starch (B.D.H.)	15·0 g.
Glycerol (B.D.H. AR)	5·0 ml.
DL alpha alanine (H. & W.)	1·0 g.
L glutamic acid (Light)	3·7 g.
Sodium fluoride (B.D.H. Puriss)	0·5 g.
Sodium chloride (H. & W. AR)	0·5 g.
Potassium chloride (H. & W. AR)	3·3 g.
Tap water	to 1070·0 ml.

    * (or 1·3 times amount normally used in 5 per cent blood agar base)

Add agar and starch to about 600 ml. tap water and steam to dissolve, shaking occasionally. (With 10-litre lots this takes three hours and needs shaking every half-hour.)

Add all other ingredients to about 100 ml. tap water. Dissolve with heat. Neutralize with 10 per cent potassium hydroxide (AR), using phenol red as external indicator. Add to agar and starch. Make up to 1070 ml. Adjust pH to 7·2. Distribute in 100 ml. amounts in 250 ml. screw-capped bottles. Autoclave at 110° C. for 10 mins.

(*b*) Cysteine-Magnesium Salt Mixture

Magnesium lactate (H. & W.)	7·6 g.
Fumaric acid (H. & W. or Light)	3·3 g.
Malonic acid (,, ,, ,,)	3·1 g.
L or DL Cysteine hydrochloride (H. & W. or Light)	2·0 g.
Distilled water	90·0 ml.
Magnesium hydroxide 8·5 suspension : qs. (approx. 44 ml.) *	

    * (Phillips' Magnesia is convenient because finely divided).

Add solids to water and bring to the boil to dissolve. Cool to about 80° C. and add magnesium hydroxide suspension until a slight but obvious excess remains after three minutes observation. Make up to 170 ml. Filter through Whatman's no. 1 paper. Bring to the boil for three minutes and transfer to a sterile screw-capped bottle which will admit a 10 ml. pipette. Store at 5°C. without further sterilization.

(*c*) 1 per cent 4 : 4 Diamido diphenylamine dihydrochloride (M. & B. 938)

Weigh out 100 mg. of M. & B. 938 on a sterile watch glass or filter paper and add to 10 ml. sterile distilled water in a screw-capped bottle. Store at 5° C. Use for up to 6 weeks provided no deposit is formed. Dilute 1 : 10 before adding to the medium.

(*d*) 50 unit per ml. Penicillin solution.

A 50 unit per ml. solution in 0·25 per cent sodium citrate ($Na_3C_6H_5O_7$, $2H_2O$, B.D.H. AR) is prepared fortnightly from the solid sodium or potassium crystalline benzyl penicillinate (e.g. Buffered Penicillin G, B.D.H.) and kept in the refrigerator.[1]

(*e*) Defibrinated horse blood.

Use blood less than 7 days old.

Preparation of Plates. Allow melted base to cool for half an hour in the water bath at 55° C. and a further 5 to 10 minutes at room temperature. To 100 ml. add, in order, with mixing after each addition, the following :

> 8 ml. cysteine magnesium salt mixture
> 1·5 ml. 0·1 per cent M. & B. 938
> 0·75 ml. 50 unit/ml. penicillin
> 60·0 ml. defibrinated horse blood (well mixed and at room temp.)

Pour five plates. Leave undisturbed overnight. Store at 5° C. without drying. Use up to 10 days old. Dry for a minimum time.

Penicillin Streptomycin Blood Agar (for *Acinetobacter parapertussis* and fungi)

Melt 100 ml. of nutrient agar (see page 320). Cool to 45° C. and add :

> 1 ml. of 600 μg/ml. (Glaxo) streptomycin
> 2·5 ml. of 50 unit/ml. penicillin
> 10·0 ml. defibrinated horse blood at room temperature.

Pour five plates for *Acin. parapertussis* culture or distribute in 1 oz. screw-capped bottles and slope for isolation of fungi.

[1] B.D.H. : British Drug Houses, Ltd., Graham Street, London, N.1.
Davis : Davis Gelatine, Ltd., 29 Mitre Street, London, E.C. 3.
H. & W. : Hopkin & Williams Ltd., Freshwater Road, Chadwell Heath, Essex, England.
Light : L. Light & Co. Ltd., Poyle Trading Estate, Colnbrook, Bucks., England.
M. & B. : May and Baker Ltd., Dagenham, Essex, England.
Phillips : Phillips, Scott and Turner, St. Mark's Hill, Surbiton, Surrey, England.

Cystine-Lactose-Electrolyte-Deficient medium (CLED), (Mackey and Sandys, 1966) obtainable dried from Oxoid Ltd CM 301.

Phenolphthalein phosphate agar (Barber and Kuper, 1951) Dissolve 0·5 g. sodium phenolphthalein phosphate in 100 ml. distilled water. Sterilize by Seitz filtration. Melt 200 ml. nutrient agar, pH 7·4, cool to 50° C. and add 2 ml. of the solution. Mix well and pour about 2 mm. deep into Petri dishes. Store the solution at 4° C.

The solution will hydrolyse after some weeks. It can be tested by removing a little and adding alkali. A pale pink colour indicates hydrolysis and the solution must be renewed. Some batches of sodium phenolphthalein phosphate are slightly hydrolysed and are unsuitable. The medium can be made selective for staphylococci by the addition of Polymyxin B in a concentration of 25 units per ml.

Salt medium for the selection of staphylococci (Maitland and Martyn, 1948)

Add 10 per cent sodium chloride to Robertson's cooked meat medium. If faeces or other heavily contaminated material is incubated in this medium overnight and then plated it will yield an almost pure growth of staphylococci if they are present.

Rappaport's medium (Rappaport *et al.*, 1956)

Solution A: Bactotryptone   0·5 g.
              NaCl   0·8 g.
              KH_2PO_4 0·16 g.
              Bi-distilled water 100 ml.
Solution B: Dissolve 40 g. $MgCl_2$ (C.P.) from a newly opened container (because it is hygroscopic) into 100 ml. water.
Solution C: 0·4 per cent solution malachite green in distilled water.

For use : to 100 ml. solution A add 10 ml. solution B and 3 ml. solution C. Mix well, distribute in 5 ml. quantities and autoclave. Store in a refrigerator.

Cetrimide agar (Brown and Lowbury, 1965)

Proteose peptone No. 3 (Difco)	20 g.
New Zealand agar[1]	15 g.
Glycerol	10 g.
Distilled water	1000 ml.

[1] See Note, page 363.

Adjust this basal medium to pH 7·2 and autoclave at 15 lb. for 15 minutes. To 100 ml. of the molten base add aseptically the following solutions, each made up in distilled water and seitz filtered:

K₂HPO₄ (anhydrous)	15 per cent solution	1 ml.
MgSO₄7H₂O	15 per cent solution	1 ml.
Cetrimide (BP)	2 per cent solution	1·5 ml.

Mix well by inversion several times and pour into Petri dishes.

Selectivity can be improved by the addition of nalidixic acid at a concentration of 15 μg per ml. A stock solution containing 10,000 μg per ml. is prepared by dissolving nalidixic acid powder in alkalinised water. After adding the weighed powder to the measured distilled water shake well and add one drop 10 N sodium hydroxide and shake again. Continue adding a drop at a time and shaking until all the powder is dissolved, store at about 4° C.

Tarshis blood agar for tubercle bacilli (Tarshis *et al.*, 1953)

Prepare the agar base by steaming 30 g. New Zealand agar and 10 ml. glycerol in 1 litre distilled water for 2 hours. Filter through paper pulp, adjust to pH 7·3 to 7·4, bottle in 500 ml. amounts and sterilize by autoclaving at 15 lb. for 15 minutes.

To prepare the medium melt the base and cool to 50° C. Add 25 per cent human blood and 50 units per ml. penicillin. Mix well. Distribute with sterile precautions in 15 ml. screw-capped bottles and slope.

The human blood is obtained by pooling 3 bottles of recently expired (3 weeks old) transfusion blood of the same group. This is done to dilute any inhibiting substance which might be present in blood from one donor.

Selective medium for isolation of *Myco. tuberculosis* (Mitchison *et al.*, 1973)

MATERIAL

Middlebrook 7H10 Agar (Difco)	
Glycerol	
Bacto-Middlebrook OADC enrichment	
Antibiotic solutions Polymyxin	100,000 units/ml.
Carbenicillin	100,000 μg/ml.
Trimethoprim	10,000 μg/ml.
Amphotericin B	2000 μg/ml.

METHOD. Suspend 19 g. agar in 1 litre distilled water. Add 5 ml. glycerol. When dissolved distribute in 200 ml. volumes and sterilize by autoclaving at 15 lb. (121° C.) for 10 minutes. Make

40 ml. of a stock antibiotic solution by mixing together 4 ml. polymyxin and 2 ml. each of carbenicillin and trimethoprim and 32 ml. sterile distilled water, this will keep for several months frozen. To each 200 ml. molten agar cooled to 50° C. add asepticcally 4 ml. of the antibiotic mixture and 1 ml. amphotericin solution Mix very well, distribute in 5 ml. volumes in screw-capped bottles, slope.

Kirchner Medium (P.H.L.S. Working Party, 1958)

> Kirchner medium base Oxoid (CM193)
> Glycerol
> Sterile horse serum

To 1 litre of base medium in distilled water add 20 ml. glycerol and mix well. Distribute 9 ml. volumes to 30 ml. (1 oz.) screw-capped bottles and sterilize by autoclaving at 115° C. for 10 minutes. To each bottle when cold add aseptically 1 ml. sterile horse serum containing 100 units penicillin per ml.

MEDIA FOR SENSITIVITY TESTS AND ASSAY

Lysed blood agar (for sulphonamide sensitivity tests). To lyse the horse blood cells, remove the serum from sedimented, defibrinated blood and replace it with sterile distilled water; mix well and store the mixture frozen solid in the refrigerator.

Melt 100 ml. of Oxoid diagnostic sensitivity test (DST) agar or Wellcotest agar and cool it to 50° C. Warm the frozen lysed horse red cells and some fresh whole horse-blood as described for the preparation of blood agar. Add 5 ml. of each to the nutrient agar, mix well and pour 15–20 ml. per 8·5-cm. Petri dish.

Blood plates left overnight to give the lysed blood time to neutralize sulphonamide inhibitors, can be " chocolated " for tests with haemophilus by heating in an inspissator in a moist atmosphere.

Wellcotest agar is free of sulphonamide and trimethoprim inhibitors and can be used without blood for tests on non-fastidious organisms.

Indicator serum peptone water for tube titration assay and sensitivity tests

To 60 ml. peptone water add 15 ml. of 10 per cent glucose solution in distilled water, and 10 ml. of a saturated solution of phenol red. Steam for one hour and allow to cool.

Add aseptically 25 ml. of sterile horse serum.

Distribute in 30 ml. screw-capped bottles. Test for sterility by incubation overnight. Store at about 4° C.

Solid medium for testing Mycobacteria

Prepare Lowenstein-Jensen medium (Mackie and McCartney, 1959). Add drugs to measured volumes of the medium in 15 ml. screw-capped bottles before inspissation.

Medium for rapid sensitivity tests of *Myco tuberculosis* in sputum by Slide Culture (Rubbo and Morris, 1951)

The basic medium is prepared in distilled water and contains the following:

	%
Asparagin	0·5
KH_2PO_4	0·5
K_2SO_4	0·5
Glycerol	2·0
Magnesium citrate	0·15

Adjust to pH 7 and sterilize in the autoclave at 15 lb. for 15 minutes. Cool.

Add aseptically 10 per cent sterile ox serum or antibiotic-free human serum (horse serum can not be successfully substituted). Mix well and distribute aseptically into sterile plugged tubes (150 mm. × 15 mm.) in 5 ml. amounts.

Note. It is wise to check that the tubes are sufficiently wide to contain the thin slides used in the test.

Streptomycin assay agar (Mitchison and Spicer, 1949)

2 per cent nutrient agar[1]	1 part
1 per cent peptone in distilled water	1 ,,

Melt the agar, mix with the peptone solution, and adjust pH to 7·8–8·0. Distribute in 19 ml. amounts in screw-capped bottles and sterilize in the autoclave.

MEDIA FOR IDENTIFICATION

Aesculin-bile medium. Add 0·1 g. aesculin to 100 ml. nutrient agar and autoclave at 15 lb. for 15 minutes.

Dissolve 20 g. Bacto dehydrated ox-gall powder in 100 ml. distilled water and sterilize by steaming for 2 hours.

Sterilize 0·5 per cent ferric citrate in distilled water by steaming for 2 hours and store both these solutions at about 4° C.

[1] See Note, page 363.

To prepare the medium melt the base and cool to 50° C. Add 5 ml. horse serum, 10 ml. 0·5 per cent ferric citrate solution and 20 ml. 20 per cent ox-gall solution. Mix well and pour into small Petri dishes.

Kligler's iron agar

The agar base is obtainable in powder or tablet form from Oxoid Ltd. (CM33, CM34). Instructions for making the medium are followed, Oxoid Manual (1973) London, see also Kligler (1918).

Nagler plate for identification of *Cl. welchi* and *Cl. bifermentans* (Hayward, 1941)

Add 20 per cent human serum or plasma and 5 per cent Fildes' extract (peptic digest of blood) to molten nutrient agar at 50° C. Pour the plate, dry off the water of condensation and mark the plate into halves with a grease pencil. On one half spread a few drops of *Cl. welchi* antitoxin of British pharmacopoeial strength (Polyvalent therapeutic antitoxin will do) and allow to dry. Seed the swab over both halves of the plate, spreading the inoculum in the same pattern on each side. Incubate in an anaerobic jar. The reaction is positive if cloudy zones are seen round colonies on the half of the medium without antitoxin and if similar colonies on the antitoxin half show no zones (see page 160).

Taschdjian's medium for chlamydospores

Sprinkle 10 g. cream of rice into 1,000 ml. boiling tap water and boil for 30 seconds. Filter through cotton; reconstitute to 1,000 ml. Add 10 g. New Zealand agar[1] and 10 ml. Tween 80. Autoclave at 15 lb. for 15 minutes; pour into plates.

Seed heavily on quarter plates, cover with a sterile coverslip and incubate at room temperature, 18–25° C.

Elek's medium for diphtheria plate virulence test (Elek, 1949)

Dissolve 20 g. of proteose-peptone (Difco) or Evans peptone, 3 g. maltose and 0·7 ml. lactic acid B.P. in 500 ml. of distilled water. Add 1·5 ml. of 40 per cent NaOH, shake well and heat to boiling point. Remove the deposit by filtration through filter paper and adjust the reaction to pH 7·8 with normal HCl. Prepare 500 ml. of 3 per cent agar[1] (powder) in 1 per cent NaCl and adjust the reaction

[1] See Note, page 363.

to pH 7·8.   Filter through paper pulp and add to the fluid base. Distribute in 10 ml. quantities and sterilize by autoclaving for 10 minutes at 10 lb.   Add 2 ml. warmed horse serum before pouring, see page 108.

Different batches of horse sera and lactic acid affect the result. Therefore when a new batch is prepared and found satisfactory large quantities of these two ingredients should be stored for future use.

Oxidation of gluconate (Shaw and Clarke, 1955)

MEDIUM

Evans' peptone	1·5 g.
Yeastrel	1·0 g.
K₂HPO₄	1·0 g.
Potassium gluconate	40·0 g.
Distilled water	1 litre

Adjust to pH 7·0, distribute in 1 ml. volumes in screwcapped bottles and autoclave, 10 lb. for 10 minutes.

METHOD : incubate inoculated medium for 48 hours at 37° C. Add 1 ml. Benedict's reagent, or 1 clinitest tablet for detecting reducing sugars and place in a boiling water bath for 10 minutes.

RESULT : A yellowish-brown precipitate of cuprous oxide shows that gluconate has been oxidized.

Malonate and Phenyl-pyruvic acid tests combined (Shaw and Clarke, 1955)

MEDIUM

(NH₄)₂SO₄	2·0 g.
K₂HPO₄	0·6 g.
KH₂PO₄	0·4 g.
NaCl	2·0 g.
Sodium malonate	3·0 g.
DL-phenylalanine	2·0 g.
Yeastrel	1·0 g.
0·5% (w/v) ethanolic bromthymol blue	5 ml.
Distilled water	1 litre

Mix well, steam for about 5 minutes, mix again and distribute in 1 ml. quantities in screw-capped bottles.   Sterilize by autoclaving at 10 lb. for 10 minutes.

METHOD.   Incubate inoculated medium overnight.   Malonate utilizers produce alkali and turn the indicator blue, non-malonate producers leave the green medium unchanged.   Add N/10 HCl until the medium turns yellow.   Then add 4 to 5 drops 10 per cent (w/v) ferric chloride solution, shake and record the colour immediately.

RESULTS. Dark green indicates a positive result and shows that phenyl-pyruvic acid has been produced from phenylalanine. A negative result remains a yellow buff colour.

Hydrocele fluid sugar agar slopes (for the identification of Neisseria)

Nutrient agar	100 ml.
Sterile hydrocele fluid	10 ml.
Sterile buffered indicator	4 ml.
10% sugar solution	10 ml.

Cool the molten nutrient agar to 50° C. and warm the other solutions in the incubator. Add them to the agar, mix well, distribute aseptically and slope.

Antibiotic-free ascitic fluid can be substituted for the hydrocele fluid or human serum inactivated at 56° C. but hydrocele fluid gives the best results. Three batches of medium are required, one containing glucose, one maltose and one sucrose.

Indicator.[1] To a 0·5 per cent solution of acid fuchsin add 2·5 per cent of N/1 sodium hydroxide. Then add about 15 ml. of activated charcoal and boil for two minutes. Filter through paper. The filtrate should be straw-coloured. Repeated boiling with charcoal may be required to obtain the correct colour.

Urea slope (Christensen) for identification of Proteus and other urease producing bacteria (Cook, 1948)

Difco bacto peptone	1 g.
Sodium chloride	5 g.
Potassium dihydrogen phosphate	2 g.
Agar, Japanese or Davis	20 g.
Distilled water	1000 ml.

Steam until the solids are dissolved, filter through lint and add 1 g. of glucose and about 3 ml. of 0·4 per cent phenol red solution. Adjust pH to 6·8 or 7·0, bottle in 250 ml. quantities and autoclave at 10 lb. for 15 minutes. Add 25 ml. of a 20 per cent solution of urea, sterilized by filtration, to each bottle of melted base at 50° C. to make a final concentration of 2 per cent urea in the medium. Mix well, distribute and slope in sterilized tubes. Test for sterility by overnight incubation at 37° C.

[1] Method from the Department of Bacteriology, London School of Hygiene and Tropical Medicine, W.C.1.

CB—N*

The following media mentioned in the text are prepared according to the formulae in *Identification of Medical Bacteria*, by Cowan and Steel (1974) :

MacConkey's bile-salt agar
Desoxycholate-citrate agar
Wilson and Blair's bismuth sulphite medium
Selenite F.
Peptone water sugar media
Litmus milk
Koser's citrate
Simmons' citrate agar
Glucose phosphate broth (for M.R. & V.P. tests)
Nutrient gelatine
Cooked meat broth *
Fildes' extract (peptic digest of sheep or horse blood)
Bordet-Gengou medium
Loeffler's medium
Dorset's egg medium
Lowenstein-Jensen medium
Sabouraud's medium for fungi
Fletcher's medium for leptospira

* The prepared broth should contain finely divided meat at least one inch in depth.

IDENTIFICATION TESTS

Niacin test

REAGENTS. (*a*) 10 per cent aqueous cyanogen bromide.[1] Since it is volatile, approximately the right volume is transferred to a weighed, brown, stoppered bottle and the amount measured by re-weighing. Distilled water is added to make a w/v solution (= saturated solution). Store at 4° C, and renew monthly.
(*b*) 3 per cent benzidine[2] in 96 per cent ethyl alcohol. Store at 4° C. and renew monthly.

METHOD.[3] Take a well-grown culture of mycobacterium on an egg medium slope ; if there is little or no water of condensation, add 0·5 ml. distilled water. Autoclave at 15 lb. for 20 minutes to sterilize. Using a 50 dropping pipette deliver 2 drops of the condensate on to a white tile. Add 2 drops cyanogen bromide solution and mix by shaking the tile. Add 2 drops benzidine solution, mix and read after 5 minutes. Positive and negative controls should be included.

[1] Eastman Chemical Co. (Kodak) Ltd., Kirkby, Liverpool.
[2] Analar grade, British Drug Houses Ltd.
[3] From Bacteriology Department, Royal Postgraduate Medical School, London.

RESULT. Positive reaction, pink ; negative reaction, colourless or gray.

Catalase test for Mycobacteria (from Middlebrook, 1954)

Add equal volumes, about 3 ml. each, of a 10 per cent aqueous solution of Tween 80 and of 100 volume solution of hydrogen peroxide to a well grown Lowenstein culture in a 15 ml screw-capped bottle so that the entire growth is submerged. Bubbles collect as foam on the top of the fluid in catalase-positive cultures.

Reading : +++ = big head of foam in 30 seconds
± = collection of a few gas bubbles over a period of 3 minutes.

Plate test for nitrate reduction (Cook, 1950)

Soak sheets of Ford 428 Mill blotting paper in warm 40 per cent potassium nitrate (Analar) solution. Dry in the incubator, cut in strips about 16 mm. wide and autoclave 10 lb. for 10 minutes. Stab 2 cultures to be tested on either side of a blood agar plate, place the strip between them and incubate overnight. Include positive and negative controls. (Stab far from each other but not more than 1 cm. from the strip. Do not attempt to test more than two unknown strains and two controls per plate.)

RESULTS. Greenish-brown discoloration round the inoculated area indicates a positive test and is due to oxidation of haemoglobin to methaemoglobin by nitrites.

Oxidase test (Kovacs, 1956, modified)

REAGENT. *p*-aminodimethylalanine oxalate (Difco). Weigh 0·2 g. into each of a series of 30 ml. (1 oz.) screw-capped bottles and store at 4° C. Add 20 ml. distilled water to a bottle when the test is required. The solution can be used repeatedly during the day and is then discarded.

METHOD. Moisten a strip of blotting paper with the reagent and place it on a microscope slide or a Petri dish. Smear it with a little growth from an overnight culture on solid medium, using a glass rod or platinum loop to avoid traces of iron oxide. When oxidation occurs indophenol is produced and a deep pinkish-purple colour is seen on the paper. No change is seen when the test is negative. Positive (*Ps. aeruginosa*) and negative (*Esch. coli*) controls should be included.

X and V growth factors for the differentiation of haemophilus

X factor = haemin. Dissolve 1 g. of haematin hydrochloride in 100 ml. of 1 per cent Na_2CO_3. Sterilize through a Seitz filter and incubate for 48 hours to test for sterility. The preparation will remain stable in the refrigerator for many months. For use add 1 to 2 drops, just enough to colour the medium, before inoculation.

V factor = co-enzyme I, which is synthesized by many bacteria and can be extracted in crude form from yeast as follows :

Emulsify 250 g. of brewer's yeast in 250 ml. of distilled water. Adjust to pH 4·6 with N/1 HCl using bromocresol green indicator. Boil for 10 minutes at 100° C. and allow the sediment to settle. Filter through a Seitz filter and test for sterility. This acid extract will keep for several months in the refrigerator. It is neutralized before use.

Neutralization : Sterilize a solution of N/10 NaOH by filtration and titrate it against 1 ml. of the acid extract using phenol red as indicator. Mark on the bottle the volume necessary to neutralize 1 ml. of extract. To 5 ml. of peptone water or peptone water sugar medium add 1 ml. of sterile extract and the appropriate amount of alkali.

These extracts may be used to improve the growth of other delicate bacteria.

SPECIMENS FOR DIAGNOSTIC TESTS

In general, specimens for culture should be received in the laboratory as soon as possible after sampling. With the exception of cerebrospinal fluid, which should be incubated, specimens are best refrigerated or left at room temperature in a cool dark place if immediate transmission is impossible. When delayed more than a few hours, transport medium and a special swab should be employed (page 362). When culture of tubercle bacilli in addition to pyogenic bacteria is required send either a liquid specimen, which can be divided, or two swabs.

When antibiotic treatment is contemplated all samples required for culture must be collected before the first dose is given.

1. *Sore throat.* A swab from the tonsils or pharynx should be cultured on the day of sampling (page 74 et seq.). If delay is unavoidable swabs in transport medium should be used (page 362).

2. *Streptococcal and diphtheria carriers.* Nose and throat swabs are required. The nasal swab should be rubbed gently on the mucosa of both anterior nares. Note these organisms are penicillin sensitive ; samples taken after treatment are unlikely to be positive. Other antibiotics may also kill them.

3. *Meningococcal carriers* (page 82). A nasopharyngeal swab must be seeded immediately on warm blood agar or be sent in transport medium (page 362). The suspected carrier should attend the laboratory whenever possible.

4. *Whooping cough* (page 155). Nasopharyngeal or per-nasal swabs are seeded directly on to Bordet-Gengou or Lacey's medium screened with penicillin, and on to penicillin-streptomycin blood agar. The inoculated medium is then returned to the laboratory as soon as possible. In Britain the medium is not often required, therefore telephone the laboratory to arrange preparation of freshly poured plates.

5. *Staph. aureus carriers* (page 307). A nasal swab only is required in the first instance. For exclusion of carriage the patient should attend the laboratory for swabbing of all carrier sites.

6. *Enteric fever or pyrexia of unknown origin.*

(*a*) First week. Blood culture (page 34). The sample of blood is best taken by a member of the laboratory staff before treatment is given. A sample of clotted blood is taken at the same time for agglutination or other serological tests later.

    Faeces culture (page 91). A small sample of faeces containing any abnormal material such as pus or mucus is sent in a screw-capped bottle or waxed carton and cultured on the day on which it was voided.

(*b*) Second and subsequent weeks. Blood culture. A positive result becomes progressively less likely.

    Faeces culture (see (*a*) above).

    Agglutination tests, 5 to 10 ml. of clotted blood required (page 66).

Urine culture (page 51). A clean mid-stream specimen is sent in sterile screw-capped bottles for culture on the day of voiding.

7. *Dysentery.* Faeces or rectal swabs must be received on the day of sampling. Swabs must not be allowed to dry (page 10).

Stools for amoebae should be passed into a warm bed-pan and must arrive still warm. The laboratory staff should be warned that a specimen is on its way.

8. *Tuberculosis.* Samples of sputum are sent for microscopic examination on six consecutive days. They can be collected separately, refrigerated or kept in a cool place and sent together. The containers must never previously have been used for the purpose or else they must be specially treated. They need not be sterile (see page 72).

Samples of sputum or any other discharge for *culture* of tubercle bacilli must be received on the day of sampling and should be sent in a sterile container.

Gastric washings must either be received within a few hours or be neutralized and refrigerated.

Laryngeal swabs should be received on the day of sampling or the patient should be sent to the laboratory (page 79).

9. *Pleural effusion.* For complete examination two samples are required, one in a sterile bottle for culture, the other in a large sterile jar containing citrate for examination and culture of tubercle bacilli (page 50).

10. *Pus and infected tissue,* should arrive as soon as possible after sampling. Biopsy specimens for culture should be placed in a sterile screw-capped bottle *without added fluid.* When actinomycosis or tuberculosis is suspected and no biopsy is made, and when sampling discharge from a sinus the whole inner dressing sent in a large sterile jar is more suitable than a swab.

11. *Meningitis.* Cerebrospinal fluid should be taken with a dry-sterilized puncture needle using extremely careful aseptic precautions. It should arrive while still warm (page 48).

12. *Conjunctivitis.* Swabs should be seeded at the bedside or the patient should be sent to the laboratory (page 107).

13. *Puerperal sepsis.* Cervical swabs are either cultured within half an hour or are specially prepared and placed in transport medium (page 362). Blood for culture should be taken, preferably, by a member of the laboratory staff, before antibiotic treatment is given (page 34).

14. *Venereal disease.* The most reliable method is to take discharge direct from the cervix or urethra with a small disposable

plastic loop and spread it directly on selective medium which is incubated forthwith. Failing this urethra and cervix are sampled with specially prepared swabs which are placed in transport medium (page 98 et seq.). Fluid from a chancre is collected in a glass capillary tube and sealed with Plasticine. It should be examined as soon as possible, within a few hours. Blood for serological tests (5 to 10 ml.) is delivered into a sterile bottle and allowed to clot. A sterile sample is essential for the treponemal immobilization test and the patient must not have received antibiotics within the previous 48 hours.

15. *Leptospirosis.* Fresh blood and alkaline urine, still warm, are examined under dark ground illumination for spirochaetes and are inoculated into animals and into fluid media for culture (page 58). In the second week of the disease 5 to 10 ml. clotted blood is sent for agglutination tests. Spirochaetes are unlikely to be demonstrable in the urine before the second week.

16. *Acute lung infection.* Samples of sputum for bacterial culture should be received on the day of collection (page 86 et seq.). If virus infection is suspected throat washings, taken by gargling virus transport medium, should be sent frozen at −60° C. (in a thermos flask containing solid CO_2). (See also 18, below.)

17. *Body fluids for antibiotic assay* should be transmitted to the laboratory as soon as possible. A note of the time and amount of the last dose given before sampling is essential. Note also any other antibacterial drugs recently received. (See page 247 *et seq.*)

18. *Virus infections.* Serological tests are available for the following infections, herpes simplex, herpes zoster, cytomegalovirus influenza, parainfluenza, mumps, adenovirus, lymphogranuloma venerum, psittacosis, Q fever and Rickettsiae. Infection by enteroviruses, i.e. poliomyelitis, ECHO and coxsackie can be confirmed serologically only after successful isolation of the virus.

Varicella and herpes infections can be distinguished rapidly from vaccinia or smallpox by electron microscopy of fluid from the bleb. When smallpox is suspected expert advice should be sought before investigations are attempted. Rapid diagnosis of respiratory syncytial virus is possible by direct immunofluorescence of virus seen in nasopharyngeal aspirates.

When virus isolation is required telephone the laboratory for advice before taking a specimen. A sample of clotted blood should be sent with the culture specimen.   If isolation succeeds a second specimen of blood will be needed to demonstrate either a rise or fall of antibody titre which is important in assessing the significance of isolation.

REFERENCES

ALWEN, J. and SMITH, D. G. (1967). *J. appl. Bact.*, **30**, 389.
BARBER, M. and KUPER, S. W. A. (1951). *J. Path. Bact.*, **63**, 65.
BROWN, V. I., and LOWBURY, E. J. L. (1965). *J. clin. Path.*, **18**, 752.
COLLEE, J. G., WATT, B., FOWLER, E. B. and BROWN, R. (1972). *J. appl. Bact.* **35**, 71.
COOK, G. T. (1948). *J. Path. Bact.*, **60**, 171.
—— (1950). *J. clin. Path.*, **3**, 359.
COOPER, G. N. (1957). *J. clin. Path.*, **10**, 226.
COWAN, S. T. (1974). *Identification of Medical Bacteria.* University Press, Cambridge.
DENSON, K. W. E., DOWNER, S. J. and JEFFRIES, L. J. (personal communication).
ELEK, S. (1949). *J. clin. Path.*, **2**, 250.
HARPER, G. J. and CAWSTON, W. C. (1945). *J. Path. Bact.*, **57**, 59.
HAYWARD, N. J. (1941). *Brit. med. J.*, **i**, 811.
—— (1947). *Recent Advances in Clinical Pathology*, ed. S. C. DYKE. Churchill, London.
KLIGLER, I. J. (1918). *J. exper. Med.*, **28**, 319.
KOVACS, N. (1956). *Nature, Lond.*, **178**, 703.
LACEY, B. W. (1951). *J. gen. Microbiol.*, **5**, vi.
—— (1954). *J. Hyg. Camb.*, **52**, 273.
LACK, C. H. and TOWERS, A. G. (1962). *Brit. Med. J.*, **ii**, 1227.
LUCAS, D. R. (personal communication).
MACKEY, J. P. and SANDYS, G. H. (1966). *Brit. med. J.*, **i**, 1173.
MACKIE, T. J. and MCCARTNEY, J. E. (1959). *Practical Bacteriology*, 9th edn. Livingston, Edinburgh and London.
MAITLAND, H. B. and MARTYN, G. (1948). *J. Path. Bact.*, **60**, 553.
MEDICAL RESEARCH COUNCIL, " A System of Bacteriology in Relation to Medicine ", (1931), **9**, 174.
MIDDLEBROOK, G. (1954). *Amer. Rev. Tuberc.*, **69**, 471.
MILES, A. A. and MISRA, S. S. (1938). *J. Hyg. Camb.*, **38**, 732.
MITCHISON, D. A., ALLEN, B. W. and LAMBERT, R. A. (1973). *J. clin. Path.*, **26**, 250.
MITCHISON, D. A. and SPICER, C. C. (1949). *J. gen. Microbiol.*, **3**, 184.
MOFFET, M., YOUNG, J. L. and STUART, R. D. (1948). *Brit. med. J.*, **ii**, 421.
PARISH, H. J., O'MEARA, R. A. Q. and CLARK, W. H. M. (1934). *Lancet*, **i**, 1054.
PHILLIPS, I., HUMPHREY, D., MIDDLETON, A. and NICOL., C. S. (1972). *Brit. J. ven. Dis.*, **48**, 287.

RAPPAPORT, F., KONFORTI, N. and NAVON, B. (1956). *J. clin. Path.*, **9,** 261.

Report of a Working Party (1966). *Mth. Bull. Minist. Hlth. Lab. Serv.*, **25,** 289.

RUBBO, S. D. and MORRIS, D. M. (1951). *J. clin. Path.*, **4,** 173.

SHAW, C. and CLARKE, P. H. (1955). *J. gen. Microbiol.*, **13,** 155.

STAMP, Lord (1947). *J. gen. Microbiol.*, **1,** 251.

STOKES, E. J. (1958). *Lancet*, **i,** 668.

TARSHIS, M. S., KINSELLA, P. C. and PARKER, M. V. (1953). *J. Bact.*, **66,** 448.

TOMLINSON, A. H. (1967). *Immunology*, **13,** 323.

UNIVERSITY COLLEGE HOSPITAL media department (1953). Unpublished.

von HAEBLER, T. and MILES, A. A. (1938). *J. Path. Bact.*, **46,** 245.

WATT, B. (1973.) *J. med. Microbiol.* **6,** 307.

WILSON, M. M. (1952). *Amer. Rev. Tuberc.*, **65,** 709.

Index

385

Rickettsia 46, 264, 271, 381
Rickettsial infection 272, 381, *see* various
  diseases
Ringworm 199
Robertson's cooked meat broth 161, *see*
  cooked meat broth
Rough variants, *see* various bacteria

Sabouraud's medium 103, 194, 376
Safety cabinet 171
Salivary calculi 191
Salmonella carriers 58, 337
  food poisoning 335 *et seq.*
Salmonella species 57, 69, 91, 140, 335
  agglutination tests 143
  difficulties in identification 145
  See also enteric fever
  typhi 140, 147, 269
  carriers 58
Salt medium, for selection of staphylococci 369
Saprophyte 1, 9, 68
  acid-fast 172, 180
Satellitism 154
  test for 154
Schick test 294
Schistosoma 98
Schultz-Charlton reaction 295
Selection of material for examination 12
Selective culture 13, *see* various specimens
Selective media, aminoglycoside 149,
  364
  for *Myco. tuberculosis* 175, 370
  for *N. gonorrhoeae* 101, 366
Selenite F medium 58, 92, 376
Sensitivity tests to antibiotics 208, *et seq.*
  bactericidal 240
  blotting paper strip method 121, 212,
  221
  combined action 239
  dilution methods 226
  disc method 210
  factors influencing 212
  for *Myco. tuberculosis* 232
  methicillin 222
  primary 208, 216
Serous fluids, collection of 11
  examination of 50
  tuberculous 178
Serratia marcescens 137, 141
Serum, haemolytic 274
  medium 134, 371
Shake culture 355
Shaker, mechanical 61
Shigella species 91, 93, 147, *see* dysentery
  sonnei 20, 93, 147
Simmon's medium 137, 376
Sinuses 191

Skin 104, 328
  carriers 104, 304, 310, 328
  tests 293
Slide agglutination 141
  coagulase test 114
  culture, for antibiotic sensitivity 234,
  372
Smallpox 381
Soft sore 155
Somatic antigen, *see* O antigen 267
Sonne's bacillus, *see Shigella sonnei*
Sore throat 1, 74, 378
Specimens, collection of 9, 378
  concentration of 13
  culture of, from sites normally sterile
  5, 19, 34
  for actinomyces 191
  for fungi, *see* various species
  from sites with a normal flora 5, 19,
  71
  for *Myco. tuberculosis* 174
  required for diagnosis 378
  unsuitable 12
Speed, need for 6, 208, 234
Spiramycin 258
Spirochaetal jaundice, *see* leptospirosis
  58, 271
Spirochaetes 58, 74, 104
Spore-bearers, in hospital infection 326
Spore-bearing bacilli 158 *et seq.*, *see
  also Clostridium* species and *Bacillus*
  species
Spores 159, 161, 162, 326
  in isolation of clostridia 159
Sporotrichum 201
" Spreader, " *see* swarming
Sputum 10, 86
  in tuberculosis 172, 177
  liquefaction of 88
  slide culture of 234
Stamp's method of preserving bacteria
  357
Standard agglutination 143, 265
Standard strains for antibiotic assay
  250, 359
  maintenance of 359
Staphylococci variants 116
Staphylococcus albus 45, 58, 68, 115, 117
Staphylococcus aureus 40, 68, 73, 80, 81,
  85, 89, 91, 99, 104, 105, 113, 222,
  307, 321, 329
  carriers of 310, 321 *et seq.*, 379
  in blood 40
  in food poisoning 91, 335
  in hospital infection 307 *et seq.*, 321
  methicillin resistant 222
  toxin and antitoxin 361
Starch, fermentation of 135
Statistical argument 1
Steam, sterilisation by 331